Modern Computer Architecture and Organization

Second Edition

Learn x86, ARM, and RISC-V architectures and the design of smartphones, PCs, and cloud servers

Jim Ledin

BIRMINGHAM—MUMBAI

Modern Computer Architecture and Organization
Second Edition

Copyright © 2022 Packt Publishing

All rights reserved. No part of this book may be reproduced, stored in a retrieval system, or transmitted in any form or by any means, without the prior written permission of the publisher, except in the case of brief quotations embedded in critical articles or reviews.

Every effort has been made in the preparation of this book to ensure the accuracy of the information presented. However, the information contained in this book is sold without warranty, either express or implied. Neither the author, nor Packt Publishing or its dealers and distributors, will be held liable for any damages caused or alleged to have been caused directly or indirectly by this book.

Packt Publishing has endeavored to provide trademark information about all of the companies and products mentioned in this book by the appropriate use of capitals. However, Packt Publishing cannot guarantee the accuracy of this information.

Senior Publishing Product Manager: Denim Pinto
Acquisition Editor – Peer Reviews: Saby Dsilva
Project Editor: Namrata Katare
Content Development Editor: Edward Doxey
Copy Editor: Safis Editing
Technical Editor: Tejas Mhasvekar
Proofreader: Safis Editing
Indexer: Subalakshmi Govindhan
Presentation Designer: Ganesh Bhadwalkar

First published: April 2020
Second edition: May 2022

Production reference: 2260422

Published by Packt Publishing Ltd.
Livery Place
35 Livery Street
Birmingham
B3 2PB, UK.

ISBN 978-1-80323-451-9

www.packt.com

Foreword

I am a software developer, not a hardware engineer. I have spent my career building software of all different kinds to solve lots of different kinds of problems. However, as a quirk, accident or fate, I have spent a fair amount of my software development career closer to the hardware than many, maybe most, software developers do these days.

In the early years of my fascination with computers I quickly discovered that the, by today's standards, incredibly crude devices that I had access to, couldn't really do anything very interesting unless I learned how to program them in assembler. So, I learned to program them in Z80, and later 6502 and 80x86 assembler.

Programming in assembler is different in lots of ways to programming in higher level languages. It immediately puts you next to the hardware. You can't ignore how memory is laid out, you need to adjust your code for it. You can't ignore the registers at your disposal, they are your variables and you need to marshal them carefully. You also learn how to communicate with other devices through I/O ports, which is, ultimately, the only way that digital devices communicate with each other. Once, when working on a particularly tricky problem, I woke up in the middle of the night and realised that I had been dreaming in 80x86 assembly language.

My career, and more importantly the hardware I had access to, developed. I got my dream job, at the time, working in the R&D division of a computer manufacturer. I worked on enhancing operating systems to work with our hardware and built device drivers to take advantage of some of the unique features of our PCs. Again, it was essential in this kind of work to have a good working knowledge of how the hardware worked.

Software development evolved. The languages that we used became more abstract, the operating systems, virtual machines, containers and public cloud infrastructure increasingly hid the details of the underlying hardware from us software developers. I recently spoke to a LISP programmer on social media who didn't realise that ultimately his lovely functional declarative structures got translated to opcodes and values in the registers of a CPU. He seemingly had no working model for how the computers that he relied upon worked. He didn't have to, but I think he would be a better programmer if he did.

In the latter part of my career I worked on some world-class high performance systems. A team I led was tasked with building one the world's highest performance financial exchanges.

In order to do so, once again we needed to dig in and really understand how the underlying hardware of our system worked. This allowed us to take full advantage of the staggering levels of performance that modern hardware is capable of. During this time we stole a term from motor-racing to try to describe our approach. In the 1970's the best Formula 1 racing driver was Jackie Stewart. He was interviewed and asked, "do you need to be an engineer to be a great racing driver?". Jackie responded, "no, but you must have Mechanical Sympathy for the car." In essence, you need to understand the capabilities of the underlying hardware to take advantage of them.

We adopted this idea of Mechanical Sympathy and applied it to our work. For example, the biggest cost in our trading system was a cache-miss. If the data we wanted to process wasn't in the appropriate cache when it was needed, we'd see orders of magnitude wiped off the performance of our system. So we needed to design our code, even though it was written in a high-level language running in a virtual machine, to maximise our chances that the data was in the cache. We needed to understand and manage the concurrency in our multi-core processors, and recognise and take advantage of things like processor cache lines and the essentially block storage nature of memory and other storage devices. The result was levels of performance that some people didn't think possible. Modern hardware is very impressive when you take advantage of it.

This interest in the hardware isn't just for high-performance computing. Estimates vary, but all agree that a significant fraction of the carbon that we emit as a species comes from powering the data-centres where our code lives. I can't think of any field of human endeavor that is as inefficient as software—for most systems, a speed increase of up to 100 times is easy if you do just a bit more work to manage the flow of information through your hardware. Nearly all systems can attain a 1000-fold increase with some more focused work, however if we could gain even a 10x improvement by better understanding how our code works and how it uses the hardware that it operates on, we could reduce the carbon footprint of computing by a factor of 10 too. That is an idea that is much more important than performance for performance's sake.

Ultimately, there is a degree to which you must understand how your computer works, and there are risks to losing touch with how the hardware we all depend upon functions. I confess that I am a nerd. I love to understand how things work. Not everyone needs to push hardware to its limits, but it is a bad idea to treat it like magic, because it's not magic. It is engineering, and engineering is always about trade-offs. You will be surprised how often the fundamentals of how your hardware works leaks out and affects the behaviour of your software, however far up the stack it sits—even if we are writing cloud-based systems in LISP.

For someone like me, Jim Ledin's *Modern Computer Architecture and Organization*, Second Edition, is a delight.

I am not a hardware engineer, and I don't want to be. For me though, a vital part of my skills as a software developer includes having a good working model for how the hardware that I rely upon, actually works. I want to maintain, and build, mechanical sympathy.

This book takes us from the basic concepts of computation, looking at the first computers and the first CPUs, to the potential of quantum computing and other near-future directions that our hardware will probably exploit. You might want to understand how a modern processor works, and get to grips with their staggering efficiency and their ability to keep themselves fed with data from stores that are hundreds of times slower than they are. You may also be interested in complex ideas that extend beyond the confines of only the hardware, such as how cryptocurrency mining works, or what the architecture of a modern self-driving car, looks like. This book can answer those questions and many, many more.

I think that it is not just computer scientists and engineers, but indeed every software developer, who will be better at their job when they have some understanding of how the devices that they use in their everyday work. When trying to understand something big and complicated in software, I still, frequently think, "well it's all just bits, bytes and opcodes really, so what is going on here?" This is the equivalent of a chemist understanding molecules and compounds and being able to go back to first principles to solve something tricky. These are the real building blocks, and it can help us all to understand them better.

I know that I will be dipping into this book on a regular basis for years to come, and I hope that you enjoy doing the same.

Dave Farley
Independent Software Engineering Consultant and Founder of Continuous Delivery Ltd

Contributors

About the author

Jim Ledin is the CEO of Ledin Engineering, Inc. Jim is an expert in embedded software and hardware design and testing. He is also an expert in system cybersecurity assessment and penetration testing. He has a B.S. degree in aerospace engineering from Iowa State University and an M.S. degree in electrical and computer engineering from the Georgia Institute of Technology. Jim is a registered professional electrical engineer in California, a Certified Information System Security Professional (CISSP), a Certified Ethical Hacker (CEH), and a Certified Penetration Tester (CPT).

I would like to thank my wife Lynda and daughter Emily for their patience and support as I focused on this project. I love you, my sweeties!

I would also like to thank Dr. Sarah M. Neuwirth and Iztok Jeras for their diligent work reviewing each of these chapters. Your input has helped create a much better book!

Thank you as well to Dave Farley for providing such an eloquent foreword.

About the reviewers

Dr. Sarah M. Neuwirth is a postdoctoral research associate in the Modular Supercomputing and Quantum Computing Group at Goethe-University Frankfurt, Germany. She also holds a visiting researcher position at the Jülich Supercomputing Centre, Germany. Sarah has more than 9 years of experience in the academic field. Her research interests include high-performance storage systems, parallel I/O and file systems, modular supercomputing (i.e., resource disaggregation and virtualization), standardized cluster benchmarking, high performance computing and networking, distributed systems, and communication protocols.

In addition, Sarah has designed the curriculum for, and taught, courses in parallel computer architecture, high performance interconnection networks, and distributed systems for the past 9 years. In 2018, Sarah was awarded her Ph.D. in computer science from the Heidelberg University, Germany. She defended her degree with highest honors (summa cum laude) and was the recipient of the ZONTA Science Award 2019 for her outstanding dissertation. Sarah also holds a M.Sc. degree (2012) and B.Sc. degree (2010) in computer science and mathematics from the University of Mannheim, Germany. She has served as a technical reviewer for several prestigious HPC-related conferences and journals, including the IEEE/ACM SC Conference Series, ACM ICPP, IEEE IDPDS, IEEE HPCC, the PDSW workshop at IEEE/ACM SC, the PERMAVOST workshop at ACM HPDC, ACM TOCS, IEEE Access, and Elsevier's FGCS.

Iztok Jeras obtained a bachelor's degree in electrical engineering and a master's in computer science at the University of Ljubljana. He worked at several Slovenian companies on micro-controller, FPGA, and ASIC designs with some embedded software and Linux IT work mixed in. In his spare time, he researches cellular automata and contributes to digital design-related open source projects. Recently he has been focusing on the RISC-V ISA.

Join our community Discord space

Join the book's Discord workspace for a monthly *Ask me Anything* session with the author:
https://discord.gg/7h8aNRhRuY

Table of Contents

Preface .... **xxix**

Chapter 1: Introducing Computer Architecture — 1

Technical requirements 2
The evolution of automated computing devices 2
  Charles Babbage's Analytical Engine • 2
  ENIAC • 4
  IBM PC • 5
    The Intel 8088 microprocessor • 6
    The Intel 80286 and 80386 microprocessors • 8
    The iPhone • 9
Moore's law 11
Computer architecture 14
  Representing numbers with voltage levels • 14
  Binary and hexadecimal numbers • 14
  The 6502 microprocessor • 18
  The 6502 instruction set • 22
Summary 25
Exercises 25

Chapter 2: Digital Logic — 27

Technical requirements 28
Electrical circuits 28
The transistor 29
Logic gates 30
Latches 35

Flip-flops 38

Registers 40

Adders 41

    Propagation delay • 43

Clocking 44

Sequential logic 45

Hardware description languages 46

    VHDL • 47

Summary 51

Exercises 52

Chapter 3: Processor Elements   53

Technical requirements 54

A simple processor 54

    Control unit • 55

        Executing an instruction – a simple example • 57

    Arithmetic logic unit • 58

    Registers • 64

The instruction set 65

Addressing modes 66

    Immediate addressing mode • 66

    Absolute addressing mode • 67

    Absolute indexed addressing mode • 68

    Indirect indexed addressing mode • 70

Instruction categories 71

    Memory load and store instructions • 72

    Register-to-register data transfer instructions • 72

    Stack instructions • 72

    Arithmetic instructions • 73

    Logical instructions • 74

    Branching instructions • 74

Subroutine call and return instructions • 75

Processor flag instructions • 75

Interrupt-related instructions • 75

No operation instruction • 76

Interrupt processing 76

$\overline{\text{IRQ}}$ processing • 76

$\overline{\text{NMI}}$ processing • 78

BRK instruction processing • 79

Input/output operations 80

Programmed I/O • 82

Interrupt-driven I/O • 82

Direct memory access • 83

Summary 84

Exercises 84

Chapter 4: Computer System Components   87

Technical requirements 88

Memory subsystem 88

Introducing the MOSFET 89

Constructing DRAM circuits with MOSFETs 92

The capacitor • 92

The DRAM bit cell • 94

DDR5 SDRAM • 96

Graphics DDR • 99

Prefetching • 99

I/O subsystem 100

Parallel and serial data buses • 100

PCI Express • 102

SATA • 104

M.2 • 105

USB • 105

Thunderbolt • 106

Graphics displays 106

VGA • 108

DVI • 108

HDMI • 109

DisplayPort • 109

Network interface 110

Ethernet • 110

Wi-Fi • 111

Keyboard and mouse 112

Keyboard • 112

Mouse • 113

Modern computer system specifications 114

Summary 116

Exercises 116

Chapter 5: Hardware-Software Interface — 117

Technical requirements 118

Device drivers 118

The parallel port • 119

PCIe device drivers • 121

Device driver structure • 122

BIOS 124

UEFI • 126

The boot process 127

BIOS boot • 128

UEFI boot • 128

Trusted boot • 130

Embedded devices • 131

Operating systems 131

Processes and threads 133
  Scheduling algorithms and process priority • 136
Multiprocessing 141
Summary 142
Exercises 142

Chapter 6: Specialized Computing Domains 145

Technical requirements 146
Real-time computing 146
  Real-time operating systems • 148
Digital signal processing 152
  ADCs and DACs • 152
  DSP hardware features • 155
  Signal processing algorithms • 157
    Convolution • 157
    Digital filtering • 158
    Fast Fourier transform (FFT) • 159
GPU processing 162
  GPUs as data processors • 164
    Big data • 164
    Deep learning • 165
Examples of specialized architectures 167
Summary 169
Exercises 169

Chapter 7: Processor and Memory Architectures 173

Technical requirements 173
The von Neumann, Harvard, and modified Harvard architectures 174
  The von Neumann architecture • 174
  The Harvard architecture • 176
  The modified Harvard architecture • 177

Physical and virtual memory 178
    Paged virtual memory • 182
    Page status bits • 186
    Memory pools • 187
Memory management unit 189
Summary 192
Exercises 192

Chapter 8: Performance-Enhancing Techniques   195

Technical requirements 196
Cache memory 196
    Multilevel processor caches • 198
    Static RAM • 199
    Level 1 cache • 200
    Direct-mapped cache • 201
    Set associative cache • 205
    Processor cache write policies • 207
    Level 2 and level 3 processor caches • 209
Instruction pipelining 211
    Superpipelining • 214
    Pipeline hazards • 215
    Micro-operations and register renaming • 217
    Conditional branches • 218
Simultaneous multithreading 220
SIMD processing 221
Summary 223
Exercises 224

Chapter 9: Specialized Processor Extensions   225

Technical requirements 225
Privileged processor modes 226

Handling interrupts and exceptions • 226

Protection rings • 230

Supervisor mode and user mode • 232

System calls • 233

Floating-point arithmetic **234**

The 8087 floating-point coprocessor • 237

The IEEE 754 floating-point standard • 239

Power management **240**

Dynamic voltage frequency scaling • 241

System security management **243**

Trusted Platform Module • 245

Thwarting cyberattackers • 246

Summary **247**

Exercises **248**

Chapter 10: Modern Processor Architectures and Instruction Sets   249

Technical requirements **249**

x86 architecture and instruction set **250**

The x86 register set • 252

x86 addressing modes • 255

Implied addressing • 256

Register addressing • 256

Immediate addressing • 256

Direct memory addressing • 256

Register indirect addressing • 256

Indexed addressing • 257

Based indexed addressing • 257

Based indexed addressing with scaling • 257

x86 instruction categories • 258

Data movement • 258

Stack manipulation • 259

    Arithmetic and logic • 259

    Conversions • 261

    Control flow • 261

    String manipulation • 261

    Flag manipulation • 262

    Input/output • 262

    Protected mode • 262

    Miscellaneous instructions • 262

    Other instruction categories • 263

    Common instruction patterns • 263

  x86 instruction formats • 264

  x86 assembly language • 264

x64 architecture and instruction set 268

  The x64 register set • 269

  x64 instruction categories and formats • 270

  x64 assembly language • 270

32-bit ARM architecture and instruction set 273

  The ARM register set • 275

  ARM addressing modes • 277

    Immediate • 277

    Register direct • 278

    Register indirect • 278

    Register indirect with offset • 278

    Register indirect with offset, pre-incremented • 278

    Register indirect with offset, post-incremented • 278

    Double register indirect • 279

    Double register indirect with scaling • 279

  ARM instruction categories • 279

    Load/store • 279

    Stack manipulation • 279

    Register movement • 280

    Arithmetic and logic • 280

    Comparisons • 280

    Control flow • 281

    Supervisor mode • 281

    Breakpoint • 281

    Conditional execution • 281

    Other instruction categories • 282

  32-bit ARM assembly language • 282

64-bit ARM architecture and instruction set 284

  64-bit ARM assembly language • 286

Summary 288

Exercises 288

Chapter 11: The RISC-V Architecture and Instruction Set   293

Technical requirements 294

The RISC-V architecture and applications 294

The RISC-V base instruction set 298

  Computational instructions • 298

  Control flow instructions • 299

  Memory access instructions • 299

  System instructions • 300

  Pseudo-instructions • 302

  Privilege levels • 304

RISC-V extensions 305

  The M extension • 306

  The A extension • 306

  The C extension • 307

  The F and D extensions • 308

  Other extensions • 308

RISC-V variants 309

64-bit RISC-V 310

Standard RISC-V configurations 311
RISC-V assembly language 312
Implementing RISC-V in an FPGA 313
Summary 317
Exercises 318

Chapter 12: Processor Virtualization 319

Technical requirements 319
Introducing virtualization 320
  Types of virtualization • 320
    Operating system virtualization • 321
    Application virtualization • 322
    Network virtualization • 323
    Storage virtualization • 323
  Categories of processor virtualization • 324
    Trap-and-emulate virtualization • 324
    Paravirtualization • 327
    Binary translation • 327
    Hardware emulation • 328

Virtualization challenges 329
  Unsafe instructions • 330
  Shadow page tables • 330
  Security • 331

Virtualizing modern processors 331
  x86 processor virtualization • 331
    x86 hardware virtualization • 333
  ARM processor virtualization • 334
  RISC-V processor virtualization • 334

Virtualization tools 335
  VirtualBox • 335
  VMware Workstation • 336

VMware ESXi • 336

KVM • 337

Xen • 337

QEMU • 337

Virtualization and cloud computing **338**

Electrical power consumption • 339

Summary **339**

Exercises **340**

Chapter 13: Domain-Specific Computer Architectures   341

Technical requirements **341**

Architecting computer systems to meet unique requirements **342**

Smartphone architecture **343**

iPhone 13 Pro Max • 343

Personal computer architecture **347**

Alienware Aurora Ryzen Edition R10 gaming desktop • 347

    Ryzen 9 5950X branch prediction • 348

    Nvidia GeForce RTX 3090 GPU • 348

    Aurora subsystems • 350

Warehouse-scale computing architecture **351**

WSC hardware • 352

Rack-based servers • 354

Hardware fault management • 357

Electrical power consumption • 358

The WSC as a multilevel information cache • 358

Deploying a cloud application • 359

Neural networks and machine learning architectures **363**

Intel Nervana neural network processor • 363

Summary **367**

Exercises **367**

Chapter 14: Cybersecurity and Confidential Computing Architectures   369

Technical requirements   370
Cybersecurity threats   370
Cybersecurity threat categories • 370

Cyberattack techniques • 372

Types of malware • 373

Post-exploitation actions • 376

Features of secure hardware   377
Identify what needs to be protected • 378

Anticipate all types of attacks • 378

Features of secure system design • 381

Secure key storage • 381

Encryption of data at rest • 381

Encryption of data in transit • 381

Cryptographically secure key generation • 382

Secure boot procedure • 382

Tamper-resistant hardware design • 383

Confidential computing   383
Designing for security at the architectural level   386
Avoid security through obscurity • 386

Comprehensive secure design • 387

The principle of least privilege • 388

Zero trust architecture • 388

Ensuring security in system and application software   390
Common software weaknesses • 390

Buffer overflow • 390

Cross-site scripting • 391

SQL injection • 391

Path traversal • 392

Source code security scans • 393

Summary 394
Exercises 394

Chapter 15: Blockchain and Bitcoin Mining Architectures   395

Technical requirements 395
Introduction to blockchain and bitcoin 396
   The SHA-256 hash algorithm • 400
   Computing SHA-256 • 402
   Bitcoin core software • 403
The bitcoin mining process 404
   Bitcoin mining pools • 405
   Mining with a CPU • 408
   Mining with a GPU • 408
Bitcoin mining computer architectures 409
   Mining with FPGAs • 411
   Mining with ASICs • 413
   Bitcoin mining economics • 416
Alternative types of cryptocurrency 417
Summary 419
Exercises 419

Chapter 16: Self-Driving Vehicle Architectures   421

Technical requirements 421
Overview of self-driving vehicles 422
   Driving autonomy levels • 422
Safety concerns of self-driving vehicles 424
Hardware and software requirements for self-driving vehicles 426
   Sensing vehicle state and the surroundings • 427
      GPS, speedometer, and inertial sensors • 427
      Video cameras • 429
      Radar • 430

Lidar • 430

  Sonar • 431

 Perceiving the environment • 431

  Convolutional neural networks • 431

  Example CNN implementation • 433

  CNNs in autonomous driving applications • 435

  Lidar localization • 440

  Object tracking • 441

 Decision processing • 442

  Lane keeping • 442

  Complying with the rules of the road • 442

  Avoiding objects • 443

  Planning the vehicle path • 443

Autonomous vehicle computing architecture 443

 Tesla HW3 Autopilot • 444

Summary 446

Exercises 446

Chapter 17: Quantum Computing and Other Future Directions in Computer Architectures   449

Technical requirements 450

The ongoing evolution of computer architectures 450

Extrapolating from current trends 452

 Moore's law revisited • 452

 The third dimension • 453

 Increased device specialization • 454

Potentially disruptive technologies 455

 Quantum physics • 455

 Spintronics • 456

 Quantum computing • 458

 Quantum code-breaking • 459

Adiabatic quantum computation • 460

The future of quantum computing • 461

Carbon nanotubes • 462

Building a future-tolerant skill set 464

Continuous learning • 464

College education • 466

Conferences and literature • 467

Summary 468

Exercises 471

Appendix   473

Answers to Exercises 473
Chapter 1: Introducing Computer Architecture 473

Exercise 1 • 473

Answer • 473

Exercise 2 • 475

Answer • 476

Exercise 3 • 479

Answer • 479

Exercise 4 • 483

Answer • 484

Exercise 5 • 485

Answer • 485

Exercise 6 • 486

Answer • 487

Chapter 2: Digital Logic 488

Exercise 1 • 488

Answer • 488

Exercise 2 • 489

Answer • 489

Exercise 3 • 490

Answer • 490

Exercise 4 • 491

Answer • 492

Exercise 5 • 492

Answer • 492

Exercise 6 • 495

Answer • 495

Chapter 3: Processor Elements 498

Exercise 1 • 498

Answer • 498

Exercise 2 • 499

Answer • 499

Exercise 3 • 499

Answer • 499

Exercise 4 • 500

Answer • 500

Exercise 5 • 502

Answer • 502

Exercise 6 • 504

Answer • 504

Chapter 4: Computer System Components 510

Exercise 1 • 510

Answer • 511

Exercise 2 • 511

Answer • 511

Chapter 5: Hardware-Software Interface 511

Exercise 1 • 511

Answer • 512

Exercise 2 • 512

Answer • 512

Chapter 6: Specialized Computing Domains 513

Exercise 1 • 513

Answer • 514

Exercise 2 • 514

Answer • 515

Exercise 3 • 516

Answer • 516

Chapter 7: Processor and Memory Architectures 517

Exercise 1 • 517

Answer • 518

Exercise 2 • 518

Answer • 518

Exercise 3 • 518

Answer • 519

Chapter 8: Performance-Enhancing Techniques 521

Exercise 1 • 521

Answer • 521

Exercise 2 • 521

Answer • 521

Exercise 3 • 522

Answer • 522

Chapter 9: Specialized Processor Extensions 522

Exercise 1 • 522

Answer • 522

Exercise 2 • 525

Answer • 525

Exercise 3 • 529

Answer • 529

Exercise 4 • 529

Answer • 529

Exercise 5 • 530

Answer • 530

Exercise 6 • 530

Answer • 530

Exercise 7 • 530

Answer • 530

Exercise 8 • 531

Answer • 531

Chapter 10: Modern Processor Architectures and Instruction Sets 531

Exercise 1 • 531

Answer • 532

Exercise 2 • 535

Answer • 535

Exercise 3 • 540

Answer • 541

Exercise 4 • 544

Answer • 544

Exercise 5 • 549

Answer • 550

Exercise 6 • 552

Answer • 552

Exercise 7 • 557

Answer • 558

Exercise 8 • 560

Answer • 560

Chapter 11: The RISC-V Architecture and Instruction Set 566

Exercise 1 • 566

Answer • 566

Exercise 2 • 567

Answer • 567

Exercise 3 • 568

Answer • 568

Chapter 12: Processor Virtualization 571

Exercise 1 • 571

Answer • 572

Exercise 2 • 573

Answer • 573

Exercise 3 • 575

Answer • 575

Chapter 13: Domain-Specific Computer Architectures 577

Exercise 1 • 577

Answer • 577

Exercise 2 • 577

Answer • 578

Chapter 14: Cybersecurity and Confidential Computing Architectures 579

Exercise 1 • 579

Answer • 579

Exercise 2 • 580

Answer • 580

Exercise 3 • 580

Answer • 581

Chapter 15: Blockchain and Bitcoin Mining Architectures 581

Exercise 1 • 581

Answer • 581

Exercise 2 • 583

Answer • 583

Chapter 16: Self-Driving Vehicle Architectures 584

Exercise 1 • 584

Answer • 584

Exercise 2 • 585

Answer • 585

Exercise 3 • 587

Answer • 588

Exercise 4 • 591

Answer • 592

Chapter 17: Future Directions in Computer Architectures 597

Exercise 1 • 597

Answer • 597

Exercise 2 • 597

Answer • 598

Exercise 3 • 598

Answer • 598

Exercise 4 • 600

Answer • 600

Other Books You May Enjoy   605

Index   609

Preface

Welcome to the second edition of *Modern Computer Architecture and Organization*. It has been my pleasure to receive a great deal of feedback and comments from readers of the first edition. Of course, I appreciate all input from readers, especially those who bring any errors and omissions to my attention.

This book presents the key technologies and components employed in modern processor and computer architectures and discusses how various architectural decisions result in computer configurations optimized for specific needs.

To understate the situation quite drastically, modern computers are complicated devices. Yet, when viewed in a hierarchical manner, the functions of each level of complexity become clear. We will cover a great many topics in these chapters and will only have space to explore each of them to a limited degree. My goal is to provide a coherent introduction to each important technology and subsystem you might find in a modern computing device and explain its relationship to other system components.

This edition includes updates on technologies that have advanced since the publication of the first edition and adds significant new content in several important areas related to computer architecture. New chapters cover the topics of cybersecurity, blockchain and bitcoin mining, and self-driving vehicle computing architectures.

While the security of computing systems has always been important, recent exploitations of major vulnerabilities in widely used operating systems and applications have resulted in substantial negative impacts felt in countries around the world. These cyberattacks have accentuated the need for computer system designers to incorporate cybersecurity as a foundational element of system architecture.

I will not be providing a lengthy list of references for further reading. The internet is your friend in this regard.

If you can manage to bypass the clamor of political and social media argumentation on the internet, you will find yourself in an enormous, cool, quiet library containing a vast quantity of accumulated human knowledge. Learn to use the advanced features of your favorite search engine. Also, learn to differentiate high-quality information from uninformed opinion. Check multiple sources if you have any doubts about the information you're finding. Consider the source: if you are looking for information about an Intel processor, search for documentation published by Intel.

By the end of this book, you will have gained a strong grasp of the computer architectures currently used in a wide variety of digital systems. You will also have developed an understanding of the relevant trends in architectural technology currently underway, as well as some possible disruptive advances in the coming years that may drastically influence the architectural development of computing systems.

Who this book is for

This book is intended for software developers, computer engineering students, system designers, computer science professionals, reverse engineers, and anyone else seeking to understand the architecture and design principles underlying all types of modern computer systems, from tiny embedded devices to smartphones to warehouse-sized cloud server farms. Readers will also explore the directions these technologies are likely to take in the coming years. A general understanding of computer processors is helpful but is not required.

What this book covers

Chapter 1, Introducing Computer Architecture, begins with a brief history of automated computing devices and describes the significant technological advances that drove leaps in capability. This is followed by a discussion of Moore's law, with an assessment of its applicability over previous decades and the implications for the future. The basic concepts of computer architecture are introduced in the context of the 6502 microprocessor.

Chapter 2, Digital Logic, introduces transistors as switching elements and explains their use in constructing logic gates. We will then see how flip-flops and registers are developed by combining simple gates. The concept of sequential logic, meaning logic that contains state information, is introduced, and the chapter ends with a discussion of clocked digital circuits.

Chapter 3, Processor Elements, begins with a conceptual description of a generic processor. We will examine the concepts of the instruction set, register set, and instruction loading, decoding, execution, and sequencing.

Memory load and store operations are also discussed. The chapter includes a description of branching instructions and their use in looping and conditional processing. Some practical considerations are introduced that lead to the necessity for interrupt processing and I/O operations.

Chapter 4, Computer System Components, discusses computer memory and its interface to the processor, including multilevel caching. I/O requirements, including interrupt handling, buffering, and dedicated I/O processors, are described. We will discuss some specific requirements for I/O devices, including the keyboard and mouse, the video display, and the network interface. The chapter ends with descriptive examples of these components in modern computer applications, including smart mobile devices, personal computers, gaming systems, cloud servers, and dedicated machine learning systems.

Chapter 5, Hardware-Software Interface, discusses the implementation of the high-level services a computer operating system must provide, including disk I/O, network communications, and interactions with users. This chapter describes the software layers that implement these features, starting at the level of the processor instruction set and registers. Operating system functions, including booting, multiprocessing, and multithreading, are also described.

Chapter 6, Specialized Computing Domains, explores domains of computing that tend to be less directly visible to most users, including real-time systems, digital signal processing, and GPU processing. We will discuss the unique requirements associated with each of these domains and look at examples of modern devices implementing these features.

Chapter 7, Processor and Memory Architectures, takes an in-depth look at modern processor architectures, including the von Neumann, Harvard, and modified Harvard variants. The chapter discusses the implementation of paged virtual memory. The practical implementation of memory management functionality within the computer architecture is introduced and the functions of the memory management unit are described.

Chapter 8, Performance-Enhancing Techniques, discusses a number of performance-enhancing techniques used routinely to reach peak execution speed in real-world computer systems. The most important techniques for improving system performance, including the use of cache memory, instruction pipelining, instruction parallelism, and SIMD processing, are the subjects of this chapter.

Chapter 9, Specialized Processor Extensions, focuses on extensions commonly implemented at the processor instruction set level to provide additional system capabilities beyond generic data processing requirements. The extensions presented include privileged processor modes, floating-point mathematics, power management, and system security management.

Chapter 10, Modern Processor Architectures and Instruction Sets, examines the architectures and instruction set features of modern processor designs, including the x86, x64, and ARM processors. One challenge that arises when producing a family of processors over several decades is the need to maintain backward compatibility with code written for earlier-generation processors. The need for legacy support tends to increase the complexity of the later-generation processors. This chapter will examine some of the attributes of these processor architectures that result from supporting legacy requirements.

Chapter 11, The RISC-V Architecture and Instruction Set, introduces the exciting new RISC-V (pronounced risk five) processor architecture and its instruction set. RISC-V is a completely open source, free-to-use specification for a reduced instruction set computer architecture. A complete instruction set specification has been released and a number of hardware implementations of this architecture are currently available. Work is ongoing to develop specifications for a number of instruction set extensions. This chapter covers the features and variants available in the RISC-V architecture and introduces the RISC-V instruction set. We will also discuss the applications of the RISC-V architecture in mobile devices, personal computers, and servers.

Chapter 12, Processor Virtualization, introduces the concepts involved in processor virtualization and explains the many benefits resulting from the use of virtualization. The chapter includes examples of virtualization based on open source tools and operating systems. These tools enable the execution of instruction set-accurate representations of various computer architectures and operating systems on a general-purpose computer. We will also discuss the benefits of virtualization in the development and deployment of real-world software applications.

Chapter 13, Domain-Specific Computer Architectures, brings together the topics discussed in previous chapters to develop an approach for architecting a computer system design to meet unique user requirements. We will discuss some specific application categories, including mobile devices, personal computers, gaming systems, internet search engines, and neural networks.

Chapter 14, Cybersecurity and Confidential Computing Architectures, focuses on the security needs of critical application areas like national security systems and financial transaction processing. These systems must be resilient against a broad range of cybersecurity threats, including malicious code, covert channel attacks, and attacks enabled by physical access to computing hardware. Topics addressed in this chapter include cybersecurity threats, encryption, digital signatures, and secure hardware and software design.

The explosion of interest in cryptocurrencies and their growing acceptance by mainstream financial institutions and retailers demonstrate that this area of computing is on a continued growth path. This edition adds a chapter on blockchain and the computational demands of bitcoin mining.

Chapter 15, *Blockchain and Bitcoin Mining Architectures*, introduces the concepts associated with blockchain, a public, cryptographically secured ledger recording a sequence of transactions. We continue with an overview of the process of bitcoin mining, which appends transactions to the bitcoin blockchain and rewards those who complete this task with payment in the form of bitcoin. Bitcoin processing requires high-performance computing hardware, which is illustrated in terms of a current-generation bitcoin mining computer architecture.

The continuing growth in the number of automobiles with partial or full self-driving capabilities demands robust, highly capable computing systems that meet the requirements for safe autonomous vehicle operation on public roadways.

Chapter 16, *Self-Driving Vehicle Architectures*, describes the capabilities required in self-navigating vehicle processing architectures. It begins with a discussion of the requirements for ensuring the safety of the autonomous vehicle and its occupants, as well as for other vehicles, pedestrians, and stationary objects. We continue with a discussion of the types of sensors and data a self-driving vehicle receives as input while driving and a description of the types of processing required for effective vehicle control. The chapter concludes with an overview of an example self-driving computer architecture.

Chapter 17, *Quantum Computing and Other Future Directions in Computer Architectures*, looks at the road ahead for computer architectures. This chapter reviews the significant advances and ongoing trends that have resulted in the current state of computer architectures and extrapolates these trends in possible future directions. Potentially disruptive technologies are discussed that could alter the path of future computer architectures. In closing, I will propose some approaches for professional development for the computer architect that should result in a future-tolerant skill set.

As in the other chapters, each of the three new chapters contains end-of-chapter exercises designed to broaden your understanding of the chapter topic and cement the information from the chapter within your knowledge base.

I hope you enjoy this updated edition as much as I have enjoyed developing it. Happy reading!

To get the most out of this book

Each chapter in this book includes a set of exercises at the end. To get the most from the book, and to cement some of the more challenging concepts in your mind, I recommend you try to work through each exercise. Complete solutions to all exercises are provided in the book and are available online at `https://github.com/PacktPublishing/Modern-Computer-Architecture-and-Organization-Second-Edition`.

In case there is a need to update the code examples and answers to the exercises, updates will appear at this GitHub repository.

Download the example code files

The code bundle for the book is hosted on GitHub at `https://github.com/PacktPublishing/Modern-Computer-Architecture-and-Organization-Second-Edition`. We also have other code bundles from our rich catalog of books and videos available at `https://github.com/PacktPublishing/`. Check them out!

Download the color images

We also provide a PDF file that has color images of the screenshots/diagrams used in this book. You can download it here: `https://static.packt-cdn.com/downloads/9781803234519_ColorImages.pdf`.

Conventions used

There are a number of text conventions used throughout this book.

`CodeInText`: Indicates code words in the text, database table names, folder names, filenames, file extensions, pathnames, dummy URLs, user input, and Twitter handles. Here is an example: "Subtraction using the `SBC` instruction tends to be a bit more confusing to novice 6502 assembly language programmers."

A block of code is set as follows:

```
; Add four bytes together using immediate addressing mode
LDA #$04
CLC
ADC #$03
ADC #$02
ADC #$01
```

Any command-line input or output is written as follows:

```
C:\>bcdedit
Windows Boot Manager
--------------------
identifier  {bootmgr}
```

Bold: Indicates a new term, an important word, or words that you see on the screen, for example, in menus or dialog boxes, also appear in the text like this. For example: "Because there are now four sets, the **Set** field in the physical address reduces to two bits and the **Tag** field increases to 24 bits."

Warnings or important notes appear like this.

Tips and tricks appear like this.

Get in touch

Feedback from our readers is always welcome.

General feedback: Email feedback@packtpub.com, and mention the book's title in the subject of your message. If you have questions about any aspect of this book, please email us at questions@packtpub.com.

Errata: Although we have taken every care to ensure the accuracy of our content, mistakes do happen. If you have found a mistake in this book we would be grateful if you would report this to us. Please visit, http://www.packtpub.com/submit-errata, selecting your book, clicking on the Errata Submission Form link, and entering the details.

Piracy: If you come across any illegal copies of our works in any form on the Internet, we would be grateful if you would provide us with the location address or website name. Please contact us at copyright@packtpub.com with a link to the material.

If you are interested in becoming an author: If there is a topic that you have expertise in and you are interested in either writing or contributing to a book, please visit http://authors.packtpub.com.

Share your thoughts

Once you've read *Modern Computer Architecture and Organization, Second Edition*, we'd love to hear your thoughts! Scan the QR code below to go straight to the Amazon review page for this book and share your feedback.

https://packt.link/r/1803234512

Your review is important to us and the tech community and will help us make sure we're delivering excellent quality content.

1

Introducing Computer Architecture

The architectures of automated computing systems have evolved from the first mechanical calculators constructed nearly two centuries ago to the broad array of modern electronic computer technologies we use directly and indirectly every day. Along the way, there have been stretches of incremental technological improvement interspersed with disruptive advances that drastically altered the trajectory of the industry. We can expect these trends to continue in the coming years.

In the 1980s, during the early days of personal computing, students and technical professionals eager to learn about computer technology had a limited range of subject matter available for this purpose. If they had a computer of their own, it was probably an IBM PC or an Apple II. If they worked for an organization with a computing facility, they might have used an IBM mainframe or a Digital Equipment Corporation VAX minicomputer. These examples, and a limited number of similar systems, encompassed most people's exposure to the computer systems of the time.

Today, numerous specialized computing architectures exist to address widely varying user needs. We carry miniature computers in our pockets and purses that can place phone calls, record video, and function as full participants on the internet. Personal computers remain popular in a format outwardly similar to the PCs of past decades. Today's PCs, however, are orders of magnitude more capable than the early generations in terms of computing power, memory size, disk space, graphics performance, and communication ability. These capabilities enable modern PCs to easily perform tasks that would have been inconceivable on early PCs, such as the real-time generation of high-resolution 3D images.

Companies offering web services to hundreds of millions of users construct vast warehouses filled with thousands of tightly coordinated computer systems capable of responding to a constant stream of user requests with extraordinary speed and precision. Machine learning systems are trained through the analysis of enormous quantities of data to perform complex activities such as driving automobiles.

This chapter begins with a presentation of some key historical computing devices and the leaps in technology associated with them. We will then examine some significant modern-day trends related to technological advances and introduce the basic concepts of computer architecture, including a close look at the 6502 microprocessor and its instruction set. The following topics will be covered in this chapter:

- The evolution of automated computing devices
- Moore's law
- Computer architecture

Technical requirements

Files for this chapter, including answers to the exercises, are available at `https://github.com/PacktPublishing/Modern-Computer-Architecture-and-Organization-Second-Edition`.

The evolution of automated computing devices

This section reviews some classic machines from the history of automated computing devices and focuses on the major advances each embodied. Babbage's Analytical Engine is included here because of the many leaps of genius represented in its design. The other systems are discussed because they embodied significant technological advances and performed substantial real-world work over their lifetimes.

Charles Babbage's Analytical Engine

Although a working model of the **Analytical Engine** was never constructed, the detailed notes **Charles Babbage** developed from 1834 until his death in 1871 described a computing architecture that appeared to be both workable and complete. The Analytical Engine was intended to serve as a general-purpose programmable computing device. The design was entirely mechanical and was to be constructed largely of brass. The Analytical Engine was designed to be driven by a shaft powered by a steam engine.

Borrowing from the punched cards of the Jacquard loom, the rotating studded barrels used in music boxes, and the technology of his earlier Difference Engine (also never completed in his lifetime, and more of a specialized calculating device than a computer), the Analytical Engine's design was, otherwise, Babbage's original creation.

Unlike most modern computers, the Analytical Engine represented numbers in signed decimal form. The decision to use base-10 numbers rather than the base-2 logic of most modern computers was the result of a fundamental difference between mechanical technology and digital electronics. It is straightforward to construct mechanical wheels with 10 positions, so Babbage chose the human-compatible base-10 format because it was not significantly more technically challenging than using some other number base. Simple digital circuits, on the other hand, are not capable of maintaining 10 different states with the ease of a mechanical wheel.

All numbers in the Analytical Engine consisted of 40 decimal digits. The large number of digits was likely chosen to reduce problems with numerical overflow. The Analytical Engine did not support floating-point mathematics.

Each number was stored on a vertical axis containing 40 wheels, with each wheel capable of resting in 10 positions corresponding to the digits 0-9. A 41st number wheel contained the sign: any even number on this wheel represented a positive sign, and any odd number represented a negative sign. The Analytical Engine axis was somewhat analogous to the register used in modern processors, except the readout of an axis was destructive—reading an axis would set it to 0. If it was necessary to retain an axis's value after it had been read, another axis had to store a copy of the value during the readout. Numbers were transferred from one axis to another, or used in computations, by engaging a gear with each digit wheel and rotating the wheel to extract the numerical value. The set of axes serving as system memory was referred to as the *store*.

The addition of two numbers used a process somewhat similar to the method of addition taught to schoolchildren. Assume a number stored on one axis, let's call it the addend, was to be added to a number on another axis that we will call the accumulator. The machine would connect each addend digit wheel to the corresponding accumulator digit wheel through a train of gears. It would then simultaneously rotate each addend digit downward to 0 while driving the accumulator digit an equivalent rotation in the increasing direction. If an accumulator digit wrapped around from 9 to 0, the next most significant accumulator digit would increment by 1. This *carry* operation would propagate across as many digits as needed (think of adding 1 to 999,999). By the end of the process, the addend axis would hold the value 0 and the accumulator axis would hold the sum of the two numbers. The propagation of carries from one digit to the next was the most mechanically complex part of the addition process.

Operations in the Analytical Engine were sequenced by music box-like rotating barrels in a construct called the **mill**, which is analogous to the control unit of a modern CPU.

Each Analytical Engine instruction was encoded in a vertical row of locations on the barrel, where the presence or absence of a stud at a particular location either engaged a section of the Engine's machinery or left the state of that section unchanged. Based on Babbage's estimate of the Engine's execution speed, the addition of two 40-digit numbers, including the propagation of carries, would take about 3 seconds.

Babbage conceived several important concepts for the Engine that remain relevant to modern computer systems. His design supported a degree of parallel processing consisting of simultaneous multiplication and addition operations that accelerated the computation of series of values intended to be output as numerical tables. Mathematical operations such as addition supported a form of pipelining, in which sequential operations on different data values overlapped in time.

Babbage was well aware of the difficulties associated with complex mechanical devices, such as friction, gear backlash, and wear over time. To prevent errors caused by these effects, the Engine incorporated mechanisms called **lockings** that were applied during data transfers across axes. The lockings forced the number wheels into valid positions and prevented the accumulation of small errors from allowing a wheel to drift to an incorrect value. The use of lockings is analogous to the amplification of potentially weak input signals to produce stronger outputs by the digital logic gates in modern processors.

The Analytical Engine was to be programmed using punched cards and supported branching operations and nested loops. The most complex program intended for execution on the Analytical Engine was developed by Ada Lovelace to compute the Bernoulli numbers, an important sequence in number theory. The Analytical Engine code to perform this computation is recognized as the first published computer program of substantial complexity.

Babbage constructed a trial model of a portion of the Analytical Engine mill, which is currently on display at the Science Museum in London.

ENIAC

ENIAC, the **Electronic Numerical Integrator and Computer**, was completed in 1945 and was the first programmable general-purpose electronic computer. The system consumed 150 kilowatts of electricity, occupied 1,800 square feet of floor space, and weighed 27 tons.

The design was based on vacuum tubes, diodes, and relays. ENIAC contained over 17,000 vacuum tubes that functioned as switching elements.

Similar to the Analytical Engine, it used base-10 representation of 10-digit decimal numbers implemented using 10-position ring counters (the ring counter will be discussed in *Chapter 2, Digital Logic*).

Input data was received from an IBM punch-card reader and the output of computations was delivered by a card punch machine.

The ENIAC architecture was capable of complex sequences of processing steps including loops, branches, and subroutines. The system had 20 10-digit accumulators that functioned like registers in modern computers. It did not initially have any memory beyond the accumulators. If intermediate values were required for use in later computations, the data had to be written to punch cards and read back in when needed. ENIAC could perform about 385 multiplications per second.

ENIAC programs consisted of plugboard wiring and switch-based function tables. Programming the system was an arduous process that often took the team of talented female programmers weeks to complete. Reliability was a problem, as vacuum tubes failed regularly, requiring troubleshooting on a day-to-day basis to isolate and replace failed tubes.

In 1948, ENIAC was improved by adding the ability to program the system via punch cards rather than plugboards. This greatly enhanced the speed with which programs could be developed. As a consultant for this upgrade, John von Neumann proposed a processing architecture based on a single memory region holding program instructions and data, a processing component with an arithmetic logic unit and registers, and a control unit that contained an instruction register and a program counter. Many modern processors continue to implement this general structure, now known as the **von Neumann architecture**. We will discuss this architecture in detail in *Chapter 3, Processor Elements*.

Early applications of ENIAC included analyses related to the development of the hydrogen bomb and the computation of firing tables for long-range artillery.

IBM PC

In the years following the construction of ENIAC, several technological breakthroughs resulted in remarkable advances in computer architectures:

- The invention of the transistor in 1947 by John Bardeen, Walter Brattain, and William Shockley delivered a vast improvement over the vacuum tube technology prevalent at the time. Transistors were faster, smaller, consumed less power, and, once production processes had been sufficiently optimized, were much more reliable than the failure-prone tubes.

- The commercialization of integrated circuits in 1958, led by Jack Kilby of Texas Instruments, began the process of combining large numbers of formerly discrete components onto a single chip of silicon.
- In 1971, Intel began production of the first commercially available microprocessor, the Intel 4004. The 4004 was intended for use in electronic calculators and was specialized to operate on 4-bit binary-coded decimal digits.

From the humble beginnings of the Intel 4004, microprocessor technology advanced rapidly over the ensuing decade by packing increasing numbers of circuit elements onto each chip and expanding the capabilities of the microprocessors implemented on those chips.

The Intel 8088 microprocessor

IBM released the IBM PC in 1981. The original PC contained an Intel 8088 microprocessor running at a clock frequency of 4.77 MHz and featured 16 KB of **Random Access Memory (RAM)**, expandable to 256 KB. It included one or, optionally, two floppy disk drives. A color monitor was also available. Later versions of the PC supported more memory, but because portions of the address space had been reserved for video memory and **Read-Only Memory (ROM)**, the architecture could support a maximum of 640 KB of RAM.

The 8088 contained 14 16-bit registers. Four were general-purpose registers (AX, BX, CX, and DX). Four were memory segment registers (CS, DS, SS, and ES) that extended the address space to 20 bits. Segment addressing functioned by adding a 16-bit segment register value, shifted left by 4 bit positions, to a 16-bit offset contained in an instruction to produce a physical memory address within a 1 MB range.

The remaining 8088 registers were the **Stack Pointer (SP)**, the **Base Pointer (BP)**, the **Source Index (SI)**, the **Destination Index (DI)**, the **Instruction Pointer (IP)**, and the **Status Flags (FLAGS)**. Modern x86 processors employ an architecture remarkably similar to this register set (*Chapter 10, Modern Processor Architectures and Instruction Sets*, will cover the details of the x86 architecture). The most obvious differences between the 8088 and x86 are the extension of the register widths to 32 bits in x86 and the addition of a pair of segment registers (FS and GS) that are used today primarily as data pointers in multithreaded operating systems.

The 8088 had an external data bus width of 8 bits, which meant it took two bus cycles to read or write a 16-bit value. This was a performance downgrade compared to the earlier 8086 processor, which employed a 16-bit external bus. However, the use of the 8-bit bus made the PC more economical to produce and provided compatibility with lower-cost 8-bit peripheral devices. This cost-sensitive design approach helped reduce the purchase price of the PC to a level accessible to more potential customers.

Program memory and data memory shared the same address space and the 8088 accessed memory over a single bus. In other words, the 8088 implemented the von Neumann architecture. The 8088 instruction set included instructions for data movement, arithmetic, logical operations, string manipulation, control transfer (conditional and unconditional jumps, and subroutine call and return), input/output, and additional miscellaneous functions. The processor required about 15 clock cycles per instruction on average, resulting in an execution speed of 0.3 **million instructions per second (MIPS)**.

The 8088 supported nine distinct modes for addressing memory. This variety of modes was needed to efficiently access a single item at a time as well as for iterating over sequences of data.

The segment registers in the 8088 architecture provided a seemingly clever way to expand the range of addressable memory without increasing the length of most instructions referencing memory locations. Each segment register allowed access to a 64-kilobyte block of memory beginning at a physical memory address defined at a multiple of 16 bytes. In other words, the 16-bit segment register represented a 20-bit base address with the lower four bits set to zero. Instructions could then reference any location within the 64-kilobyte segment using a 16-bit offset from the address defined by the segment register.

The CS register selected the code segment location in memory and was used in fetching instructions and performing jumps and subroutine calls and returns. The DS register defined the data segment location for use by instructions involving the transfer of data to and from memory. The SS register set the stack segment location, which was used for local memory allocation within subroutines and for storing subroutine return addresses.

Programs that required less than 64 kilobytes in each of the code, data, and stack segments could ignore the segment registers entirely because those registers could be set once at program startup (programming language compilers would do this automatically) and remain unchanged through execution. Easy!

Things got quite a bit more complicated when a program's data size increased beyond 64 kilobyte. Though the use of segment registers resulted in a clean hardware design, using those registers caused many headaches for software developers. Compilers for the 8088 architecture distinguished between *near* and *far* references to memory. A near pointer represented a 16-bit offset from the current segment register base address. A far pointer contained 32 bits of addressing information: a 16-bit segment register value and a 16-bit offset. Far pointers consumed an additional 16 bits of data memory and they also required additional processing time.

Making single memory access using a far pointer involved the following steps:

1. Save the current segment register contents to a temporary location
2. Load the new segment value into the register
3. Access the data (reading or writing as needed) using an offset from the segment base
4. Restore the original segment register value

When using far pointers, it was possible to declare data objects (for example, an array of characters representing a document in a text editor) up to 64 KB in size. If you needed a larger structure, you had to work out how to break it into chunks no larger than 64 KB and manage them yourself. As a result of such segment register manipulations, programs that required extensive access to data items larger than 64 KB became quite complex and were susceptible to code size bloat and slower execution.

The IBM PC motherboard contained a socket for an optional Intel 8087 floating-point coprocessor. The designers of the 8087 invented data formats and processing rules for 32-bit and 64-bit floating-point numbers that became enshrined in 1985 as the IEEE 754 floating-point standard, which remains in near-universal use today. The 8087 could perform about 50,000 floating-point operations per second. We will look at floating-point processing in detail in *Chapter 9, Specialized Processor Extensions*.

The Intel 80286 and 80386 microprocessors

The second generation of the IBM PC, the PC AT, was released in 1984. **AT** stood for **Advanced Technology**, which referred to several significant enhancements over the original PC that mostly resulted from the use of the Intel 80286 processor.

Like the 8088, the 80286 was a 16-bit processor, and it maintained backward compatibility with the 8088: 8088 code could run unmodified on the 80286. The 80286 had a 16-bit data bus and 24 address lines supporting a 16-megabyte address space. The external data bus width was 16 bits, improving data access performance over the 8-bit bus of the 8088. The instruction execution rate (instructions per clock cycle) was about double the 8088 in many applications. This meant that at the same clock speed, the 80286 would be twice as fast as the 8088. The original PC AT clocked the processor at 6 MHz and a later version operated at 8 MHz. The 6 MHz variant of the 80286 achieved an instruction execution rate of about 0.9 MIPS, roughly three times that of the 8088.

The 80286 implemented a protected virtual address mode intended to support multiuser operating systems and multitasking.

In protected mode, the processor enforced memory protection to ensure one user's programs could not interfere with the operating system or with other users' programs. This groundbreaking technological advance in personal computing remained little used for many years, mainly because of the prohibitive cost of adding sufficient memory to a computer system to make it useful in a multiuser, multitasking context.

Following the 80286, the next generation of the x86 processor line was the 80386, introduced in 1985. The 80386 was a 32-bit processor with support for a flat 32-bit memory model in protected mode. The flat memory model allowed programmers to address up to 4 GB directly, without the need to manipulate segment registers. Compaq introduced an IBM PC-compatible personal computer based on the 80386 called the DeskPro in 1986. The DeskPro shipped with a version of Microsoft Windows targeted to the 80386 architecture.

The 80386 maintained substantial backward compatibility with the 80286 and 8088 processors. The processor architecture implemented in the 80386 remains the current standard x86 architecture. We will examine this architecture in detail in *Chapter 10, Modern Processor Architectures and Instruction Sets*.

The initial version of the 80386 was clocked at 33 MHz and achieved about 11.4 MIPS. Modern implementations of the x86 architecture run several hundred times faster than the original as the result of higher clock speeds, performance enhancements, including the extensive use of multilevel cache memory, and more efficient instruction execution at the hardware level. We will examine the benefits of cache memory in *Chapter 8, Performance-Enhancing Techniques*.

The iPhone

In 2007, Steve Jobs introduced the iPhone to a world that had no idea it had any use for such a device. The iPhone built upon previous revolutionary advances from Apple Computer, including the Macintosh computer, released in 1984, and the iPod music player of 2001. The iPhone combined the functions of the iPod, a mobile telephone, and an internet-connected computer.

The iPhone did away with the hardware keyboard that was common on smartphones of the time and replaced it with a touchscreen capable of displaying an onscreen keyboard, or any other type of user interface. In addition to touches for selecting keyboard characters and pressing buttons, the screen supported multi-finger gestures for actions such as zooming a photo.

The iPhone ran the OS X operating system, the same OS used on the flagship Macintosh computers of the time.

This decision immediately enabled the iPhone to support a vast range of applications already developed for Macs and empowered software developers to rapidly introduce new applications tailored to the iPhone, after Apple began allowing third-party application development.

The iPhone 1 had a 3.5" screen with a resolution of 320x480 pixels. It was 0.46 inches thick (thinner than other smartphones), had a built-in 2-megapixel camera, and weighed 4.8 oz. A proximity sensor detected when the phone was held to the user's ear and turned off screen illumination and touchscreen sensing during calls. It had an ambient light sensor to automatically set the screen brightness and an accelerometer that detected whether the screen was being held in portrait or landscape orientation.

The iPhone 1 included 128 MB of RAM and 4 GB, 8 GB, or 16 GB of flash memory, and supported **Global System for Mobile communications (GSM)** cellular communication, Wi-Fi (802.11b/g), and Bluetooth.

In contrast to the abundance of openly available information about the IBM PC, Apple was notoriously reticent about releasing the architectural details of the iPhone's construction. Apple released no information about the processor or other internal components of the first iPhone, simply referring to it as a **closed system**.

Despite the lack of official information from Apple, other parties enthusiastically tore down the various iPhone models and attempted to identify the phone's components and how they interconnected. Software sleuths have devised various tests that attempt to determine the specific processor model and other digital devices implemented within the iPhone. These reverse engineering efforts are subject to error, so descriptions of the iPhone architecture in this section should be taken with a grain of salt.

The iPhone 1 processor was a 32-bit ARM11 manufactured by Samsung running at 412 MHz. The ARM11 was an improved variant of previous-generation ARM processors and included an 8-stage instruction pipeline and support for **Single Instruction-Multiple Data (SIMD)** processing to improve audio and video performance. The ARM processor architecture will be discussed further in *Chapter 10, Modern Processor Architectures and Instruction Sets*.

The iPhone 1 was powered by a 3.7 V lithium-ion polymer battery. The battery was not intended to be replaceable, and Apple estimated it would lose about 20 percent of its original capacity after 400 charge and discharge cycles. Apple quoted up to 250 hours of standby time and 8 hours of talk time on a single charge.

Six months after the iPhone was introduced, *Time* magazine named the iPhone the "Invention of the Year" for 2007. In 2017, *Time* ranked the *50 Most Influential Gadgets of All Time*. The iPhone topped the list.

In the next section, we will examine the interplay of technological advances in computing over time and the underlying physical limits of silicon-based integrated circuits.

Moore's law

For those working in the rapidly advancing field of computer technology, it is a significant challenge to make plans for the future. This is true whether the goal is to plot your own career path or for a giant semiconductor corporation to identify optimal R&D investments. No one can ever be completely sure what the next leap in technology will be, what effects from it will ripple across the industry and its users, or when it will happen. One approach that has proven useful in this difficult environment is to develop a rule of thumb, or empirical law, based on experience.

Gordon Moore co-founded Fairchild Semiconductor in 1957 and was later the chairman and CEO of Intel. In 1965, Moore published an article in *Electronics* magazine in which he offered his prediction of the changes that would occur in the semiconductor industry over the next 10 years. In the article, he observed that the number of formerly discrete components, such as transistors, diodes, and capacitors, that could be integrated onto a single chip had been doubling approximately yearly and the trend was likely to continue over the next 10 years. This doubling formula came to be known as **Moore's law**. This was not a scientific law in the sense of the law of gravity. Rather, it was based on an observation of historical trends, and he believed this formulation had some ability to predict the future.

Moore's law turned out to be impressively accurate over those 10 years. In 1975, he revised the predicted growth rate for the following 10 years to double the number of components per integrated circuit every 2 years, rather than yearly. This pace continued for decades, up until about 2010. In more recent years, the growth rate has appeared to decline slightly. In 2015, Brian Krzanich, Intel CEO, stated that the company's growth rate had slowed to doubling about every two and a half years.

Even though the time to double integrated circuit density is increasing, the current pace represents a phenomenal rate of growth that can be expected to continue into the future, just not quite as rapidly as it once progressed.

Moore's law has proven to be a reliable tool for evaluating the performance of semiconductor companies over the decades.

Companies have used it to set goals for the performance of their products and to plan their investments. By comparing the integrated circuit density increases for a company's products against prior performance, and against other companies, semiconductor executives and industry analysts can evaluate and score company performance. The results of these analyses have fed directly into decisions to invest in enormous new fabrication plants and to push the boundaries of ever-smaller integrated circuit feature sizes.

The decades since the introduction of the IBM PC have seen tremendous growth in the capabilities of single-chip microprocessors. Current processor generations are hundreds of times faster, operate natively on 32-bit and 64-bit data, have far more integrated memory resources, and unleash vastly more functionality, all packed into a single integrated circuit.

The increasing density of semiconductor features, as predicted by Moore's law, has enabled these improvements. Smaller transistors run at higher clock speeds due to the shorter connection paths between circuit elements. Smaller transistors also, obviously, allow more functionality to be packed into a given amount of die area. Being smaller and closer to neighboring components allows the transistors to consume less power and generate less heat.

There was nothing magical about Moore's law. It was an observation of the trends in progress at the time. One trend was the steadily increasing size of semiconductor dies. This was the result of improving production processes that reduced the density of defects, which allowed acceptable production yield with larger integrated circuit dies. Another trend was the ongoing reduction in the size of the smallest components that could be reliably produced in a circuit. The final trend was what Moore referred to as the "cleverness" of circuit designers in making increasingly efficient and effective use of the growing number of circuit elements placed on a chip.

Traditional semiconductor manufacturing processes have begun to approach physical limits that will eventually put the brakes on growth under Moore's law. The smallest features on current commercially available integrated circuits are around 5 **nanometers (nm)**. For comparison, a typical human hair is about 50,000 nm thick, and a water molecule (one of the smallest molecules) is 0.28 nm across. There is a point beyond which it is simply not possible for circuit elements to become smaller as the sizes approach atomic scale.

In addition to the challenge of building reliable circuit components from a small number of molecules, other physical effects with names such as *Abbe diffraction limit* become significant impediments to single-digit nanometer-scale circuit production.

We won't get into the details of these phenomena; it's sufficient to know the steady increase in integrated circuit component density that has proceeded for decades under Moore's law is going to become a lot harder to continue over the coming years.

This does not mean we will be stuck with processors essentially the same as those that are now commercially available. Even as the rate of growth in transistor density slows, semiconductor manufacturers are pursuing several alternative methods to continue growing the power of computing devices. One approach is specialization, in which circuits are designed to perform a specific category of tasks extremely well rather than performing a wide variety of tasks merely adequately.

Graphics Processing Units (GPUs) are an excellent example of specialization. The original generation of GPUs focused exclusively on improving the speed at which three-dimensional graphics scenes could be rendered, mostly for use in video gaming. The calculations involved in generating a three-dimensional scene are well defined and must be applied to thousands of pixels to create a single frame. The process is repeated for each subsequent frame, and frames must be redrawn at a 60 Hz or higher rate to provide a satisfactory user experience. The computationally demanding and repetitive nature of this task is ideally suited for acceleration via hardware parallelism. Multiple computing units within a GPU simultaneously perform essentially the same calculations on different input data to produce separate outputs. Those outputs are combined to generate the entire scene. Modern GPU architectures have been enhanced to support other computing domains, such as training neural networks on massive amounts of data. GPU architectures will be covered in detail in *Chapter 6, Specialized Computing Domains*.

As Moore's law shows signs of fading over the coming years, what advances might take its place to kick off the next round of innovations in computer architectures? We don't know for sure today, but some tantalizing options are currently under intense study. Quantum computing is one example of these technologies. We will cover that technology in *Chapter 17, Quantum Computing and Other Future Directions in Computer Architectures*.

Quantum computing takes advantage of the properties of subatomic particles to perform computations in a manner that traditional computers cannot. A basic element of quantum computing is the **qubit**, or quantum bit. A qubit is similar to a regular binary bit, but in addition to representing the states 0 and 1, qubits can attain a state that is a superposition (or mixture) of the 0 and 1 states. When measured, the qubit output will always be 0 or 1, but the probability of producing either output is a function of the qubit's quantum state prior to being read. Specialized algorithms are required to take advantage of the unique features of quantum computing.

Another future possibility is that the next great technological breakthrough in computing devices will be something that we either haven't thought of or, if we have thought about it, we may have dismissed the idea out of hand as unrealistic. The iPhone, discussed in the preceding section, is an example of a category-defining product that revolutionized personal communication and enabled the use of the internet in new ways. The next major advance may be a new type of product, a surprising new technology, or some combination of product and technology. Right now, we don't know what it will be or when it will happen, but we can say with confidence that such changes are coming.

The next section introduces some fundamental digital computing concepts that must be understood before we delve into digital circuitry and the details of modern computer architecture in the coming chapters.

Computer architecture

The descriptions of a number of key architectures from the history of computing presented in the previous sections of this chapter included some terms that may or may not be familiar to you. This section will introduce the conceptual building blocks that are used to construct modern-day processors and related computer subsystems.

Representing numbers with voltage levels

One ubiquitous feature of modern computers is the use of voltage levels to indicate data values. In general, only two voltage levels are recognized: a low level and a high level. The low level is often assigned the value 0, and the high level is assigned the value 1.

The voltage at any point in a circuit (digital or otherwise) is analog in nature and can take on any voltage within its operating range. When changing from the low level to the high level, or vice versa, the voltage must pass through all voltages in between. In the context of digital circuitry, the transitions between low and high levels happen quickly and the circuitry is designed to not react to voltages between the high and low levels.

Binary and hexadecimal numbers

The circuitry within a processor does not work directly with numbers, in any sense. Processor circuit elements obey the laws of electricity and electronics and simply react to the inputs provided to them. The inputs that drive these actions result from the code developed by programmers and from the data provided as input to the program. The interpretation of the output of a program as, say, numbers in a spreadsheet, or characters in a word processing program, is a purely human interpretation that assigns meaning to the result of the electronic interactions within the processor. The decision to assign 0 to the low voltage and 1 to the high voltage is the first step in the interpretation process.

The smallest unit of information in a digital computer is a binary digit, called a **bit**, which represents a discrete data element containing the value 0 or 1. Multiple bits can be placed together to enable the representation of a greater range of values. A **byte** is composed of 8 bits placed together to form a single value. A byte is the smallest unit of information that can be read from or written to memory by most modern processors. Some computers, past and present, use a different number of bits for the smallest addressable data item, but the 8-bit byte is the most common size.

A single bit can take on two values: 0 and 1. Two bits placed together can take on four values: 00, 01, 10, and 11. Three bits can take on eight values: 000, 001, 010, 011, 100, 101, 110, and 111. In general, a group of n bits can take on 2^n values. An 8-bit byte, therefore, can represent 2^8, or 256, unique values.

The binary number format is not most people's first choice when it comes to performing arithmetic. Working with numbers such as 11101010 can be confusing and error-prone, especially when dealing with 32- and 64-bit values. To make working with these numbers somewhat easier, **hexadecimal** numbers are often used instead. The term *hexadecimal* is often shortened to *hex*.

In the hexadecimal number system, binary numbers are separated into groups of 4 bits. With 4 bits in the group, the number of possible values is 2^4, or 16. The first 10 of these 16 numbers are assigned the digits 0-9, and the last 6 are assigned the letters A-F. *Table 1.1* shows the first 16 binary values starting at 0, along with the corresponding hexadecimal digit and the decimal equivalent to the binary and hex values:

Binary	Hexadecimal	Decimal
0000	0	0
0001	1	1
0010	2	2
0011	3	3
0100	4	4
0101	5	5
0110	6	6
0111	7	7
1000	8	8
1001	9	9

1010	A	10
1011	B	11
1100	C	12
1101	D	13
1110	E	14
1111	F	15

Table 1.1: Binary, hexadecimal, and decimal numbers

The binary number 11101010 can be represented more compactly by breaking it into two 4-bit groups (1110 and 1010) and writing them as the hex digits EA. A 4-bit grouping is sometimes referred to as a *nibble*, meaning it is half a byte. Because binary digits can take on only two values, binary is a base-2 number system. Hex digits can take on 16 values, so hexadecimal is base-16. Decimal digits can have 10 values, and therefore decimal is base-10.

When working with these different number bases, it is easy for things to become confusing. Is a number written as 100 a binary, hexadecimal, or decimal value? Without additional information, you can't tell. Various programming languages and textbooks have taken different approaches to remove this ambiguity. In most cases, decimal numbers are unadorned, so the number 100 is usually decimal. In programming languages such as C and C++, hexadecimal numbers are prefixed by *0x*, so the number 0x100 is 100 hex. In assembly languages, either the prefix character *$* or the suffix *h* might be used to indicate hexadecimal numbers. The use of binary values in programming is less common, mostly because hexadecimal is preferred due to its compactness. Some compilers support the use of *0b* as a prefix for binary numbers.

HEXADECIMAL NUMBER REPRESENTATION

This book uses either the prefix $ or the suffix h to represent hexadecimal numbers, depending on the context. The suffix b will represent binary numbers, and the absence of a prefix or suffix indicates decimal numbers.

Bits are numbered individually within a binary number, with bit 0 as the rightmost, least significant bit. Bit numbers increase in magnitude leftward, up to the most significant bit at the far left.

Some examples should make this clear. In *Table 1.1*, the binary value 0001b (1 decimal) has bit number 0 set to 1 and the remaining three bits are cleared to 0. For 0010b (2 decimal), bit 1 is set and the other bits are cleared. For 0100b (4 decimal), bit 2 is set and the other bits are cleared.

SET VERSUS CLEARED

A bit that is set has the value 1. A bit that is cleared has the value 0.

An 8-bit byte can take on values from $00h to $FF, equivalent to the decimal range 0-255. When performing addition at the byte level, the result can exceed 8 bits. For example, adding $01 to $FF results in the value $100. When using 8-bit registers, this represents a carry into the 9th bit, which must be handled appropriately by the processor hardware and by the software performing the addition.

In unsigned arithmetic, subtracting $01 from $00 results in a value of $FF. This constitutes a wraparound to $FF. Depending on the computation being performed, this may or may not be the desired result. Once again, the processor hardware and the software must handle this situation to arrive at the desired result.

When appropriate, negative values can be represented using binary numbers. The most common signed number format in modern processors is **two's complement**. In two's complement, 8-bit signed numbers span the range from -128 to 127. The most significant bit of a two's complement data value is the sign bit: a 0 in this bit represents a positive number and a 1 represents a negative number. A two's complement number can be negated (multiplied by -1) by inverting all the bits, adding 1, and ignoring the carry. Inverting a bit means changing a 0 bit to 1 and a 1 bit to 0. See *Table 1.2* for some step-by-step examples negating signed 8-bit numbers:

Decimal value	Binary value	Invert the bits	Add one	Negated result
0	00000000b	11111111b	00000000b	0
1	00000001b	11111110b	11111111b	-1
-1	11111111b	00000000b	00000001b	1
127	01111111b	10000000b	10000001b	-127
-127	10000001b	01111110b	01111111b	127

Table 1.2: Negation operation examples

Negating 0 returns a result of 0, as you would expect mathematically.

> **TWO'S COMPLEMENT ARITHMETIC**
>
> Two's complement arithmetic is identical to unsigned arithmetic at the bit level. The manipulations involved in addition and subtraction are the same whether the input values are intended to be signed or unsigned. The interpretation of the result as signed or unsigned depends entirely on the intent of the user.

Table 1.3 shows how the binary values 00000000b to 11111111b correspond to signed values over the range -128 to 127, and unsigned values from 0 to 255:

Binary	Signed Decimal	Unsigned Decimal
00000000b	0	0
00000001b	1	1
00000010b	2	2
⋮	⋮	⋮
01111110b	126	126
01111111b	127	127
10000000b	-128	128
10000001b	-127	129
10000010b	-126	130
⋮	⋮	⋮
11111101b	-3	253
11111110b	-2	254
11111111b	-1	255

Table 1.3: Signed and unsigned 8-bit numbers

Signed and unsigned representations of binary numbers extend to larger integer data types. 16-bit values can represent unsigned integers from 0 to 65,535, and signed integers in the range -32,768 to 32,767. 32-bit, 64-bit, and even larger integer data types are commonly available in modern processors and programming languages.

The 6502 microprocessor

This section introduces a processor architecture that is relatively simple compared to more powerful modern processors.

The intent here is to provide a whirlwind introduction to some basic concepts shared by processors spanning the spectrum from very low-end microcontrollers to sophisticated multi-core 64-bit processors.

The 6502 processor was introduced by MOS Technology in 1975. The 6502 found widespread use in video game consoles from Atari and Nintendo and in computers marketed by Commodore and Apple. Versions of the 6502 continue to be in widespread use today in embedded systems, with estimates of between 5 and 10 billion (yes, *billion*) units produced as of 2018. In popular culture, both Bender, the robot in *Futurama*, and the T-800 robot in *The Terminator* appear to have employed the 6502, based on onscreen evidence.

Like many early microprocessors, the 6502 was powered by 5 **volts** (**V**) **direct current** (**DC**). In these circuits, a low signal level is any voltage between 0 and 0.8 V. A high signal level is any voltage between 2 and 5 V. Voltages between these ranges occur only during transitions from low to high and from high to low. The low signal level is defined as logical 0, and the high signal level is defined as logical 1. *Chapter 2, Digital Logic*, will delve further into the electronic circuits used in digital electronics.

The **word length** of a processor defines the size of the fundamental data element the processor operates upon. The 6502 has a word length of 8 bits. This means the 6502 reads and writes memory 8 bits at a time and stores data internally in 8-bit wide registers.

Program memory and data memory share the same address space and the 6502 accesses its memory over a single bus. Like the Intel 8088, the 6502 implements the von Neumann architecture. The 6502 has a 16-bit address bus, enabling the addressing of 64 kilobytes of memory.

1 KB is defined as 2^{10}, or 1,024 bytes. The number of unique binary combinations of the 16 address lines is 2^{16}, which permits access to 65,536 byte-wide memory locations. Note that just because a device can address 64 KB, it does not mean there must be memory at each of those locations. The Commodore VIC-20, based on the 6502, contained just 5 KB of RAM and 20 KB of ROM.

The 6502 contains internal storage areas called registers. A **register** is a location in a logical device in which a word of information can be stored and acted upon during computation. A typical processor contains a small number of registers for temporarily storing data values and performing operations such as addition or address computations.

Figure 1.1 shows the 6502 register structure. The processor contains five 8-bit registers (A, X, Y, P, and S) and one 16-bit register (PC). The numbers above each register indicate the bit numbers at each end of the register:

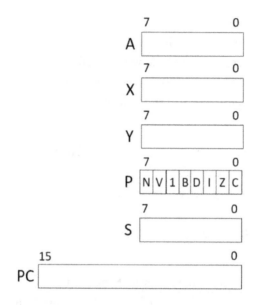

Figure 1.1: 6502 register set

Each of the A, X, and Y registers can serve as a general-purpose storage location. Program instructions can load a value into one of those registers and, some instructions later, use the saved value for some purpose if the intervening instructions did not modify the register contents. The A register is the only register capable of performing arithmetic operations. The X and Y registers, but not the A register, can be used as index registers in calculating memory addresses.

The P register contains processor flags. Each bit in this register has a unique purpose, except for the bit labeled 1. The 1 bit is unused and can be ignored. Each of the remaining bits in this register is called a **flag** and indicates a specific condition that has occurred or represents a configuration setting. The 6502 flags are as follows:

- N: Negative sign flag: This flag is set when the result of an arithmetic operation sets bit 7 in the result. This flag is used in signed arithmetic.
- V: Overflow flag: This flag is set when a signed addition or subtraction results in overflow or underflow outside the range -128 to 127.

- **B: Break flag**: This flag indicates a Break (BRK) instruction has executed. This bit is not present in the P register itself. The B flag value is only relevant when examining the P register contents as stored on the stack by a BRK instruction or by an interrupt. The B flag is set to distinguish a software interrupt resulting from a BRK instruction from a hardware interrupt during interrupt processing.
- **D: Decimal mode flag**: If set, this flag indicates processor arithmetic will operate in **Binary-Coded Decimal (BCD)** mode. BCD mode is rarely used and won't be discussed here, other than to note that this base-10 computation mode evokes the architectures of the Analytical Engine and ENIAC.
- **I: Interrupt disable flag**: If set, this flag indicates that interrupt inputs (other than the non-maskable interrupt) will not be processed.
- **Z: Zero flag**: This flag is set when an operation produces a result of 0.
- **C: Carry flag**: This flag is set when an arithmetic operation produces a carry.

The N, V, Z, and C flags are the most important flags in the context of general computing involving loops, counting, and arithmetic.

The S register is the **stack pointer**. In the 6502, the stack is the region of memory from addresses $100 to $1FF. This 256-byte range is used for the temporary storage of parameters within subroutines and holds the return address when a subroutine is called. At system startup, the S register is initialized to point to the top of this range. Values are "pushed" onto the stack using instructions such as PHA, which pushes the contents of the A register onto the stack.

When a value is pushed onto the stack, the 6502 stores the value at the address indicated by the S register, after adding the fixed $100 offset, and then decrements the S register. Additional values can be placed on the stack by executing more push instructions. As additional values are pushed, the stack grows downward in memory. Programs must take care not to exceed the fixed 256-byte size of the stack when pushing data onto it.

Data stored on the stack must be retrieved in the reverse of the order from which it was pushed onto the stack. The stack is a **Last-In, First-Out (LIFO)** data structure, meaning when you "pop" a value from the stack, it is the byte most recently pushed onto it. The PLA instruction increments the S register by 1 and then copies the value at the address indicated by the S register (plus the $100 offset) into the A register.

The PC register is the program counter. This register contains the memory address of the next instruction to be executed. Unlike the other registers, the PC is 16 bits long, allowing access to the entire 6502 address space.

Each instruction consists of a 1-byte operation code, called **opcode** for short, and may be followed by 0 to 2 operand bytes, depending on the type of instruction. After each instruction executes, the PC updates to point to the next instruction following the one that just completed. In addition to automatic updates during sequential instruction execution, the PC can be modified by jump instructions, branch instructions, and subroutine call and return instructions.

The 6502 instruction set

We will now examine the 6502 instruction set. Instructions are individual processor commands that, when strung together sequentially, execute the algorithm coded by the programmer. An instruction contains a binary number called an operation code (or **opcode**) that tells the processor what to do when that instruction executes.

If they wish, programmers can write code directly using processor instructions. We will see examples of this later in this section. Programmers can also write code in a so-called high-level language. The programmer then uses a software tool called a **compiler** that translates the high-level code into a (usually much longer) sequence of processor instructions.

In this section, we are working with code written as sequences of processor instructions. This form of source code is called **assembly language**.

Each of the 6502 instructions has a three-character mnemonic. In assembly language source files, each line of code contains an instruction mnemonic followed by any operands associated with the instruction. The combination of the mnemonic and the operands defines the **addressing mode**. The 6502 supports several addressing modes providing a great deal of flexibility in accessing data in registers and memory. For this introduction, we'll only work with the **immediate** addressing mode, in which the operand itself contains a value rather than indicating a register or memory location containing the value. An immediate value is preceded by a # character.

In 6502 assembly, decimal numbers have no adornment (48 means 48 decimal), while hexadecimal values are preceded by a $ character ($30 means 30 hexadecimal, equivalent to 00110000b and to 48 decimal). An immediate decimal value looks like #48 and an immediate hexadecimal value looks like #$30.

Some assembly code examples will demonstrate the 6502 arithmetic capabilities. Five 6502 instructions are used in the following examples:

- LDA loads register A with a value
- ADC performs addition using Carry (the C flag in the P register) as an additional input and output

- SBC performs subtraction using the C flag as an additional input and output
- SEC sets the C flag directly
- CLC clears the C flag directly

Since the C flag is an input to the addition and subtraction instructions, it is important to ensure it has the correct value prior to executing the ADC or SBC instructions. Before initiating an addition operation, the C flag must be clear to indicate there is no carry from a prior addition. When performing multi-byte additions (say, with 16-bit, 32-bit, or 64-bit numbers), the carry, if any, will propagate from the sum of one byte pair to the next as you add the more significant bytes together. If the C flag is set when the ADC instruction executes, the effect is to add 1 to the result. After the ADC completes, the C flag serves as the ninth bit of the result: a C flag result of 0 means there was no carry, and a 1 indicates there was a carry from the 8-bit register.

Subtraction using the SBC instruction tends to be a bit more confusing to novice 6502 assembly language programmers. Schoolchildren learning subtraction use the technique of borrowing when subtracting a larger digit from a smaller digit. In the 6502, the C flag represents the opposite of Borrow. If C is 1, then Borrow is 0, and if C is 0, Borrow is 1. Performing a simple subtraction with no incoming Borrow requires setting the C flag before executing the SBC command.

The following examples employ the 6502 as a calculator using inputs defined as immediate values in the code and with the result stored in the A register. The **Results** columns show the final value of the A register and the states of the N, V, Z, and C flags:

Instruction Sequence	Description	Results				
		A	N	V	Z	C
CLC LDA #1 ADC #1	8-bit addition with no Carry input: Clear the Carry flag, then load an immediate value of 1 into the A register and add 1 to it.	$02	0	0	0	0
SEC LDA #1 ADC #1	8-bit addition with a Carry input: Set the Carry flag, then load an immediate value of 1 into the A register and add 1 to it.	$03	0	0	0	0
SEC LDA #1 SBC #1	8-bit subtraction with no Borrow input: Set the Carry flag, then load an immediate value of 1 into the A register then subtract 1 from it. C = 1 indicated no Borrow occurred.	$00	0	0	1	1

CLC LDA #1 SBC #1	8-bit subtraction with a Borrow input: Clear the Carry flag, then load an immediate value of 1 into the A register and subtract 1 from it. C = 0 indicates a Borrow occurred.	$FF	1	0	0	0
CLC LDA $FF ADC #1	Unsigned overflow: Add 1 to $FF. C = 1 indicates a Carry occurred.	$00	0	0	1	1
SEC LDA #0 SBC #1	Unsigned underflow: Subtract 1 from 0. C = 0 indicates a Borrow occurred.	$FF	1	0	0	0
CLC LDA #$7F ADC #1	Signed overflow: Add 1 to $7F. V = 1 indicates signed overflow occurred.	$80	1	1	0	0
SEC LDA #$80 SBC #1	Signed underflow: Subtract 1 from $80. V = 1 indicates signed underflow occurred.	$7F	0	1	0	1

Table 1.4: 6502 arithmetic instruction sequences

If you don't happen to have a 6502-based computer with an assembler and debugger handy, there are several free 6502 emulators available online that you can run in your web browser. One excellent emulator is available at `https://skilldrick.github.io/easy6502/`. Visit the website and scroll down until you find a default code listing with buttons for assembling and running 6502 code. Replace the default code listing with a group of three instructions from *Table 1.4* and then assemble the code.

To examine the effect of each instruction in the sequence, use the debugger controls to single-step through the instructions and observe the result of each instruction on the processor registers.

This section has provided a very brief introduction to the 6502 processor and a small subset of its capabilities. One point of this analysis was to illustrate the challenge of dealing with the issue of carries when performing addition and borrows when doing subtraction. From Charles Babbage to the designers of the 6502 to the developers of modern computer systems, computer architects have developed solutions to the problems of computation and implemented them using the best technology available to them.

Summary

This chapter began with a brief history of automated computing devices and described significant technological advances that drove leaps in computational capability. A discussion of Moore's law followed with an assessment of its applicability over past decades and its implications for the future. The basic concepts of computer architecture were introduced through a discussion of the 6502 microprocessor registers and instruction set. The history of computer architecture is fascinating, and I encourage you to explore it further.

The next chapter will introduce digital logic, beginning with the properties of basic electrical circuits and proceeding through the design of digital subsystems used in modern processors. You will learn about logic gates, flip-flops, and digital circuits including multiplexers, shift registers, and adders. The chapter includes an introduction to hardware description languages, which are specialized computer languages used in the design of complex digital devices such as computer processors.

Exercises

1. Using your favorite programming language, develop a simulation of a single-digit decimal adder that operates in the same manner as in Babbage's Analytical Engine. First, prompt the user for two digits in the range 0-9: the addend and the accumulator. Display the addend, the accumulator, and the carry, which is initially 0. Perform a series of cycles as follows:

    1. If the addend is 0, display the values of the addend, accumulator, and carry and terminate the program
    2. Decrement the addend by 1 and increment the accumulator by 1
    3. If the accumulator is incremented from 9 to 0, increment the carry
    4. Go back to *step 1*
    5. Test your code with these sums: 0+0, 0+1, 1+0, 1+2, 5+5, 9+1, and 9+9

2. Create arrays of 40 decimal digits each for the addend, accumulator, and carry. Prompt the user for two decimal integers of up to 40 digits each. Perform the addition digit by digit using the cycles described in *Exercise 1* and collect the carry output from each digit position in the carry array. After the cycles are complete, insert carries, and, where necessary, ripple them across digits to complete the addition operation. Display the results after each cycle and at the end. Test with the same sums as in *Exercise 1* and also test the sums 99+1, 999999+1, 49+50, and 50+50.

3. Modify the program of *Exercise 2* to implement the subtraction of 40-digit decimal values. Perform borrowing as required. Test with 0-0, 1-0, 1000000-1, and 0-1. What is the result for 0-1?

4. 6502 assembly language references data in memory locations using an operand value containing the address (without the # character, which indicates an immediate value). For example, the `LDA $00` instruction loads the byte at memory address $00 into A. `STA $01` stores the byte in A into address $01. Addresses can be any value in the range of 0 to $FFFF, assuming memory exists at the address and the address is not already in use for some other purpose. Using your preferred 6502 emulator, write 6502 assembly code to store a 16-bit value in addresses $00-$01, store a second value in addresses $02-$03, then add the two values and store the result in $04-$05. Be sure to propagate any carry between the two bytes. Ignore any carry from the 16-bit result. Test with $0000+$0001, $00FF+$0001, and $1234+$5678.

5. Write 6502 assembly code to subtract two 16-bit values in a manner similar to *Exercise 4*. Test with $0001-$0000, $0001-$0001, $0100-$00FF, and $0000-$0001. What is the result for $0000-$0001?

6. Write 6502 assembly code to store two 32-bit integers to addresses $00-03 and $04-$07, and then add them, storing the results in $08-$0B. Use a looping construct, including a label and a branch instruction, to iterate over the bytes of the two values to be added. Search the internet for the details of the 6502 decrement and branch instructions and the use of labels in assembly language. Hint: The 6502 zero-page indexed addressing mode works well in this application.

Join our community Discord space

Join the book's Discord workspace for a monthly *Ask me Anything* session with the author: `https://discord.gg/7h8aNRhRuY`

2
Digital Logic

This chapter builds upon the introductory topics presented in *Chapter 1, Introducing Computer Architecture*, and provides a firm understanding of the digital building blocks used in the design of modern processors and other sophisticated electronic circuits. We begin with a discussion of basic electrical circuit elements. Next, we introduce transistors and examine their use as switching components in simple logic gates. We then construct latches, flip-flops, and ring counters from logic gates. More complex processor components, including registers and adders, are developed by combining the devices introduced earlier. The concept of sequential logic, which means logic that contains state information that varies over time, is developed. The chapter ends with an introduction to hardware description languages, which represent the design method of choice for complex digital devices.

The following topics will be covered in this chapter:

- Electrical circuits
- The transistor
- Logic gates
- Latches
- Flip-flops
- Registers
- Adders
- Clocking
- Sequential logic
- Hardware description languages

Technical requirements

Files for this chapter, including answers to the exercises, are available at `https://github.com/PacktPublishing/Modern-Computer-Architecture-and-Organization-Second-Edition`.

Electrical circuits

We begin this chapter with a brief review of the properties of electrical circuits.

Conductive materials, such as copper, exhibit the ability to easily produce an electric current in the presence of an electric field. Nonconductive materials, for example, glass, rubber, and **polyvinyl chloride** (**PVC**), inhibit the flow of electricity so thoroughly that they are used as insulators to protect electrical conductors against short circuits. In metals, electrical current consists of electrons in motion. Materials that permit some electrical current to flow, while predictably restricting the amount allowed to flow, are used in the construction of resistors.

The relationship between electrical current, voltage, and resistance in a circuit is analogous to the relationship between flow rate, pressure, and flow restriction in a hydraulic system. Consider a kitchen water tap: pressure in the pipe leading to the tap forces water to flow when the valve is opened. If the valve is opened just a tiny bit, the flow from the faucet is a trickle. If the valve is opened further, the flow rate increases. Increasing the valve opening is equivalent to reducing the resistance to water flow through the faucet.

In an electrical circuit, voltage corresponds to the pressure in the water pipe. Electrical current, measured in **amperes** (often shortened to **amps**), corresponds to the rate of water flow through the pipe and faucet. Electrical resistance corresponds to the flow restriction resulting from a partially opened valve.

The quantities of voltage, current, and resistance are related by the formula $V = IR$, where V is the voltage (in volts), I is the current (in amperes), and R is the resistance (in ohms). In other words, the voltage across a resistive circuit element equals the product of the current through the element and its resistance. This is **Ohm's law**, named in honor of Georg Ohm, who first published the relationship in 1827.

Figure 2.1 shows a simple circuit representation of this relationship. The stacked horizontal lines to the left indicate a voltage source, such as a battery or a computer power supply. The zig-zag shape to the right represents a resistor. The lines connecting the components are wires, which are assumed to be perfect conductors. A perfect conductor allows electrical current to flow with no resistance.

The current, denoted by the letter *I*, flows around the circuit clockwise, out the positive side of the battery, through the resistor, and back into the negative side of the battery. The negative side of the battery is defined in this circuit as the voltage reference point, with a voltage of zero volts:

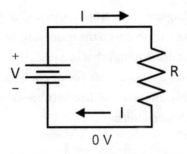

Figure 2.1: Simple resistive circuit

Using the water pipe analogy, the wire at zero volts represents a pool of water. A "pump" (the battery in the diagram) draws water from the pool and pushes it out of the "pump" at the top of the battery symbol into a pipe at a higher pressure. The water flows as current *I* to the faucet, represented by resistor *R* to the right. After passing through the flow-restricted faucet, the water ends up in the pool where it is available to be drawn into the pump again.

If we assume the battery voltage, or pressure rise across the water pump, is constant, then any increase in resistance *R* will reduce the current *I* by an inversely proportional amount. Doubling the resistance cuts the current in half, for example. Doubling the voltage, perhaps by placing two batteries in series, as is common in flashlights, will double the current through the resistor.

In the next section, we introduce the transistor, which serves as the basis for all modern digital electronic devices.

The transistor

A **transistor** is a semiconductor device that, for the purpose of this discussion, functions as a digital switch. A **semiconductor** is a material that exhibits properties between those of good conductors (like copper wire) and good insulators (like glass or plastic). In a suitable circuit configuration, the conductivity of a semiconductor device can be varied by a control input. A transistor used in this manner becomes a digital switching element.

The transistor switching operation is electrically equivalent to changing between very high and very low resistance based on the state of an input signal. One important feature of switching transistors is that the switching input does not need to be very strong.

This means that a very small current at the switching input can turn on and turn off a much larger current passing through the transistor. A single transistor's output current can drive many other transistor inputs. This characteristic is vital to the development of complex digital circuits.

Figure 2.2 shows the schematic diagram of the NPN transistor. NPN refers to the construction of the interconnected silicon regions that make up the transistor. An *N* region of silicon has material added to it (using a process called **doping**) that increases the number of available electrons present. A *P* region is doped to have a reduced number of available electrons. An NPN transistor contains two *N* sections, with a *P* section sandwiched between them. The three terminals of the device are connected to each of these regions:

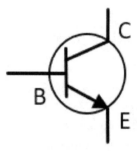

Figure 2.2: NPN transistor schematic symbol

The collector, labeled *C* in *Figure 2.2*, is connected to one of the *N* regions, and the emitter, *E*, is connected to the other *N* region. The base, *B*, connects to the *P* region between the two *N* regions. The collector "collects" current and the emitter "emits" current, as indicated by the arrow. The base terminal is the control input. By changing the voltage applied to the base terminal, and thus altering the amount of current flowing into the base, current entering via the collector and exiting via the emitter can be adjusted.

Logic gates

Figure 2.3 is a schematic diagram of a transistor NOT gate. This circuit is powered by a 5 V supply. The input signal might come from a pushbutton circuit that produces 0 V when the button is not pressed and 5 V when it is pressed. *R1* limits the current flowing from the input terminal to the transistor base terminal when the input is high (near 5 V). In a typical circuit, *R1* has a value of about 1,000 ohms. *R2* might have a value of 5,000 ohms. *R2* limits the current flowing from the collector to the emitter when the transistor is switched on:

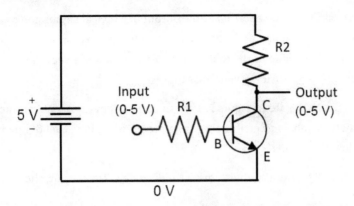

Figure 2.3: Transistor NOT gate

The input terminal accepts voltage inputs over the range 0 to 5 V, but since we are interested in digital circuit operation, we are only interested in signals that are either near 0 V (low) or near 5 V (high). We will assume that all voltage levels between the low and high states are transient during near-instantaneous transitions between the low and high states.

A typical NPN transistor has a switching voltage of about 0.7 V. When the input terminal is held at a low voltage, 0.2 V for example, the transistor is effectively switched off and has a very large resistance between the collector and emitter. This allows *R2*, connected to the 5 V power supply, to pull the output signal to a high state near 5 V.

When the input signal voltage rises above 0.7 V and into the 2 to 5 V range, the transistor switches on and the resistance between the collector and the emitter becomes very small. This, in effect, connects the output terminal to 0 V through a resistance that is much smaller than *R2*. This pulls the output terminal to a low voltage, typically around 0.2 V.

To summarize the behavior of this circuit, when the input terminal is high, the output terminal is low. When the input terminal is low, the output terminal is high. This function describes a NOT gate, in which the output is the inverse of the input. Assigning the low signal level the binary value 0 and the high signal level the value 1, the behavior of this gate is summarized in the truth table of *Table 2.1*:

Input	Output
0	1
1	0

Table 2.1: NOT gate truth table

A **truth table** is a tabular representation of the output of a logical expression as a function of all possible combinations of inputs. Each column represents one input or output, with the output(s) shown on the right-hand side of the table. Each row presents one set of input values together with the output of the expression given those inputs.

Circuits such as the NOT gate in *Figure 2.3* are so common in digital electronics that they are assigned schematic symbols to enable the construction of higher-level diagrams representing more complex logic functions.

The symbol for a NOT gate is a triangle with a small circle at the output, shown in *Figure 2.4*:

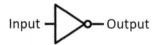

Figure 2.4: NOT gate schematic symbol

The triangle represents an amplifier, which means this is a device that turns a weaker input signal into a stronger output signal. The circle represents the inversion operator, which converts the signal to its binary opposite.

Next, we will look at some more complex logical operations that can be developed by building upon the NOT gate circuit. The circuit in *Figure 2.5* uses two transistors to perform an AND operation on the inputs $Input_1$ and $Input_2$. An AND operation has an output of 1 when both inputs are 1, otherwise the output is 0. Resistor R2 pulls the *Output* signal low unless both transistors have been switched on by high levels at the $Input_1$ and $Input_2$ signals:

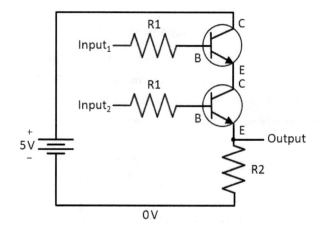

Figure 2.5: Transistor AND gate

Table 2.2 is the truth table for the AND gate. In simple terms, the *Output* signal is true (at the 1 level) when both the *Input₁* and *Input₂* inputs are true, and false (0) otherwise:

Input₁	Input₂	Output
0	0	0
1	0	0
0	1	0
1	1	1

Table 2.2: AND gate truth table

The AND gate has its own schematic symbol, shown in *Figure 2.6*:

Figure 2.6: AND gate schematic symbol

An OR gate has an output of 1 when either the *A* or *B* input is 1, and when both inputs are 1. Here is the truth table for the OR gate:

A	B	Output
0	0	0
1	0	1
0	1	1
1	1	1

Table 2.3: OR gate truth table

The OR gate schematic symbol is shown in *Figure 2.7*:

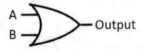

Figure 2.7: OR gate schematic symbol

The exclusive-OR, or XOR, operation produces an output of 1 when just one of the A and B inputs is 1. The output is 0 when both inputs are 0 and when both are 1. This is the XOR truth table:

A	B	Output
0	0	0
1	0	1
0	1	1
1	1	0

Table 2.4: XOR gate truth table

The XOR gate schematic symbol is shown in *Figure 2.8*:

Figure 2.8: XOR gate schematic symbol

Each of the AND, OR, and XOR gates can be implemented with an inverting output. The function of the gate is the same as described in the preceding section, except the output is inverted (0 is replaced with 1 and 1 is replaced with 0 in the *Output* column in *Table 2.2*, *Table 2.3*, and *Table 2.4*). The schematic symbol for an AND, OR, or XOR gate with inverted output has a small circle added on the output side of the symbol, just as on the output of the NOT gate. The names of the gates with inverted outputs are NAND, NOR, and XNOR. The letter *N* in each of these names indicates NOT. For example, NAND means NOT AND, which is functionally equivalent to an AND gate followed by a NOT gate.

Simple logic gates can be combined to produce more complex functions. A multiplexer is a circuit that selects one of multiple inputs to pass through to its output based on the state of a selector input. *Figure 2.9* is the diagram of a two-input multiplexer:

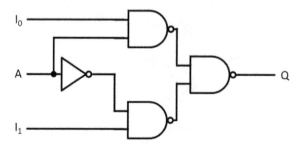

Figure 2.9: Two-input multiplexer circuit

The two single-bit data inputs are I_0 and I_1. The selector input A passes the value of I_0 through to the output Q when A is high. It passes I_1 to the output when A is low. One use of a multiplexer in processor design is to select input data from one of multiple sources when loading an internal register.

The truth table representation of the two-input multiplexer is shown in *Table 2.5*. In this table, the value X indicates "don't care," meaning it does not matter what value that signal has in determining the Q output:

A	I_0	I_1	Q
1	0	X	0
1	1	X	1
0	X	0	0
0	X	1	1

Table 2.5: Two-input multiplexer truth table

The logic gates presented in this section, and circuits constructed from them, are referred to as **combinational logic** when the output at any moment depends only on the current state of the inputs. For the moment, we're ignoring propagation delay and assuming that the circuit output responds immediately to changes in its inputs. In other words, given these assumptions, the output does not depend on prior input values. Combinational logic circuits have no memory of past inputs or outputs.

In the next section, we will look at some circuits that can retain memory of past operations.

Latches

Combinational logic does not directly permit the storage of data as is needed for digital functions such as processor registers. Logic gates can be used to create data storage elements by using feedback from a gate output to the input of a gate preceding that point in the signal chain.

A **latch** is a single-bit memory device constructed from logic gates. *Figure 2.10* shows a simple type of latch called the **Set-Reset**, or **SR, latch**. The feature that provides memory in this circuit is the feedback from the output of the AND gate to the input of the OR gate:

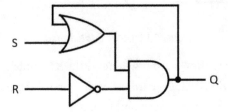

Figure 2.10: SR latch circuit

Based on the inputs S and R, the circuit can either set the output Q to high, reset Q to low, or cause the output Q to be held at its last value. In the hold state, both S and R are low, and the state of the output Q is retained. Pulsing S high (going from low to high then back to low) causes the output Q to go high and remain at that level. Pulsing R high causes Q to go low and stay low. If both S and R are set high, the R input overrides the S input and forces Q low.

The truth table for the SR latch is shown in *Table 2.6*. The output Q_{prev} represents the most recent value of Q selected through the actions of the S and R inputs:

S	R	Action	Q
0	0	Hold	Q_{prev}
1	0	Set	1
X	1	Reset	0

Table 2.6: SR latch truth table

One thing to be aware of with this latch circuit, and with volatile memory devices in general, is that the initial state of the Q output upon power-up is not well defined. The circuit startup behavior and the resulting value of Q depend on the characteristics and timing of the individual gates as they come to life. After power-on, and prior to beginning use of this circuit for productive purposes, it is necessary to pulse the S or R input to place Q into a known state.

The **gated D latch**, in which **D** stands for **data**, has many uses in digital circuits. The term **gated** refers to the use of an additional input that enables or inhibits the passage of data through the circuit. *Figure 2.11* shows an implementation of the gated D latch:

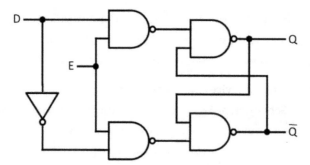

Figure 2.11: Gated D latch circuit

The D input passes through to the Q output whenever the E (enable) input is high. When E is low, the Q output retains its previous value regardless of the state of the D input. The $\bar{Q}$ output always holds the inverse of the Q output (the horizontal bar above $\bar{Q}$ means NOT):

D	E	Q	$\bar{Q}$
0	1	0	1
1	1	1	0
X	0	Q_{prev}	$\bar{Q}_{prev}$

Table 2.7: Gated D latch truth table

It is worth taking a moment to trace the logical flow of this circuit to understand its operation. The left half of *Figure 2.11*, consisting of the D input, the NOT gate, and the two leftmost NAND gates, is a combinational logic circuit, meaning the output is always a direct function of the input.

First, consider the case when the E input is low. With E low, one of the inputs to each of the two left-hand NAND gates is low, which forces the output of both gates to 1 (refer to *Table 2.2* and the AND gate truth table and remember that the NAND gate is equivalent to an AND gate followed by a NOT gate). In this state, the value of the D input is irrelevant, and one of Q or $\bar{Q}$ must be high and the other must be low, because of the cross-connection of the outputs of the two rightmost NAND gates feeding back to the gate inputs. This state will be retained as long as E is low.

When E is high, depending on the state of D, one of the two leftmost NAND gates will have a low output and the other will have a high output. The one with the low output will drive the connected rightmost NAND gate to a high output. This output will feed back to the input of the other right-hand side NAND gate and, with both inputs high, will produce a low output. The result is that the input D will propagate through to the output Q and the inverse of D will appear at output $\bar{Q}$.

It is important to understand that Q and $\bar{Q}$ cannot both be high or low at the same time because this would represent a conflict between the outputs and inputs of the two rightmost NAND gates. If one of these conditions happens to arise fleetingly, such as during power-up, the circuit will self-adjust to a stable configuration, with Q and $\bar{Q}$ holding opposite states. As with the SR latch, the result of this self-adjustment is not predictable, so it is important to initialize the gated D latch to a known state before using it in any operations. Initialization is performed by setting E high, setting D to the desired initial Q output, and then setting E low.

The gated D latch described previously is a **level-sensitive device**, meaning the output Q changes to follow the D input while the E input is held high. In more complex digital circuits, it becomes important to synchronize multiple circuit elements connected in series without the need to carefully account for propagation delays across the individual devices. The use of a **shared clock signal** as an input to multiple elements enables this type of synchronization.

In a shared-clock configuration, components update their outputs based on clock signal edges (edges are the moments of transition from low to high or high to low) rather than responding continuously to high or low input signal levels.

Edge triggering is useful because the clock signal edges identify precise moments at which device inputs must be stable and valid. After the clock edge has passed, the device's inputs are free to vary in preparation for the next active clock edge without the possibility of altering the circuit outputs. The flip-flop circuit, discussed next, responds to clock edges, providing this desirable characteristic for complex digital designs.

Flip-flops

A device that changes its output state only when a clock signal makes a specified transition (either low to high or high to low) is referred to as an **edge-sensitive device**. **Flip-flops** are similar to latches, with the key difference being that the output of a flip-flop changes in response to a clock edge rather than responding continuously to its input signal while enabled.

The **positive edge-triggered D flip-flop** is a popular digital circuit component used in a wide variety of applications. The D flip-flop typically includes set and reset input signals that perform the same functions as in the SR latch. This flip-flop has a D input that functions just like the D input of the gated D latch. Instead of an enable input, the D flip-flop has a clock input that triggers the transfer of the D input to the Q output and, with inversion, to the $\bar{Q}$ output on the clock's rising edge. Other than within a very narrow time window surrounding the rising edge of the clock signal, the flip-flop does not respond to the value of the D input. When active, the S and R inputs override any activity on the D and clock inputs.

Figure 2.12 presents the schematic symbol for the D flip-flop. The clock input is indicated by the small triangle on the left-hand side of the symbol:

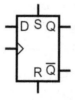

Figure 2.12: D flip-flop schematic symbol

Consider the following table. The upward-pointing arrows in the *CLK* column indicate the rising edge of the clock signal. The Q and $\bar{Q}$ outputs shown in the table rows with upward-pointing arrows represent the state of the outputs following the rising clock edge.

S	R	D	CLK	Q	$\bar{Q}$
0	0	1	↑	1	0
0	0	0	↑	0	1
0	0	X	Stable	Q_{prev}	$\bar{Q}_{prev}$
1	0	X	X	1	0
0	1	X	X	0	1

Table 2.8: D flip-flop truth table

Flip-flops can be connected in series to enable the transfer of data bits from one flip-flop to the next on sequential clock cycles. This is achieved by connecting the Q output of the first flip-flop to the *D* input of the second one, and so on for any number of stages. This structure, called a **shift register**, has many applications, two of which are serial-to-parallel conversion and parallel-to-serial conversion.

If the Q output at the end of a shift register is connected to the *D* input at the other end of the register, the result is a **ring counter**. Ring counters are used for tasks such as the construction of **finite state machines**. Finite state machines implement a mathematical model that is always in one of a set of well-defined states. Transitions between states occur when inputs satisfy the requirements to transition to a different state.

The ring counter in *Figure 2.13* has four positions. The counter is initialized by pulsing the *RST* input high and then low. This sets the Q output of the first (leftmost) flip-flop to 1 and the remaining flip-flop Q outputs to 0. After that, each rising edge of the *CLK* input transfers the 1 bit to the next flip-flop in the sequence. The fourth *CLK* pulse transfers the 1 back to the leftmost flip-flop. At all times, each of the flip-flops has a Q output of 0 except for one that has a 1 output.

The flip-flops are edge-sensitive devices and are all driven by a common clock signal, making this a **synchronous circuit**:

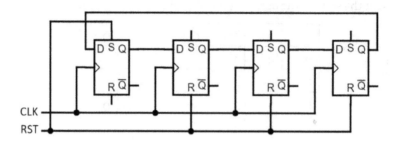

Figure 2.13: Four-position ring counter circuit

This circuit contains four ring counter states. Adding six more flip-flops would bring the number of states to 10. As we discussed in *Chapter 1, Introducing Computer Architecture*, the ENIAC used vacuum tube-based 10-position ring counters to maintain the state of decimal digits. A 10-state ring counter based on the circuit in *Figure 2.13* can perform the same function.

In the next section, we will construct registers for data storage from flip-flops.

Registers

Processor registers temporarily store data values and serve as input to and output from a variety of instruction operations, including data movement to and from memory, arithmetic, and bit manipulation. Most general-purpose processors include instructions for shifting binary values stored in registers to the left or right and for performing rotation operations in which data bits shifted out one end of the register are inserted at the opposite end. The rotation operation is similar to the ring counter, except the bits in a rotation can hold arbitrary values, while a ring counter typically transfers a single 1 bit through the sequence of locations. Circuits performing these functions are constructed from the low-level gates and flip-flops presented earlier in this chapter.

Registers within a processor are usually written and read in parallel, meaning all the bits are written or read on individual signal lines simultaneously under the control of a common clock edge. The examples presented in this section use 4-bit registers for simplicity, but it is straightforward to extend these designs to 8, 16, 32, or 64 bits.

Figure 2.14 shows a simple 4-bit register with parallel input and output. This is a synchronous circuit, in which data bits provided on inputs D_0-D_3 are loaded into the flip-flops on the rising edge of the *CLK* signal. The data bits appear immediately at the Q_0-Q_3 outputs and retain their state until new data values are loaded on a subsequent rising clock edge:

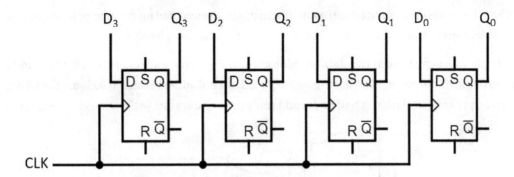

Figure 2.14: 4-bit register circuit

To perform useful functions beyond simply storing data in a register, it must be possible to load data from multiple sources into the register, perform operations on the register contents, and write the resulting data value to one of potentially many destinations.

In general-purpose processors, a data value can usually be loaded into a register from a memory location, from an input port, or transferred from another register. Operations performed on the register contents might include incrementing, decrementing, arithmetic operations, shifting, rotating, and bit manipulations such as AND, OR, and XOR. Note that incrementing or decrementing an integer is equivalent to the addition or subtraction of an operand with a second implied operand of 1. Once a register contains the result of a computation, its contents can be written to a memory location, to an output port, or to another register.

Figure 2.9 presented a circuit for a two-input multiplexer. It is straightforward to extend this circuit to support a larger number of inputs, any of which can be selected by control signals. The single-bit multiplexer can be replicated to support simultaneous operation across all the bits in a processor word. Such a circuit is used to select among a variety of sources when loading a register with data. When implemented in a processor, logic triggered by instruction opcodes sets the multiplexer control inputs to route data from the selected source to the specified destination register. *Chapter 3, Processor Elements*, will expand on the use of multiplexers for data routing to registers and to other units within the processor.

The next section will introduce circuits for adding binary numbers.

Adders

General-purpose processors usually support the addition operation for performing calculations on data values and, separately, to manage the instruction pointer. Following the execution of each instruction, the instruction pointer increments to the next instruction location.

When the processor supports multi-word instructions, the updated instruction pointer must be set to its current value plus the number of words in the just-completed instruction.

A simple adder circuit adds two data bits plus an incoming carry and produces a 1-bit sum and a carry output. This circuit, shown in *Figure 2.15*, is called a **full adder** because it includes the incoming carry in the calculation. A **half adder** adds only the two data bits without an incoming carry:

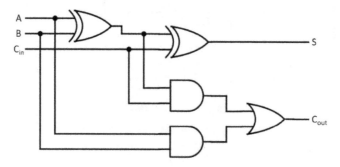

Figure 2.15: Full adder circuit

The full adder uses logic gates to produce its output as follows. The sum bit S is 1 only if the total number of bits with a value of 1 in the collection A, B, C_{in} is an odd number. Otherwise, S is 0. The two XOR gates perform this logical operation. C_{out} is 1 if both A and B are 1, or if just one of A and B is 1 and C_{in} is also 1. Otherwise, C_{out} is 0.

The circuit in *Figure 2.15* can be condensed to a schematic block that has three inputs and two outputs for use in higher-level diagrams. *Figure 2.16* is a 4-bit adder with four blocks representing copies of the full adder circuit of *Figure 2.15*. The inputs are the two words to be added, A_0-A_3 and B_0-B_3, and the incoming carry, C_{in}. The output is the sum, S_0-S_3, and the outgoing carry, C_{out}:

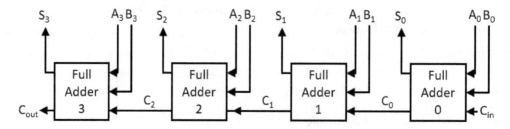

Figure 2.16: 4-bit adder circuit

It is important to note that this circuit is a combinational circuit, meaning that once the inputs have been set, the outputs will be generated directly. This includes the carry propagation from bit to bit, no matter how many bits are affected by carries. Because the carry flows across bit by bit, this configuration is referred to as a **ripple carry adder**.

It takes some time for the carries to propagate across all the bit positions and for the outputs to stabilize at their final value.

Since we are now discussing a circuit that has a signal path passing through a significant number of devices, it is appropriate to discuss the implications of the time required for signals to travel from end to end through multiple components.

Propagation delay

When the input of a logic device changes, the output does not change instantly. There is a time lag between a change of state at the input and when the result appears at the output. This is called **propagation delay**. The propagation delay through a circuit places an upper limit on the clock frequency at which the circuit can operate. In a microprocessor, the clock speed determines the speed at which the device can execute instructions.

Placing multiple combinational circuits in series results in an overall propagation delay equal to the sum of the delays of the individual devices. A gate may have a different propagation delay for a low-to-high transition than for a high-to-low transition, so the larger of these two values should be used when estimating the worst-case delay through a circuit path.

As shown in *Figure 2.15*, the longest path (in terms of the number of gates in series) from input to output for the full adder is from the *A* and *B* inputs to the C_{out} output: a total of three sequential gates. If all the 4-bit adder input signals in *Figure 2.16* are set simultaneously, the three-gate delay related to the *A* and *B* inputs will take place simultaneously across all four of the adders. However, the C_0 output from full adder 0 is only guaranteed to be stable after the three-gate delay across full adder 0. Once C_0 is stable, there is an additional two-gate delay across full adder 1 (note that in *Figure 2.15*, C_{in} only passes through two sequential levels of gates).

The overall propagation delay for the circuit in *Figure 2.16* is therefore three gate delays across full adder 0 followed by two gate delays across each of the remaining three full adders, a total of nine gate delays. This may not seem like a lot, but consider a 32-bit adder: the propagation delay for this adder is three gate delays for full adder 0 plus two gate delays for each of the remaining 31 adders, a total of 65 gate delays.

The path with the maximum propagation delay through a combinational circuit is referred to as the **critical path**. The critical path delay places an upper limit on the clock frequency that can be used to drive the circuit.

Logic gates from the **Advanced Schottky Transistor-Transistor Logic** family, abbreviated to **(AS) TTL**, are among the fastest individually packaged gates available today.

An (AS) TTL NAND gate has a propagation delay of 2 **nanoseconds** (**ns**) under typical load conditions. For comparison, light in a vacuum travels just under 2 feet in 2 ns.

In the 32-bit ripple carry adder, 65 propagation delays through (AS) TTL gates result in a delay of 130 ns between setting the inputs and receiving final, stable outputs. To form a rough estimate, let's assume this is the worst-case propagation delay through an entire processor integrated circuit. We'll also ignore any additional time required to hold inputs stable before and after an active clock edge. This adder, then, cannot perform sequential operations on input data more often than once every 130 ns.

When performing 32-bit addition with a ripple carry adder, the processor uses a clock edge to transfer the contents of two registers (each consisting of a set of D flip-flops) plus the processor C flag to the adder inputs. The subsequent clock edge loads the results of the addition into a destination register. The C flag receives the value of C_{out} from the adder.

A clock with a period of 130 ns has a frequency of (1/130 ns), which is 7.6 MHz. This certainly does not seem very fast, especially when considering that many low-cost processors are available today with clock speeds greater than 4 GHz. Part of the reason for this discrepancy is the inherent speed advantage of integrated circuits containing massive numbers of tightly packed transistors, and the other part is the result of the cleverness of designers, as referenced by Gordon Moore, which was discussed in *Chapter 1, Introducing Computer Architecture*. To perform the adder function efficiently, many design optimizations have been developed to substantially reduce the worst-case propagation delay. *Chapter 8, Performance-Enhancing Techniques*, will discuss some of the methods processor architects use to wring higher speeds from their designs.

In addition to gate delays, there is also some delay resulting from signal traveling through wires and integrated circuit conductive paths. The propagation speed through a wire or other type of conductive material varies depending on the material used for conduction and on the insulating material surrounding the conductor. Depending on these and other factors, signal propagation speed in digital circuits is typically 50-90% of the speed of light in a vacuum.

The next section discusses the generation and use of clocking signals in digital circuits.

Clocking

The clock signal serves as the heartbeat of a processor. This signal is usually a square wave signal operating at a fixed frequency. A square wave is a digital signal that oscillates between high and low states, spending equal lengths of time at the high and low levels on each cycle. *Figure 2.17* shows an example of a square wave over time:

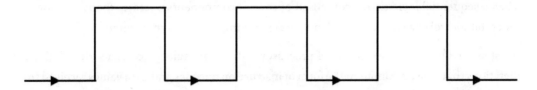
Figure 2.17: Square wave signal

The clock signal in a computer system is usually generated from a crystal oscillator providing a base frequency of a few **megahertz (MHz)**. 1 MHz is 1 million cycles per second. A crystal oscillator relies on the resonant vibration of a physical crystal, usually made of quartz, to generate a cyclic electrical signal. The mechanical vibration of the crystal is transformed into an electrical signal by the piezoelectric effect. The **piezoelectric effect** is the electrical charge that accumulates in certain crystals when placed under mechanical stress. Quartz crystals resonate at precise frequencies, which leads to their use as timing elements in computers, wristwatches, and other digital devices.

Although crystal oscillators are more accurate time references than alternative timing references that find use in low-cost devices, crystals exhibit errors in frequency that accumulate over periods of days and weeks to gradually drift by seconds and then minutes away from the correct time. To avoid this problem, most internet-connected computers access a time server periodically to reset their internal clocks to the current time as published by a precise atomic reference clock.

Phase-locked loop (PLL) frequency multiplier circuits are used to generate the high-frequency clock signals needed by multi-GHz processors. A PLL frequency multiplier generates a square wave output frequency that is an integer multiple of the input frequency provided to it from the crystal oscillator. The ratio of the PLL clock output frequency to the input frequency it receives is called the **clock multiplier**.

A PLL frequency multiplier operates by continuously adjusting the frequency of its internal oscillator to maintain the correct clock multiplier ratio relative to the PLL input frequency. Modern processors usually have a crystal oscillator clock signal input and contain several PLL frequency multipliers producing different frequencies. These PLL output frequencies then drive core processor operations at the highest possible speed while simultaneously interacting with components that require lower clock frequencies, such as system memory and peripheral devices.

Sequential logic

Digital circuitry that generates outputs based on a combination of current inputs and past inputs is called **sequential logic**. This contrasts with combinational logic, in which outputs depend only on the current state of the inputs.

When a sequential logic circuit composed of several components operates those components under the control of a shared clock signal, the circuit implements **synchronous logic**.

The steps involved in the execution of processor instructions take place as a series of discrete operations that consume input in the form of instruction opcodes and data values received from various sources. This activity takes place under the coordination of a master clock signal. The processor maintains internal state information from one clock step to the next, and from one instruction to the next.

Modern complex digital devices, including processors, are almost always implemented as synchronous sequential logic devices. Low-level internal components, such as the gates, multiplexers, registers, and adders discussed previously, are usually combinational logic circuits. These lower-level components, in turn, receive inputs under the control of synchronous logic. After allowing sufficient time for signal propagation across the combinational components, the shared clock signal transfers the outputs of those components to other portions of the architecture under the control of processor instructions and the logic circuits that carry out those instructions.

Chapter 3, Processor Elements, will introduce the higher-level processor components that implement more complex functionality, including instruction decoding, instruction execution, and arithmetic operations.

The next section introduces the idea of designing digital hardware using languages that are very similar to traditional computer programming languages.

Hardware description languages

It is straightforward to represent simple digital circuits using logic diagrams like the ones presented earlier in this chapter. When designing digital devices that are substantially more complex, however, the use of logic diagrams quickly becomes unwieldy. As an alternative to the logic diagram, several **hardware description languages** have been developed over the years. This evolution has been encouraged by **Moore's law**, which drives digital system designers to continually find new ways to quickly make the most effective use of the constantly growing number of transistors available in integrated circuits.

Hardware description languages are not the exclusive province of digital designers at semiconductor companies; even hobbyists can acquire and use these powerful tools at an affordable cost. Some are even free.

A **gate array** is a logic device containing many logic elements such as NAND gates and D flip-flops that can be connected to form arbitrary circuits.

A category of gate arrays called **field-programmable gate arrays (FPGAs)** enables end users to implement their own designs into gate array chips using just a computer, a small development board, and an appropriate software package.

A developer can define a complex digital circuit using a hardware description language and program it into a chip directly, resulting in a fully functional, high-performance custom digital device. Modern low-cost FPGAs contain enough gates to implement complex modern processor designs. As one example, an FPGA-programmable design of the RISC-V processor (discussed in detail in *Chapter 11, The RISC-V Architecture and Instruction Set*) is available in the form of open source hardware description language code.

VHDL

VHDL is one of the leading hardware description languages in use today. Development of the VHDL language began in 1983 under the guidance of the U.S. Department of Defense. The syntax and some of the semantics of VHDL are based on the **Ada programming language**. The Ada language is, incidentally, named in honor of Ada Lovelace, the programmer of Charles Babbage's Analytical Engine, discussed in *Chapter 1, Introducing Computer Architecture*. Verilog is another popular hardware design language with capabilities similar to VHDL. This book will use VHDL exclusively, but the examples can be implemented just as easily in **Verilog**.

VHDL is a multilevel acronym where the V stands for **VHSIC**, which means **very high-speed integrated circuit**, and **VHDL** stands for **VHSIC hardware description language**. The following code presents a VHDL implementation of the full adder circuit shown in *Figure 2.15*:

```
-- Load the standard libraries

library IEEE;
  use IEEE.STD_LOGIC_1164.ALL;

-- Define the full adder inputs and outputs

entity FULL_ADDER is
  port (
    A     : in   std_logic;
    B     : in   std_logic;
    C_IN  : in   std_logic;
    S     : out  std_logic;
    C_OUT : out  std_logic
```

```
    );
end entity FULL_ADDER;


-- Define the behavior of the full adder

architecture BEHAVIORAL of FULL_ADDER is

begin

    S     <= (A XOR B) XOR C_IN;
    C_OUT <= (A AND B) OR ((A XOR B) AND C_IN);

end architecture BEHAVIORAL;
```

This code is a straightforward textual description of the full adder in *Figure 2.15*. Here, the section introduced with entity FULL_ADDER is defines the inputs and outputs of the full adder component. The architecture section toward the end of the code describes how the circuit logic operates to produce the outputs *S* and *C_OUT* given the inputs *A*, *B*, and *C_IN*. The term std_logic refers to a single-bit binary data type. The <= characters represent signal assignment, which indicates a wire-like connection that drives the output on the left-hand side with the value computed on the right-hand side.

The following code references the FULL_ADDER VHDL as a component in the implementation of the 4-bit adder design presented in *Figure 2.16*:

```
-- Load the standard libraries

library IEEE;
    use IEEE.STD_LOGIC_1164.ALL;

-- Define the 4-bit adder inputs and outputs

entity ADDER4 is
    port (
        A4      : in    std_logic_vector(3 downto 0);
        B4      : in    std_logic_vector(3 downto 0);
        SUM4    : out   std_logic_vector(3 downto 0);
        C_OUT4  : out   std_logic
```

```vhdl
    );
end entity ADDER4;

-- Define the behavior of the 4-bit adder

architecture BEHAVIORAL of ADDER4 is

  -- Reference the previous definition of the full adder

  component FULL_ADDER is
    port (
      A          : in    std_logic;
      B          : in    std_logic;
      C_IN       : in    std_logic;
      S          : out   std_logic;
      C_OUT      : out   std_logic
    );
  end component;

  -- Define the signals used internally in the 4-bit adder
  signal c0, c1, c2 : std_logic;

begin

  -- The carry input to the first adder is set to 0
  FULL_ADDER0 : FULL_ADDER
    port map (
      A          => A4(0),
      B          => B4(0),
      C_IN       => '0',
      S          => SUM4(0),
      C_OUT      => c0
    );

  FULL_ADDER1 : FULL_ADDER
    port map (
      A          => A4(1),
```

```
            B          => B4(1),
            C_IN       => c0,
            S          => SUM4(1),
            C_OUT      => c1
         );

      FULL_ADDER2 : FULL_ADDER
         port map (
            A          => A4(2),
            B          => B4(2),
            C_IN       => c1,
            S          => SUM4(2),
            C_OUT      => c2
         );

      FULL_ADDER3 : FULL_ADDER
         port map (
            A          => A4(3),
            B          => B4(3),
            C_IN       => c2,
            S          => SUM4(3),
            C_OUT      => C_OUT4
         );

   end architecture BEHAVIORAL;
```

This code is a textual description of the 4-bit adder in *Figure 2.16*. Here, the section introduced with entity ADDER4 is defines the inputs and outputs of the 4-bit adder component. The phrase std_logic_vector(3 downto 0) represents a 4-bit vector data type with bit number 3 in the most significant position and bit number 0 in the least significant position.

The FULL_ADDER component is defined in a separate file, referenced here by the section beginning component FULL_ADDER is. The statement signal c0, c1, c2 : std_logic; defines the internal carry values between the full adders. The four port map sections define the connections between the 4-bit adder signals and the inputs and outputs of each of the single-bit full adders. To reference a bit in a bit vector, the bit number follows the parameter name in parentheses. For example, *A4(0)* refers to the least significant of the 4 bits in *A4*.

Note the use of hierarchy in this design. A simple component, the single-bit full adder, was first defined in a discrete, self-contained block of code. This block was then used to construct a more complex circuit, the 4-bit adder. This hierarchical approach can be extended through many levels to define an extremely complex digital device constructed from less complex components, each of which, in turn, is constructed from even simpler parts. This general approach is used routinely in the development of modern processors containing billions of transistors, while managing complexity to keep the design understandable by humans at each level of the architectural hierarchy.

The code presented in this section provides all the circuit definitions that a logic synthesis software tool suite requires to implement the 4-bit adder as a component in an FPGA device. Of course, additional circuitry is required to present meaningful inputs to the adder circuit and then to process the results of an addition operation after allowing for propagation delay.

This section provided a very brief introduction to VHDL. The intent is to make you aware that hardware description languages such as VHDL are the current state of the art in complex digital circuit design. In addition, you should know that some very low-cost options are available for FPGA development tools and devices. The exercises at the end of this chapter will introduce you to some highly capable FPGA development tools that are free. You are encouraged to search the internet and learn more about VHDL and other hardware description languages and try your hand at developing some simple (and not-so-simple) circuit designs.

Summary

This chapter began with an introduction to the properties of electrical circuits and showed how components such as voltage sources, resistors, and wires are represented in circuit diagrams. The transistor was introduced, with a focus on its use as a switching element in digital circuits. The NOT gate and the AND gate were constructed from transistors and resistors. Additional types of logic gates were defined and truth tables were presented for each device. Logic gates were used to construct more complex digital circuits, including latches, flip-flops, registers, and adders. The concept of sequential logic was introduced, and its applicability to processor design was discussed. Finally, hardware description languages were introduced and a 4-bit adder example was presented in VHDL.

You should now understand basic digital circuit concepts and the design tools used in the development of modern processors. The next chapter will expand upon these building blocks to explore the functional components of modern processors, leading to a discussion of how those components coordinate to implement the primary processor operational cycle of instruction loading, decoding, and execution.

Exercises

1. Rearrange the circuit in *Figure 2.5* to convert the AND gate to a NAND gate. Hint: there is no need to add or remove components.
2. Create a circuit implementation of an OR gate by modifying the circuit in *Figure 2.5*. Wires, transistors, and resistors can be added as needed.
3. Search the internet for free VHDL development software suites that include a simulator. Get one of these suites, set it up, and build any simple demo projects that come with the suite to ensure it is working properly.
4. Using your VHDL tool set, implement the 4-bit adder using the code listings presented in this chapter.
5. Add test driver code (search the internet for *VHDL testbench* to find examples) to your 4-bit adder to drive it through a limited set of input sets and verify that the outputs are correct.
6. Expand the test driver code and verify that the 4-bit adder produces correct results for all possible combinations of inputs.

Join our community Discord space

Join the book's Discord workspace for a monthly *Ask me Anything* session with the author:
`https://discord.gg/7h8aNRhRuY`

3
Processor Elements

This chapter begins our development of a comprehensive understanding of modern processor architectures. Building upon the basic digital circuits introduced in *Chapter 2, Digital Logic*, we discuss the functional units of a simple, generic computer processor. Concepts related to the instruction set and register set are introduced, followed by a discussion of the steps involved in instruction loading, decoding, execution, and sequencing. Addressing modes and instruction categories are discussed in the context of the 6502 processor architecture. We choose to focus on this venerable processor for its structural cleanliness and simplicity, which allows us to consider basic concepts without distractions. The requirement for processor interrupt handling is introduced, using the example of 6502 interrupt processing. The standard approaches that modern processors employ for **input/output (I/O)** operations are introduced, including **direct memory access (DMA)**.

After completing this chapter, you will understand the basic components of a processor and the structure of processor instruction sets. You will have learned the categories of processor instructions, why interrupt processing is necessary, and will have an understanding of I/O operations.

The following topics will be covered in this chapter:

- A simple processor
- The instruction set
- Addressing modes
- Instruction categories
- Interrupt processing
- Input/output operations

Technical requirements

Files for this chapter, including answers to the exercises, are available at https://github.com/PacktPublishing/Modern-Computer-Architecture-and-Organization-Second-Edition.

A simple processor

The 6502 processor architecture and a small subset of its instructions were introduced in *Chapter 1, Introducing Computer Architecture*. In this section, we will build upon that foundation to present some of the functional components universally employed in processor architectures, from the tiniest embedded controllers to the most powerful server CPUs.

The integrated circuit at the core of a computer system goes by a few different names: the **Central Processing Unit (CPU)**, microprocessor, or, simply, processor. A microprocessor is a single integrated circuit that implements all the functions of a processor. This book will refer to all categories of CPUs and microprocessors as processors.

A processor like the 6502 contains three logically distinct functional units:

- The **control unit** manages the overall operation of the device. This includes fetching the next instruction from memory, decoding the instruction to determine the operation to perform, and distributing the execution of the instruction to appropriate elements within the processor.
- The **Arithmetic Logic Unit (ALU)** is a combinational circuit that performs arithmetic and bit manipulation operations.
- The **register set** provides source and destination locations for instruction inputs and outputs. Registers are also used as temporary storage locations.

The following diagram shows the flow of control and data among the control unit, the registers, the ALU, system memory, and input/output devices:

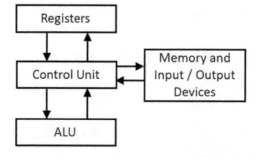

Figure 3.1: Interactions between processor functional units

The control unit directs overall processor operations to carry out each instruction. The registers, ALU, memory, and I/O devices respond to commands initiated by the control unit.

Control unit

The control unit of a modern processor is a synchronous sequential digital circuit. It interprets processor instructions and manages the execution of those instructions by interacting with the other functional units within the processor and with external components, including memory and input/output devices. The control unit is a key part of the 6502's von Neumann architecture.

In this chapter, the term *memory* refers to **Random Access Memory (RAM)** external to the processor's execution units. Cache memory, which often resides within a microprocessor integrated circuit, will be covered in later chapters.

Some examples of I/O devices are the computer keyboard, the mouse, disk storage, and graphical video displays. Other common I/O devices include network interfaces, Wi-Fi and Bluetooth® wireless interfaces, sound output to speakers, and microphone input.

When a computer system is powered on, the processor undergoes a reset process to initialize its internal components. Following a reset, the processor loads the **Program Counter (PC)** with the memory location of the first instruction to be executed. Software developers who construct the lowest-level system software components must configure their development tools to produce a code memory image that begins execution at the address required by the processor architecture.

The PC is a central component of the control unit. The PC always contains the memory address of the next instruction to be executed. At the beginning of each instruction execution cycle, the control unit reads the data word at the memory address indicated by the PC and places it in an internal register for decoding and execution. The first word of an instruction contains an **opcode**. Based on the opcode bit pattern, the control unit may read additional memory locations following the opcode to retrieve data needed by the instruction, such as a memory address or data operand.

As the control unit begins executing instructions, it performs the repetitive cycle shown in *Figure 3.2*:

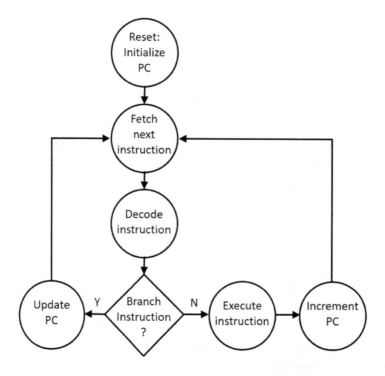

Figure 3.2: Instruction execution cycle

Following a reset, the PC contains the initial instruction location. The control unit fetches the first instruction from memory and decodes it. During decoding, the control unit determines the actions required by the instruction.

As part of the decoding process, the control unit identifies the category of instruction. The two basic instruction categories represented in *Figure 3.2* are branching instructions and all other instructions. Branching instructions are implemented directly by the control unit. These instructions cause the contents of the PC to be replaced with the memory address of the branch destination. Examples of instructions that perform branching are conditional branch instructions (when the branch is taken), subroutine calls, subroutine returns, and unconditional branching (also called *jump*) instructions.

Instructions that do not involve branching are carried out by processor circuitry under the direction of the control unit.

In a sense, the control unit manages the execution of the non-branching instructions in a manner similar to the Analytical Engine's mill (see *Chapter 1, Introducing Computer Architecture*), except, instead of using the presence of studs on a rotating barrel to engage portions of the mill machinery, the control unit uses the decoded bits of the instruction opcode to activate particular sections of digital circuitry. The selected circuit components perform the tasks required by the instruction.

The process of instruction execution may include actions such as reading or writing a register, reading or writing a memory location, directing the ALU to perform a mathematical operation, or other miscellaneous activities.

In most processors, the execution of a single instruction extends over multiple processor clock cycles. The instruction cycle count can vary significantly from simple instructions that require a small number of clock cycles to complex operations that take many cycles to complete. The control unit orchestrates all this activity.

The circuits managed by the control unit are constructed from the simple logic gates discussed in *Chapter 2, Digital Logic*, and are often composed of higher-level constructs such as multiplexers, latches, and flip-flops. Multiplexers, in particular, are commonly used by control unit logic to selectively route data to a particular destination.

Executing an instruction – a simple example

Consider a simplified example of two 6502 instructions, TXA and TYA. TXA copies the contents of register X to register A, and TYA does the same thing using the Y register as the source. If we consider these two instructions in isolation, the execution of both instructions can be implemented as shown in *Figure 3.3*:

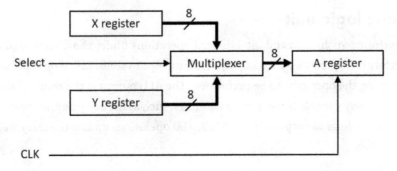

Figure 3.3: 6502 *TXA and* TYA *instructions*

The circuit in *Figure 3.3* assumes the X and Y registers are D flip-flop registers (as in *Figure 2.14*), except in the 6502 they are 8-bit registers rather than 4 bits. The multiplexer is implemented as eight copies of the two-input, single-bit multiplexer of *Figure 2.9*, all controlled by a single shared selector input. In *Figure 3.3*, thicker lines represent 8-bit data buses and thinner lines are individual logic signals. The short lines crossing the thick lines with the numeral 8 above them indicate the number of bits in the bus.

To execute the TXA instruction, the following steps are performed:

1. The control unit first sets the **Select** input to direct the X register data bits through to the output of the multiplexer. This presents the data from X at the inputs to the **A register**.
2. After the **Select** input to the multiplexer has been set, the control unit must pause to permit the propagation of the data bits to the multiplexer outputs.
3. After the multiplexer outputs have stabilized, the control unit generates a rising edge on the **CLK** signal to load the X register data bits into register A.

To execute the TYA instruction, the control unit performs the same sequence of steps, except it must first set the **Select** input to feed the Y register to the multiplexer output.

This is a very simple example of a control unit instruction operation, but it demonstrates that an individual instruction may consist of multiple steps and may involve just a small portion of the logic circuits present in the processor. Unused components within the processor must be managed by the control unit to ensure that they remain idle when not needed to execute an instruction. This prevents those components from interfering with the executing instruction and keeps power consumption to a minimum.

Arithmetic logic unit

The ALU performs arithmetic and bit-oriented operations under the direction of the control unit. To perform an operation, the ALU requires data input values, called **operands**, along with a code indicating the operation to be performed. The ALU output is the result of the operation. ALU operations may use one or more processor flags, such as the carry flag, as input, and set the states of processor flags as outputs. In the 6502, ALU operations update the carry, negative, zero, and overflow flags.

An ALU is a combinational circuit, which implies its outputs update asynchronously in response to changes at the inputs and it retains no memory of previous operations.

To execute an instruction involving the ALU, the control unit applies inputs to the ALU, pauses to allow for the propagation delay across the ALU, and then transfers the ALU output to the destination specified by the instruction.

The ALU contains an adder circuit to perform addition and subtraction operations. In a processor with two's complement arithmetic, subtraction can be implemented by first performing a two's complement negation of the right operand and adding the result to the left operand. Mathematically, when performing subtraction in this manner, the expression $A-B$ is transformed into $A+(-B)$.

As you'll recall from *Chapter 1, Introducing Computer Architecture*, the two's complement negation of a signed number is achieved by inverting all the bits in the operand and adding 1 to the result. Incorporating this operation, subtraction represented as $A+(-B)$ becomes $A+(NOT(B)+1)$.

Looking at subtraction in this form should clarify the use of the 6502 carry flag in conjunction with the SBC instruction. The C flag provides the "+1" in subtraction when there is no borrow. If there is a borrow, the sum must be reduced by 1, which is accomplished by setting the C flag to 0.

To summarize, in the 6502, subtraction logic is identical to addition logic with the single difference that the B operand in $A-B$ is routed through a set of NOT gates to invert all eight of the bits prior to feeding $NOT(B)$ to the adder input.

Figure 3.4 is a functional representation of the addition and subtraction operations in the 6502:

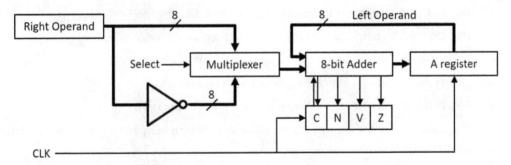

Figure 3.4: 6502 addition and subtraction operations

Similar to *Figure 3.3*, *Figure 3.4* is a highly simplified representation of the 6502 processor depicting only the components involved in the ADC and SBC instructions. The **Select** input in *Figure 3.4* chooses whether the operation is addition or subtraction. Addition requires selecting the upper multiplexer input, while the lower multiplexer input selects subtraction. In the 6502 architecture, the A register is always the left operand for subtraction.

The inputs to the adder are the left and right operands and the C flag. When executing an ADC or SBC instruction, the control unit presents the right operand to the multiplexer data inputs and sets the multiplexer select input to the appropriate state for the given instruction. After pausing for propagation through the NOT gate, the multiplexer, and the adder, the control unit generates a clock edge that latches the adder outputs into the A register and the processor flags register.

The processor flags are set as shown in the following execution of an ADC or SBC instruction:

- **C** indicates whether an addition generated in a carry (C = 1) or whether a subtraction produced a borrow (C = 0).
- **N** contains the value of bit 7 of the result.
- **V** indicates whether a signed overflow occurred (V = 1 if overflow occurred).
- **Z** is 1 if the result is zero. Z is 0 otherwise.

An ALU supports a variety of operations in addition to adding and subtracting two numbers. In the 6502, most operations with two operands use the A register as the left operand. The right operand is either read from a memory location or provided as an immediate value in the next memory location after the opcode. All 6502 ALU operands and results are 8-bit values. The 6502 ALU operations are as follows:

- **ADC, SBC**: Add or subtract two operands with carry input.
- **DEC, DEX, DEY**: Decrement a memory location or register by one.
- **INC, INX, INY**: Increment a memory location or register by one.
- **AND**: Perform a bitwise logical AND operation on two operands.
- **ORA**: Perform a bitwise logical OR operation on two operands.
- **EOR**: Perform a bitwise logical XOR operation on two operands.
- **ASL, LSR**: Shift the A register or memory location left or right by one bit position and insert 0 into the vacated bit position.
- **ROL, ROR**: Rotate the A register or memory location left or right by one bit position and insert the C flag value into the vacated bit position.
- **CMP, CPX, CPY**: Subtract two operands and discard the result, setting the N, Z, and C flags based on the result of the subtraction.
- **BIT**: Perform a bitwise logical AND between two operands and use the Z flag to indicate whether the result is 0. In addition, copy bits 7 and 6 of the left operand to the N and V flags.

The 6502 has limited ALU capabilities when compared to more complex modern processors such as the x86 family.

For example, in the 6502, the programmer must implement multiplication and division in code as repetitive addition and subtraction operations. Also, the 6502 can shift or rotate a value by just one bit position per instruction.

The x86, on the other hand, directly implements multiplication and division instructions, and the x86 shift and rotate instructions include a parameter indicating the number of bit positions to shift in a single instruction.

An ALU is a necessarily complex logic device, making it an ideal candidate for design with a hardware design language. The following listing is a VHDL implementation of a portion of a 6502-like ALU:

```
-- Load the standard libraries

library IEEE;
  use IEEE.STD_LOGIC_1164.ALL;
  use IEEE.NUMERIC_STD.ALL;

-- Define the 8-bit ALU inputs and outputs

entity ALU is
  port (
    -- Left operand
    LEFT        : in   std_logic_vector(7 downto 0);
    -- Right operand
    RIGHT       : in   std_logic_vector(7 downto 0);
    -- ALU operation
    OPCODE      : in   std_logic_vector(3 downto 0);
    -- Carry input
    C_IN        : in   std_logic;
    -- ALU output
    RESULT      : out  std_logic_vector(7 downto 0);
    -- Carry output
    C_OUT       : out  std_logic;
    -- Negative flag output
    N_OUT       : out  std_logic;
    -- Overflow flag output
    V_OUT       : out  std_logic;
    -- Zero flag output
```

```vhdl
      Z_OUT        : out    std_logic
  );
end entity ALU;

-- Define the behavior of the 8-bit ALU

architecture BEHAVIORAL of ALU is

begin

  P_ALU : process (LEFT, RIGHT, OPCODE, C_IN) is

    variable result8  : unsigned(7 downto 0);
    variable result9  : unsigned(8 downto 0);
    variable right_op : unsigned(7 downto 0);

  begin

    case OPCODE is

      when "0000" | "0001" => -- Addition or subtraction

        if (OPCODE = "0000") then
          right_op := unsigned(RIGHT);       -- Addition
        else
          right_op := unsigned(not RIGHT); -- Subtraction
        end if;

        result9 := ('0' & unsigned(LEFT)) +
                   unsigned(right_op) +
                   unsigned(std_logic_vector'(""& C_IN));
        result8 := result9(7 downto 0);

        C_OUT <= result9(8);                 -- C flag

        -- V flag
        if (((LEFT(7) XOR result8(7)) = '1') AND
```

```vhdl
                    ((right_op(7) XOR result8(7)) = '1')) then
           V_OUT <= '1';
         else
           V_OUT <= '0';
         end if;

       when "0010" =>                          -- Increment
         result8 := unsigned(LEFT) + 1;
       when "0011" =>                          -- Decrement
         result8 := unsigned(LEFT) - 1;
       when "0101" =>                          -- Bitwise AND
         result8 := unsigned(LEFT and RIGHT);
       when "0110" =>                          -- Bitwise OR
         result8 := unsigned(LEFT or RIGHT);
       when "0111" =>                          -- Bitwise XOR
         result8 := unsigned(LEFT xor RIGHT);
       when others =>
         result8 := (others => 'X');

     end case;

     RESULT <= std_logic_vector(result8);

     N_OUT <= result8(7);                      -- N flag

     if (result8 = 0) then                     -- Z flag
       Z_OUT <= '1';
     else
       Z_OUT <= '0';
     end if;

   end process P_ALU;

 end architecture BEHAVIORAL;
```

This code defines a simple ALU as a combinational circuit with a left operand, right operand, opcode, and the C flag as its inputs. The outputs are the result of the operation along with the C, N, V, and Z flags.

Next, we'll look at the purpose and functions of processor registers.

Registers

Processor registers are internal storage locations that serve as sources and destinations for instruction operations. Registers provide the quickest data access in a processor but are limited to a very small number of locations due to their high cost in terms of die area. The width of a register in bits is generally the same as the processor word size.

The 6502, as we have seen, has only three 8-bit registers: A, X, and Y. The x86 has six 32-bit registers suitable for temporary data storage: EAX, EBX, ECX, EDX, ESI, and EDI. In many processor architectures, specific registers are assigned to provide inputs required by certain instructions. For example, in the x86 architecture, a single `REP MOVSD` instruction moves a block of data with the length (in words) in ECX beginning at the source address in ESI to the destination address in EDI.

When designing a new processor architecture, it is critical to evaluate the trade-off between the number of registers and the number and complexity of instructions available to the processor. For a given integrated circuit die size and fabrication process (which together constrain the number of transistors available for the processor), adding more registers to the architecture reduces the number of transistors available for executing instructions and for performing other functions. In contrast, adding instructions with complex capabilities may limit the die space available for registers. This tension between instruction set complexity and the number of registers is expressed in the categorization of an architecture as CISC or RISC:

- **CISC (Complex Instruction Set Computer)** processors are characterized as having a rich instruction set providing a variety of features, such as the ability to load operands from memory, perform an operation, and store the result to memory, all in one instruction. In a CISC processor, an instruction may take many clock cycles to execute as the processor performs all required subtasks. The `REP MOVSD` instruction mentioned previously is an example of a single instruction with a potentially lengthy execution time. CISC processors tend to have a smaller number of registers due, in part, to the die space required for the circuitry occupied by the instruction set logic. The x86 is a classic example of CISC architecture.
- **RISC (Reduced Instruction Set Computer)** processors, on the other hand, have a smaller number of instructions that each perform simpler tasks in comparison to CISC instructions. Performing an operation on data values stored in memory might require a pair of load instructions to load two operands from memory into registers, another instruction to perform the operation, and a final instruction to store the result back to memory.

The key distinction between CISC and RISC is that RISC architectures are optimized to execute individual instructions at very high speed. Even though reading memory, performing the operation, and writing the result back to memory require several more instructions in an RISC processor than in a CISC processor, the overall start-to-finish time may be comparable or even faster for the RISC processor. Some examples of RISC architectures are ARM, discussed in *Chapter 10, Modern Processor Architectures and Instruction Sets*, and RISC-V, discussed in *Chapter 11, The RISC-V Architecture and Instruction Set*.

The reduction in instruction set complexity in RISC processors leaves more die space for registers, which means there is generally a larger number of registers in RISC processors in comparison to CISC processors. The ARM architecture, for example, has 13 general-purpose registers, while the RISC-V base 32-bit architecture has 31 general-purpose registers.

The larger number of registers in RISC architectures reduces the need to access system memory because more registers are available for storing intermediate results. This improves performance because accessing system memory is significantly more time-consuming than accessing data located in processor registers.

Think of a processor register as a set of D flip-flops in which each flip-flop contains one bit of the register's data. Each of a register's flip-flops is loaded with data by a common clock signal. Input to a register may arrive at the flip-flops after passing through a multiplexer that selects one of potentially many data sources under the control of the executing instruction.

As an alternative to using a multiplexer for this purpose, an instruction may load a register from a data bus internal to the processor. In this configuration, the control unit manages the internal bus to ensure that only the desired data source is driving the data bus lines during the clock edge that loads the register, while all other data sources on the bus are inhibited from placing data on the bus.

The following sections will introduce the full range of instructions in a processor instruction set and the addressing modes used by them.

The instruction set

Similar instructions to those discussed earlier are implemented within most general-purpose processor architectures, though more sophisticated processors augment their capabilities with additional categories of instructions. The more advanced instructions available in modern processor architectures will be introduced in later chapters.

CISC processors generally support multiple addressing modes. Addressing modes are designed to enable efficient access to sequential memory locations for use by software algorithms running on the processor. The next section describes the instruction addressing modes implemented by the 6502 processor. The section following that will introduce the categories of instructions implemented by the 6502, most of which are represented in modern processor architectures. Specialized instructions for processing interrupts and for input/output operations will then be covered, including an explanation of processor features that enable high-performance input and output operations on sizable blocks of data.

Addressing modes

CISC processors support multiple addressing modes for instructions that require transferring data between memory and registers. RISC processors have a more limited number of addressing modes. Each processor architecture defines its collection of addressing modes based on an analysis of the anticipated memory access patterns that software will use on that architecture.

To introduce the 6502 addressing modes, this section employs a simple example of 6502 code that adds together four data bytes. To avoid extraneous details, the example will ignore any carry from the 8-bit sum.

Immediate addressing mode

In immediate addressing, the operand value immediately follows the opcode in memory. For the first example, assume we are given the values of the four bytes to sum and asked to write a 6502 program to perform that task. This allows us to enter the byte values directly into our code. The bytes in this example are $01 through $04. We'll be adding the bytes together in reverse order ($04 down to $01) in anticipation of the use of a looping construct later in this section. This code uses the **immediate addressing mode** to add the four bytes together:

```
; Add four bytes together using immediate addressing mode
LDA #$04
CLC
ADC #$03
ADC #$02
ADC #$01
```

Notice that assembly language comments begin with a semicolon character. When these instructions finish execution, the A register will hold the value $0A, the sum of the four bytes listed as operands.

Recall from *Chapter 1, Introducing Computer Architecture*, that in 6502 assembly language, an immediate value is preceded by a # character and the $ character indicates the value is hexadecimal. An immediately addressed operand is read from the memory address following the instruction opcode. Immediate addressing is handy because there is no need to reserve the memory location from which to read the operand. However, immediate addressing mode is only useful when the data value is known at the time the program is written.

Absolute addressing mode

Absolute addressing mode, sometimes called **direct addressing mode**, specifies the memory location containing the value to be read or written by the instruction. The 6502 has 16 address bits, so an address field that supports accessing all available memory is two bytes long. A complete instruction to access data at an arbitrary 6502 memory location consists of three bytes: the first byte is the opcode, followed by two bytes for the address to be read or written. The two address bytes must be stored with the lower-order byte first, followed by the higher-order byte. The high-order byte of a 16-bit value contains the most significant 8 bits, and the low-order byte contains the least significant 8 bits.

The convention of storing the lower-order byte of a two-byte address at the lower memory address makes the 6502 a **little-endian** processor. The x86 is also little-endian. The ARM and RISC-V architectures allow the selection of big- or little-endian mode under software control (this is called **bi-endianness**), but most operating systems running on these architectures select little-endian mode.

For the absolute addressing mode example, we begin with some setup code to store the four bytes to be added together into addresses $200 through $203. The instructions to add the four bytes follow the setup code. This example uses absolute addressing mode to sum the four bytes:

```
; Initialize the data in memory
LDA #$04
STA $0203
LDA #$03
STA $0202
LDA #$02
STA $0201
LDA #$01
STA $0200
; Add four bytes together using absolute addressing mode
LDA $0203
```

```
CLC
ADC $0202
ADC $0201
ADC $0200
```

Unlike immediate addressing mode, absolute addressing permits summing four values that are not known until the time of program execution: the ADC instructions will add together whatever values have been stored in locations $200-$203. A limitation of this addressing mode is that the addresses of the bytes to be added must be specified when the program is written. This code cannot sum bytes located at an arbitrary location in memory.

Our simple example has the downside of unnecessarily stringing together a sequence of nearly identical instructions. To avoid this, it is usually desirable to place a repetitive sequence of code into a loop construct. The next two examples use a 6502 addressing mode that facilitates looping operations, although we won't implement a loop until the second example.

Absolute indexed addressing mode

Absolute indexed addressing mode computes a memory address by adding a base address provided in the instruction to a value contained in the X or Y register. The following example adds the bytes at addresses $0200 through $0203 using absolute indexed addressing. The X register provides an offset from the base of the byte array at address $0200:

```
; Initialize the data in memory
LDA #$04
STA $0203
LDA #$03
STA $0202
LDA #$02
STA $0201
LDA #$01
STA $0200
; Add four bytes together using absolute indexed addressing mode
LDX #$03
CLC
LDA $0200, X
DEX
ADC $0200, X
DEX
```

```
ADC $0200, X
DEX
ADC $0200, X
```

The `DEX` instruction decrements (subtracts 1 from) the X register. Although this code makes things worse in terms of increasing the number of instructions it takes to add the four bytes together, we see that the instruction sequence `DEX` followed by `ADC $0200, X` now repeats three times.

We can use conditional branching to perform the same addition sequence in a loop:

```
; Initialize the data in memory
LDA #$04
STA $0203
LDA #$03
STA $0202
LDA #$02
STA $0201
LDA #$01
STA $0200
; Add four bytes together using absolute indexed addressing mode
LDX #$03
LDA $0200, X
DEX
CLC
ADD_LOOP:
ADC $0200, X
DEX
BPL ADD_LOOP
```

The `BPL` instructions means "branch on plus," which conditionally transfers control to the instruction preceded by the `ADD_LOOP` label. `BPL` executes the branch only when the processor N flag is clear. If the N flag is set, `BPL` continues with the next instruction in memory.

The code in this example might not seem to have been worth the effort involved in constructing a loop just to add up four bytes. However, note that this version can be modified to add together 100 sequential bytes by simply changing the operand of the `LDX` instruction. Extending the previous example to add 100 bytes together in the same manner would require quite a bit more work, and the instructions would consume far more memory.

This example has the same limitation as the absolute address mode example, both of which set the start of the byte array at a memory location defined at the time the program was written. The next addressing mode removes this limitation and sums an array of bytes, starting at any address in memory.

Indirect indexed addressing mode

Indirect indexed addressing mode uses a two-byte address stored in the memory address range $00-$FF as the base address and adds the contents of the Y register to that base to produce the memory address used by the instruction. In the following example, the base address of the byte array ($0200) is first stored in little-endian order at addresses $0010 and $0011. The code uses indirect indexed addressing in a loop to add the bytes together:

```
; Initialize the data in memory
LDA #$04
STA $0203
LDA #$03
STA $0202
LDA #$02
STA $0201
LDA #$01
STA $0200
; Initialize the pointer to the byte array
LDA #$00
STA $10
LDA #$02
STA $11
; Add four bytes together using indirect indexed addressing mode
LDY #$03
LDA ($10), Y
DEY
CLC
ADD_LOOP:
ADC ($10), Y
DEY
BPL ADD_LOOP
```

With indirect indexed addressing, any memory address can be stored at addresses $10-$11 before the summing code executes. Note that indirect indexed addressing must use the Y register as the address offset. The X register is not available for use in this addressing mode.

The 6502 has some other addressing modes available: **zero-page addressing mode** provides instructions that are smaller (one byte less in length) and faster to execute for absolute and absolute indexed addressing by working only with memory addresses in the range $00-$FF. The term *zero-page* refers to the high byte of the 16-bit address, which is zero for addresses in this range. Other than exhibiting improved performance in terms of faster execution speed and reduced code memory usage, the zero-page addressing modes behave the same as the corresponding addressing modes described earlier.

Another mode is called **indexed indirect addressing mode**, which is like indirect indexed addressing except that the X register is used instead of Y, and the offset contained in X is added to the address provided in the instruction to determine the address of the pointer to the data. For example, assume that X contains the value 8. The LDA ($10, X) instruction adds the contents of X to $10, producing the result $18. The instruction then uses the 16-bit memory address read from addresses $18-$19 as the target memory address for loading the A register.

Indexed indirect addressing is not relevant to our example summing a sequence of bytes. One example application of this mode is selecting a value from a sequential list of pointers, where each pointer contains the address of a character string. The X register can reference one of the strings as an offset from the beginning of the pointer list. An instruction such as LDA ($10, X) will load the address of the selected string into A.

The addressing modes available in CISC processor architectures and, to a lesser degree, in RISC architectures are intended to support efficient methods of accessing various types of data structures in system memory.

The next section discusses the categories of instructions implemented in the 6502 architecture and how each instruction makes use of the available addressing modes.

Instruction categories

This section presents the categories of instructions available in the 6502 processor. The purpose of discussing the 6502 here is to introduce the concepts associated with the instruction set of a processor architecture that is simpler than the modern 32- and 64-bit processors we will examine in later chapters. By the time we get to those processors, the underlying instruction set concepts should be quite familiar.

Memory load and store instructions

The 6502 uses load and store instructions to read data values from system memory into processor registers and to write registers out to system memory. In the 6502 architecture, the LDA, LDX, and LDY instructions load the register identified in the instruction mnemonic with an 8-bit word from system memory. LDA supports all addressing modes available in the 6502, while LDX and LDY each support a more limited subset of addressing modes: immediate, absolute, and absolute indexed.

After each of these instructions finishes executing, the N and Z flags indicate whether the value that was loaded is negative (that is, bit 7 is set) and whether the value is zero.

STA, STX, and STY store the register identified in the instruction to memory. Each store instruction supports the same addressing modes as the load instruction for that register, except the store instructions do not support immediate addressing mode. These instructions update the N and Z flags to reflect the value stored.

Register-to-register data transfer instructions

These instructions copy an 8-bit word from one of the A, X, and Y registers to another register. These instructions use **implied addressing mode**, which means the source and destination of each instruction are indicated directly by the instruction opcode.

TAX copies the A register contents to the X register. TAY, TXA, and TYA perform similar operations between the register pairs indicated by the instruction mnemonic. These instructions update the N and Z flags.

Stack instructions

The TXS instruction copies the X register to the stack pointer (S) register. This instruction must be used to initialize the S register during system startup. TSX copies the S register to the X register. TSX updates the N and Z flags. TXS does not affect any flags.

PHA pushes the A register contents onto the stack. PHP pushes the processor flags onto the stack as an 8-bit word. These instructions do not affect the processor flags. Pushing a value onto the stack consists of writing the register to the memory address computed by adding $100 to the S register and then decrementing the S register.

PLA and PLP pop the A register and the flags register from the stack, respectively. Popping a value first increments the S register and then transfers the value at the location computed by adding $100 to the S register to the target register location.

PLA updates the N and Z flags. PLP sets or clears six of the seven processor flags based on the value popped from the stack. The B (break) flag is only meaningful in a copy of the processor flags register that has been pushed onto the stack by an interrupt or by the PHP instruction. This distinguishes a BRK instruction from a hardware interrupt request. Both the PHP and BRK instructions push the flags register with the B bit (bit 4) set.

Hardware interrupts generated via the processor $\overline{\text{IRQ}}$ (Interrupt Request) and $\overline{\text{NMI}}$ (Non-Maskable Interrupt) pins push the processor flags register with the B bit cleared. Interrupt processing and the BRK instruction will be discussed later in this chapter.

Arithmetic instructions

As we've seen, addition and subtraction are performed by the ADC and SBC instructions. The left operand of each instruction is the A register, which is also the destination for the result of the operation. All addressing modes are available for designating the right operand. The Z, C, N, and V flags are updated to reflect the result of the operation.

INC and DEC, respectively, increment or decrement the specified memory location by adding 1 to, or subtracting 1 from, the value at that location. The result is stored at the same memory location. Absolute and absolute indexed addressing modes are supported. These instructions update the N and Z flags based on the result of the operation.

The INX, DEX, INY, and DEY instructions increment or decrement the X or Y register, as indicated by the mnemonic. These instructions update the N and Z flags.

The CMP instruction performs a comparison by subtracting the operand value from the A register. The behavior of CMP is very similar to the instruction sequence SEC followed by SBC. The N, Z, and C flags are set to reflect the result of the subtraction. The differences between CMP and SBC are as follows:

- CMP discards the result of the subtraction (the value in A is unaffected by the CMP instruction)
- CMP does not use decimal mode if the D flag is set
- CMP does not affect the value of the V flag
- CMP supports all addressing modes

The CPX and CPY instructions are like CMP, except the X or Y register is used as the left operand as indicated in the mnemonic, and only absolute and absolute indexed addressing modes are supported for the right operand.

Logical instructions

The `AND`, `EOR`, and `ORA` instructions perform bitwise AND, XOR, and OR operations, respectively, between the A register and the operand. The result is stored in the A register. The Z and N flags are updated to reflect the result of the operation. All addressing modes are supported.

The `ASL` instruction shifts the operand one bit left, inserting a zero as the least significant bit. The most significant bit is shifted into the C flag. This is equivalent to multiplying the A register by two and placing the most significant bit of the 9-bit result in C.

Similar to `ASL`, `LSR` shifts the operand one bit right, inserting a zero as the most significant bit. The least significant bit is shifted into the C flag. This is equivalent to the division of an unsigned operand by two, with the remainder placed in C.

The `ROL` and `ROR` instructions shift the A register one bit to the left or right, respectively. The previous value of the C flag is shifted into the bit location vacated by the shift operation. The bit shifted out of A is stored in the C flag.

`ASL`, `LSR`, `ROL`, and `ROR` support **accumulator addressing mode**, which uses the A register as the operand. This mode is specified by using the special operand value "A," as in `ASL A`. These four instructions also support absolute and absolute indexed addressing modes.

The `BIT` instruction performs a bitwise AND between the operand and the A register, and the result is discarded. The Z flag is updated based on the result of this operation. Bits 7 and 6 from the memory location are copied to the N and V flags, respectively. Only absolute addressing mode is supported.

Branching instructions

The `JMP` instruction loads the operand into the PC and continues execution with the instruction at that location. The destination, a two-byte absolute address, can be anywhere in the 6502's address space:

- The `BCC` and `BCS` instructions perform conditional branching if the C flag is clear or set, respectively
- The `BNE` and `BEQ` instructions perform conditional branching if the Z flag is clear or set, respectively
- The `BPL` and `BMI` instructions perform conditional branching if the N flag is clear or set, respectively

- The BVC and BVS instructions perform conditional branching if the V flag is clear or set, respectively

The conditional branch instructions use **relative addressing mode**, where the target address is a signed 8-bit offset (in the range -128 to +127 bytes) from the address of the instruction following the branch instruction.

Subroutine call and return instructions

The JSR instruction pushes the address of the instruction following the JSR instruction (minus one) onto the stack, loads the address provided as the 16-bit operand into the PC, and then continues execution from the instruction at that location.

RTS is used to end a subroutine. The return PC value (minus one) is pulled from the stack and loaded into the PC. The RTS instruction increments the PC before it is used as the address of the next instruction to execute.

Processor flag instructions

The processor flag instructions operate directly on the flags, setting or clearing a single flag.

- The SEC and CLC instructions set and clear the C flag, respectively.
- The SED and CLD instructions set and clear the D flag, respectively.
- The CLV instruction clears the V flag. No instruction simply sets the V flag.

Interrupt-related instructions

Interrupt-related instructions allow the processor to manage the handling of externally generated interrupts and to generate software-triggered interrupts. External interrupts can be of two types: maskable and non-maskable. Each interrupt type is triggered by its own input pin on the 6502 processor.

Maskable interrupts can be disabled by setting the processor I flag. When masked, the processor ignores the associated input pin. Non-maskable interrupts, as the name implies, cannot be inhibited and will always initiate a processor interrupt when the appropriate signal transition occurs on the associated pin. We'll cover interrupt processing in more detail in an upcoming section.

The SEI and CLI instructions set and clear the I flag, respectively. When the I flag is set, maskable interrupts are disabled, or masked.

The BRK instruction triggers a non-maskable interrupt. The memory address two bytes after the BRK instruction is pushed onto the stack, followed by the processor flags register. The PC is loaded with the interrupt handler address, which is read from memory addresses $FFFE-$FFFF. The interrupt handler then begins to execute.

The BRK instruction does not alter any register contents (other than the stack pointer) or processor flags. The flags register pushed onto the stack has the B bit set to indicate that the interrupt is the result of a BRK instruction.

RTI returns from an interrupt handler. This instruction restores the processor flags from the stack and restores the PC. After the processor flags have been restored, the B flag is not meaningful and should be ignored.

Interrupt processing and the use of the BRK instruction will be discussed further in the *Interrupt processing* section of this chapter.

No operation instruction

The NOP instruction (often referred to as *no-op*) does nothing except advance the PC to the following instruction.

NOP instructions are sometimes used as a debugging tool during program development. For example, one or more instructions can be effectively "commented out" by filling the memory addresses for those instructions with $EA bytes. $EA is the hexadecimal value of the 6502 NOP opcode.

Interrupt processing

Processors generally support some form of interrupt handling for responding to service requests from external devices. Conceptually, interrupt handling resembles a scenario in which you are busy working on a task and your phone rings. After answering the call and perhaps jotting a note for later action ("buy bread," for example), you resume the interrupted task. We humans employ several similar mechanisms, such as doorbells and alarm clocks, which enable us to interrupt lower priority activities and respond to more immediate needs.

$\overline{\text{IRQ}}$ processing

The 6502 integrated circuit has two input signals that allow external components to notify the processor of a need for attention. The first is the interrupt request input, $\overline{\text{IRQ}}$. $\overline{\text{IRQ}}$ is an active low (meaning the signal is at its low, or 0, level; that's what the bar over the IRQ characters means) input that generates a processor interrupt when pulled low. Think of this signal as a telephone ringer notifying the processor of an incoming call.

The 6502 cannot respond instantly to a low signal level on the $\overline{\text{IRQ}}$ input. Before the 6502 can begin to process the interrupt, it must first complete the instruction already in progress. Next, it pushes the return address (the address of the next instruction that would have been executed after the instruction in progress) onto the stack, followed by the processor flags register. Since this interrupt was generated by the $\overline{\text{IRQ}}$ input, the B flag in the processor flags on the stack will be 0.

Unlike the `JSR` instruction, the return address pushed in response to the $\overline{\text{IRQ}}$ input is the actual address of the next instruction to be executed, rather than the instruction address minus 1. The interrupt return address will not be incremented to generate the return address as occurs during RTS instruction execution.

In the next stage of interrupt processing, the processor loads the address of the $\overline{\text{IRQ}}$ handler routine from memory addresses `$FFFE-$FFFF` into the PC. The 6502 then begins executing the interrupt handler code at that address. The interrupt handler is the code that identifies the peripheral that initiated the interrupt and performs the processing required to satisfy the request, and then returns control to the code that was executing prior to the interrupt.

When the interrupt handler is finished, it executes the `RTI` instruction. `RTI` pops the processor flags and the PC from the stack and resumes execution at the instruction following the instruction that was in progress when the $\overline{\text{IRQ}}$ input was driven low.

The $\overline{\text{IRQ}}$ input is a **maskable interrupt**, meaning it is possible to perform the equivalent of putting the telephone ringer on mute. When $\overline{\text{IRQ}}$ processing begins, the 6502 automatically sets the I flag, which masks (disables) the $\overline{\text{IRQ}}$ input until the I flag is cleared.

The I flag will be cleared when the `RTI` instruction restores the processor flags because the I flag could not have been set when the processor began responding to the $\overline{\text{IRQ}}$. The I flag can also be cleared by the `CLI` instruction, which means it is possible to enable $\overline{\text{IRQ}}$ interrupts while processing an $\overline{\text{IRQ}}$ interrupt. An interrupt handled while processing another interrupt is referred to as a **nested interrupt**.

The $\overline{\text{IRQ}}$ input is level-sensitive, which means any time the $\overline{\text{IRQ}}$ input is low and the I flag is cleared, the processor will initiate the interrupt processing sequence. One consequence of this is that, at the completion of processing an interrupt, the 6502's interactions with the interrupt source must ensure that the $\overline{\text{IRQ}}$ input is no longer low. If $\overline{\text{IRQ}}$ remains low when the `RTI` instruction is executed, the 6502 will immediately begin the interrupt handling process all over again.

Interrupts initiated via the $\overline{\text{IRQ}}$ input handle most routine interactions between the 6502 and peripheral devices. For example, the keyboard is an interrupt source in most computers.

Each keypress generates an $\overline{\text{IRQ}}$ interrupt. During keyboard interrupt processing, the 6502 reads the identification of the key from the keyboard interface and stores it into a queue for later processing by the currently active application. The $\overline{\text{IRQ}}$ handler code does not need to know anything about what the key press information will be used for; it just saves the data for later use.

$\overline{\text{NMI}}$ processing

The second interrupt input to the 6502 is the **non-maskable interrupt**, $\overline{\text{NMI}}$. As its name implies, the $\overline{\text{NMI}}$ input is not masked by the I flag. $\overline{\text{NMI}}$ is an edge-sensitive input that triggers on the falling edge of the signal.

The processing of $\overline{\text{NMI}}$ interrupts is similar to the processing of $\overline{\text{IRQ}}$ interrupts, except the address of the interrupt handler routine is loaded from memory addresses $FFFA-$FFFB and the I flag has no effect on this type of interrupt.

Because $\overline{\text{NMI}}$ is non-maskable, it can be triggered at any time, including when the 6502 is in the middle of handling an $\overline{\text{IRQ}}$ interrupt, or even while handling an earlier $\overline{\text{NMI}}$ interrupt.

The $\overline{\text{NMI}}$ input is normally reserved for very high-priority conditions that cannot be delayed or missed. One possible use of $\overline{\text{NMI}}$ interrupts is to trigger the incrementing of a real-time clock at regular intervals.

This example of $\overline{\text{NMI}}$ handler code increments a 32-bit clock counter located at addresses $10-$13 each time the interrupt occurs:

```
; Increment a 32-bit clock counter at each /NMI interrupt
NMI_HANDLER:
INC $10
BNE NMI_DONE
INC $11
BNE NMI_DONE
INC $12
BNE NMI_DONE
INC $13
NMI_DONE:
RTI
```

When referring to hardware signals in program source code, a leading forward slash can be used to indicate an active low signal. $\overline{\text{NMI}}$ is represented as /NMI in the preceding code comment.

BRK instruction processing

The BRK instruction triggers processing that is very similar to an $\overline{\text{IRQ}}$ interrupt. Because BRK is an instruction, there is no need to wait for the completion of an instruction in progress before initiating interrupt processing. During BRK execution, the return address (the address of the BRK instruction plus 2) and the processor flags are pushed onto the stack, much like the response to a low level on the $\overline{\text{IRQ}}$ input. Note that by adding 2 to the BRK instruction address, the return address is not pointed to the byte after BRK, but to the second byte after it.

The BRK instruction is non-maskable: the state of the I flag does not affect the execution of the BRK instruction.

The BRK handler shares the same address as the $\overline{\text{IRQ}}$ handler, which is located at memory addresses $FFFE-$FFFF. Since the BRK instruction and $\overline{\text{IRQ}}$ use the same handler, the B flag must be consulted to identify the interrupt source during processing. The B flag in the processor flags pushed onto the stack (this is *not* the B flag in the processor flags (P) register) is set in response to a BRK instruction and clear when processing an $\overline{\text{IRQ}}$ interrupt.

The BRK instruction is not present in most 6502 applications. A traditional use of this instruction is to set breakpoints when debugging a program. By temporarily replacing the opcode byte at the desired break location with a BRK instruction, the debugging program (often called a **monitor** in smaller computer systems) can gain control, allowing the user to display and modify register contents and memory locations before resuming execution.

The following example code implements a minimal handler that differentiates between $\overline{\text{IRQ}}$ interrupts and BRK instructions. It uses memory address $14 as a temporary storage location:

```
; Handle /IRQ interrupts and BRK instructions
IRQ_BRK_HANDLER:
; Save the A register
STA $14
; Retrieve the processor flags from the stack into A
PLA
PHA
; Check if the B bit is set in the flags on the stack
AND $10 ; $10 selects the B bit
; If the result is nonzero, B was set: Handle the BRK
BNE BRK_INSTR
; B was not set: Handle the /IRQ here
```

```
; …
JMP IRQ_DONE
BRK_INSTR:
; Handle the BRK instruction here
; …
IRQ_DONE:
; Restore the A register and return
LDA $14
RTI
```

This example showed how to differentiate between interrupts initiated by the processor $\overline{\text{IRQ}}$ input and those resulting from the BRK instruction in the 6502 architecture. In more sophisticated processors, including those we will discuss in later chapters, there are unique interrupt vectors (interrupt handler starting addresses) for each interrupt input signal. These architectures also contain extensive support for debugging activities such as setting breakpoints at specified instruction locations.

The preceding sections introduced the categories of instructions in the 6502 architecture and provided a brief description of each instruction within those categories. Although the 6502 is much simpler than modern 32- and 64-bit processors, this discussion presented the most common types of instructions and addressing modes used in even the most sophisticated modern processors, including instructions supporting the universal concept of interrupt processing.

The next section will introduce the fundamentals of I/O processing, which performs data transfer between the processor and peripheral devices.

Input/output operations

The goal of the I/O portion of a processor architecture is to efficiently transfer data between external peripheral devices and system memory. Input operations transfer data from the external world into memory and output operations send data from memory to an outside destination.

The format of the data on the external side of the I/O interface varies widely. Here are some examples of the external representations of computer I/O data:

- Signals on a video cable connected to a monitor
- Voltage fluctuations on the wires in an Ethernet cable
- Magnetic patterns on the surface of a disk
- Sound waves produced by computer speakers

Regardless of the form the data takes when it is outside the computer, the connection of any I/O device with the processor must comply with the processor's I/O architecture and the I/O device must be compatible with any other I/O devices that happen to be present in the computer system.

The processor uses the instruction categories, addressing modes, and interrupt processing methods described earlier in this chapter to conduct interactions with I/O devices. The difference here is that instead of reading and writing system memory, the instructions read from, and write to, locations that communicate with an I/O device.

Memory-mapped I/O and **port-mapped I/O** are the two main approaches employed in modern processors to access I/O devices. Memory-mapped I/O dedicates portions of the system address space to I/O devices. The processor accesses peripheral devices at predefined addresses using the same instructions and addressing modes it uses to read and write system memory. The 6502 employs memory-mapped I/O to communicate with its peripherals.

Processors that use port-mapped I/O implement a separate category of instructions for performing I/O operations. Port-mapped I/O devices have a dedicated address space independent from system memory. I/O devices are assigned **port numbers** as addresses. The x86 architecture employs port-mapped I/O.

One drawback of memory-mapped I/O is the need to dedicate part of the system address space to I/O devices, thereby reducing the maximum amount of memory that can be installed in the computer system. A drawback of port-mapped I/O is the requirement for the processor to implement additional instructions to perform I/O operations.

Some implementations of port-mapped I/O provide additional hardware signals to indicate when an I/O device is being addressed as opposed to system memory. Using this signal as a selector (which effectively becomes another address bit), the same address lines can be used for accessing memory and I/O devices. Alternatively, some higher-end processors implement an entirely separate bus for performing port-mapped I/O operations. This architecture allows I/O and memory access operations to proceed simultaneously.

In the simplest approach to I/O, the processor handles all steps in an I/O operation itself, using instructions to transfer data between memory and the I/O device. More complex processor architectures provide hardware features to accelerate repetitive I/O operations. We will discuss three methods of performing I/O with varying degrees of processor involvement: programmed I/O, interrupt-driven I/O, and direct memory access.

Programmed I/O

Programmed I/O simply means that the processor performs every step of the I/O data transfer operation using program instructions. Consider a keyboard that presents itself to the processor as two memory-mapped one-byte addresses in the processor's I/O address region. One of these bytes contains status information, specifically a bit indicating when a key has been pressed. The second byte contains the value of the key that was pressed.

Each time a key is pressed, the *key available* status bit is set. When using programmed I/O, the processor must periodically read the keyboard status register to see whether a key has been pressed. If the status bit indicates a key has been pressed, the processor reads the keyboard data register, which turns off the key available status bit until the next keypress occurs.

If the keyboard data register can only hold one key at a time, this keyboard status checking operation must occur frequently enough that no key presses get lost, even when a very fast typist is at the keyboard. As a result, the processor must spend a significant amount of its time checking to see whether a key has been pressed. Most of these checks will be fruitless whenever fast typing is not taking place.

It should be clear that programmed I/O is not a very efficient method for general usage. It is similar in concept to checking your phone every few seconds to see if someone is calling you.

Programmed I/O makes sense in some situations. For example, the one-time configuration of a peripheral device during system startup is a reasonable application of this technique.

Interrupt-driven I/O

An I/O device can use interrupts to notify the processor when action is needed. In the case of the simple keyboard interface, instead of merely setting a bit in a status register, the peripheral could pull the 6502's $\overline{\text{IRQ}}$ line low to initiate an interrupt each time a key is pressed. This allows the processor to go about its business without constantly checking for keypresses. The processor will only focus attention on the keyboard interface when there is work to be done, as indicated by the interrupt. Using interrupts to trigger I/O operations is analogous to adding a ringer to the phone that we had to check for incoming calls every few seconds when using programmed I/O.

The 6502 has a single maskable **interrupt input signal** ($\overline{\text{IRQ}}$) available for I/O operations. Computer systems usually contain multiple sources of I/O interrupts. This makes the task of servicing interrupts a bit more complicated in the 6502 because the processor must first identify which peripheral initiated the interrupt before it can begin transferring data.

The interrupt handler has to poll each interrupt-capable device to locate the interrupt source. In the case of the keyboard interface, this polling operation consists of reading the keyboard status register to determine if the bit is set, indicating a keypress occurred. Once the processor has identified the device responsible for the interrupt, it branches to code that interacts with the device to complete the requested I/O task. In the case of the keyboard interface, this processing performs the steps of reading the keyboard data register and clearing the key available status bit, which deactivates the $\overline{\text{IRQ}}$ input signal.

Interrupts from external devices are asynchronous events, meaning they can occur at any time. Computer system design requires careful consideration of the possibility that interrupts may be generated at potentially unexpected times, such as during system startup or while processing other interrupts. Interrupts from multiple devices may occur simultaneously, or nearly simultaneously, and in random order. Interrupt-handling hardware circuitry and interrupt-servicing code must ensure that all interrupts are detected and processed regardless of the existence of these timing peculiarities.

Interrupt-driven I/O eliminates the processor's need to periodically check I/O devices to see whether action is needed. However, handling an interrupt may consume significant processor time if it involves transferring a large block of data. This occurs frequently during operations such as reading from, or writing to, a disk drive. The next I/O method we will discuss removes the need for the processor to perform the work of transferring these large blocks of data.

Direct memory access

Direct Memory Access (DMA) permits peripheral device I/O operations to access system memory independent of the processor. When using DMA to transfer a block of data, the processor sets up the operation by configuring a DMA controller with the starting address of the data block to be transferred, the block length, and the destination address. After initiating the DMA, the processor is free to continue other work. Following completion of the operation, the DMA controller generates an interrupt to inform the processor that the transfer is complete.

Within a computer system, a DMA controller may be implemented as a separate integrated circuit managed by the processor, or a processor architecture may contain one or more integrated DMA controllers.

I/O devices that move substantial amounts of data, such as disk drives, sound cards, graphics cards, and network interfaces, generally rely on DMA to efficiently transfer data into and out of system memory. DMA is also useful for transferring blocks of data within system memory.

The 6502 architecture does not support DMA operations, but the original IBM PC included a DMA controller. Almost every 32-bit and 64-bit processor architecture provides extensive support for DMA operations.

DMA is one of many techniques that improve computer system performance by accelerating repetitive operations. In *Chapter 5, Hardware-Software Interface* and *Chapter 9, Specialized Processor Extensions*, we will see some examples of the use of DMA to accelerate I/O operations.

Summary

This chapter described the primary functional units of a simple processor: the control unit, the ALU, and the registers. An overview of processor instructions and addressing modes followed. The instruction categories implemented by the 6502 processor were introduced with the goal of demonstrating the variety and utility of instructions available in a relatively simple processor architecture.

The concepts involved in interrupt processing were introduced and demonstrated in the context of the 6502 architecture. The chapter concluded with an overview of the most common architectural approaches to I/O operations (memory-mapped I/O and port-mapped I/O) and the basic modes of performing I/O in a computer system (programmed I/O, interrupt-driven I/O, and DMA).

Having completed this chapter, you should now possess a conceptual understanding of processor functional units, instruction processing, interrupt handling, and input/output operations. This information forms the basis for the next chapter, which covers architecture at the computer system level.

Exercises

1. Consider the addition of two signed 8-bit numbers (that is, numbers in the range -128 to +127) where one operand is positive and the other is negative. Is there any pair of 8-bit numbers of different signs that, when added together, will exceed the range -128 to +127? This would constitute a signed overflow. Note: We're only looking at addition here because, as we've seen, subtraction in the 6502 architecture is the same as addition with the right operand's bits inverted.

2. If the answer to *Exercise 1* is "no," this implies the only way to create a signed overflow is to add two numbers of the same sign. If an overflow occurs, what can you say about the result of performing XOR between the most significant bit of each operand with the most significant bit of the result? In other words, what will be the result of the expressions, `left(7) XOR result(7)` and `right(7) XOR result(7)`? In these expressions, (7) indicates bit 7, the most significant bit.

3. Review the VHDL listing in the *Arithmetic logic unit* section in this chapter and determine whether the logic for setting or clearing the V flag is correct for addition and subtraction operations. Check the results of adding 126+1, 127+1, -127+(-1), and -128+(-1).

4. When transferring blocks of data over an error-prone transmission medium, it is common to use a **checksum** to determine whether any data bits were lost or corrupted during transmission. The checksum is typically appended to the transferred data record. One checksum algorithm uses these steps:

    1. Add all the bytes in the data record together, retaining only the lowest 8 bits of the sum
    2. The checksum is the two's complement of the 8-bit sum
    3. Append the checksum byte to the data record

   After receiving a data block with the appended checksum, the processor can determine whether the checksum is valid by simply adding all the bytes in the record, including the checksum, together. The checksum is valid if the lowest 8 bits of the sum are zero. Implement this checksum algorithm using 6502 assembly language. The data bytes begin at the memory location stored in addresses $10-$11 and the number of bytes (including the checksum byte) is provided as an input in the X register. Set the A register to 1 if the checksum is valid, and to 0 if it is invalid.

5. Make the checksum validation code from *Exercise 4* into a labeled subroutine that can be called with a JSR instruction and that ends with an RTS instruction.

6. Write and execute a set of tests to verify the correct operation of the checksum testing subroutine you implemented in *Exercises 4-5*. What is the shortest block of data on which your code can perform checksum validation? What is the longest block?

Join our community Discord space

Join the book's Discord workspace for a monthly *Ask me Anything* session with the author:
https://discord.gg/7h8aNRhRuY

4
Computer System Components

This chapter introduces the lower-level components used in the construction of computer systems. We begin with the **metal-oxide-semiconductor (MOS)** transistor, which is employed extensively in memory circuits and across virtually all other modern digital devices. We will examine the design of MOS transistor-based computer memory and its interface with the processor. We'll look at modern computer input/output interfaces, with a focus on the use of high-speed serial communication within the computer case, as well as data transfer over cable connections to external components. The functional requirements of system I/O devices including the graphics display, network interface, keyboard, and mouse will be discussed. The chapter ends with a descriptive example of the specifications for a modern computer motherboard.

After completing this chapter, you will have a solid understanding of the hardware components of modern computer systems, from technical specifications down to the circuit level. You will have learned how system memory is implemented, including the basics of caching. You will understand the mechanisms of efficient I/O operations and the use of **Universal Serial Bus (USB)** to connect the keyboard, mouse, and other I/O devices. You will understand the computer's network interface and will be familiar with the key technical aspects of a modern computer motherboard.

The following topics will be covered in this chapter:

- Memory subsystem
- Introducing the MOSFET
- Constructing DRAM circuits with MOSFETs
- I/O subsystem
- Graphics displays

- Network interface
- Keyboard and mouse
- Modern computer system specifications

Technical requirements

Files for this chapter, including answers to the exercises, are available at https://github.com/PacktPublishing/Modern-Computer-Architecture-and-Organization-Second-Edition.

Memory subsystem

The **memory subsystem** is an addressable sequence of storage locations containing instructions and data for use by the processor as it executes programs. Modern computer systems and digital devices often contain over a billion 8-bit storage locations in main memory, each of which can be independently read and written by the processor.

As we saw in *Chapter 1, Introducing Computer Architecture*, the design of the Babbage Analytical Engine included a collection of axes, each holding 40 decimal digit wheels, as the means of storing data during computations. Reading data from an axis was a destructive operation, resulting in zeros on each of an axis's wheels after the read was complete. This was an entirely mechanical method of data storage.

From the 1950s to the 1970s, the preferred implementation technology for digital computer memory was the magnetic core. One bit of core memory is stored in a small toroidal (donut-shaped) ceramic permanent magnet. The set of cores making up a memory array is arranged in a rectangular grid with horizontal and vertical connecting wires. Writing to a bit location involves introducing enough current in the wires connected to the bit location to flip the polarity of the core's magnetic field. A 0 bit might be defined as clockwise magnetic flux circulation within the core and a 1 bit as counterclockwise flux circulation.

Reading a bit from core memory consists of attempting to set the bit to the 0 polarity and observing the electrical response. If the selected core already contains a 0 bit, there will be no response. If the core holds a 1, a detectable voltage pulse will occur as the polarity changes. As in the Analytical Engine, a core memory read operation is destructive. After reading a bit value of 1 from memory, a subsequent write must be performed to restore the state of the bit.

Magnetic core memory is non-volatile: the contents will be retained indefinitely after power has been removed. It also has characteristics that make it valuable in applications such as spacecraft where radiation tolerance is important. In fact, the Space Shuttle computers employed core memory into the late 1980s.

Modern consumer and business computer systems use MOSFET-based DRAM circuits almost exclusively for main system memory. The next section presents the features of the MOSFET.

Introducing the MOSFET

Chapter 2, Digital Logic, described the NPN transistor, a type of **bipolar junction transistor** (BJT). The NPN transistor is called bipolar because it relies on both positive (P) and negative (N) charge carriers to function.

In semiconductors, electrons serve as the negative charge carriers. There are no physical particles with a positive charge involved in a semiconductor operation. Instead, the absence of a normally present electron in an atom exhibits the same properties as a positively charged particle. These missing electrons are referred to as **holes**. Holes function as the positive charge carriers in bipolar junction transistors.

The concept of holes is so fundamental to semiconductor operation that William Shockley, one of the inventors of the transistor, wrote a book entitled *Electrons and Holes in Semiconductors*, published in 1950. We'll next examine the behavior of positive and negative charge carriers in unipolar transistors.

As an alternative to the BJT transistor structure, the **unipolar transistor** relies on only one of the two types of charge carriers. The **metal-oxide-semiconductor field-effect transistor** (MOSFET) is a unipolar transistor suitable for use as a digital switching element. Like the NPN transistor, the MOSFET is a three-terminal device that employs a control input to turn the flow of current across the other two terminals on and off. The terminals of a MOSFET are named **gate**, **drain**, and **source**. The gate terminal controls the flow of current between the drain and source terminals.

MOSFETs are categorized as **enhancement mode** or **depletion mode** devices. An enhancement mode MOSFET blocks current flow between drain and source when the gate voltage is zero and allows current flow when the gate voltage is above the threshold voltage. A depletion mode MOSFET functions in the opposite manner, blocking current flow when the gate voltage is high and allowing current flow when the gate voltage is zero.

The following figure is the schematic representation of an n-channel enhancement mode MOSFET:

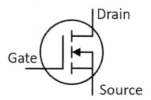

Figure 4.1: N-channel enhancement mode MOSFET

For our purposes, the n-channel enhancement mode MOSFET functions as a switch: when the gate terminal voltage is low (below the threshold voltage) relative to the source terminal, there is very high resistance between the drain and source terminals. When the gate terminal voltage is high (above the threshold voltage) relative to the source terminal, there is very little resistance between those terminals. The "n" in n-channel refers to a channel in the silicon that has been doped to provide an increased number of electrons (negative charge carriers).

The MOSFET's behavior resembles the operation of the NPN transistor discussed in *Chapter 2, Digital Logic*. There is, however, a key difference: the MOSFET is a voltage-controlled device, while the NPN transistor is a current-controlled device. The base terminal of the NPN transistor requires a small but steady current to activate the device as a switch, thereby allowing current to flow between the emitter and collector terminals. The MOSFET, on the other hand, requires only a gate-to-source voltage above a threshold to switch the current flow on between the drain and source terminals. The gate input requires almost no current flow to keep the switch open. Because of this, a MOSFET consumes significantly less power than an NPN transistor performing the equivalent digital function.

Mohamed Atalla and Dawon Kahng invented the MOSFET at Bell Telephone Laboratories in 1959. It was not until the early 1970s that production processes had matured sufficiently to support the reliable production of MOS integrated circuits. Since then, the MOSFET has been by far the most common type of transistor used in integrated circuits. In 2018, it was estimated that 13 sextillion (a **sextillion** is 1 followed by 21 zeros) transistors had been manufactured, 99.9 percent of which were MOSFETs. The MOSFET is the most frequently manufactured device in human history.

The p-channel enhancement mode MOSFET is like the n-channel enhancement mode MOSFET, except it exhibits the opposite response to the gate-to-source voltage: a gate terminal voltage lower than the source terminal beyond the threshold voltage allows current to flow between the drain and source, while a gate-to-source voltage less than the threshold inhibits current between the drain and source. The "p" in p-channel refers to channel doping that increases the number of holes (positive charge carriers). The following figure is the schematic diagram of a p-channel enhancement mode MOSFET:

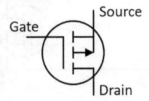

Figure 4.2: P-channel enhancement mode MOSFET

To distinguish between the schematic representations of the n-channel and p-channel MOSFETs, observe that the source terminal is connected to the center of the three internal connections in both configurations. The directional arrow on this connection points toward the gate in an n-channel MOSFET and away from the gate in a p-channel MOSFET.

Both the n-channel and p-channel enhancement mode MOSFETs can be considered normally open switches, meaning they do not conduct current when the gate-source voltage difference is small. Both n- and p-channel MOSFETs are also available in the depletion mode configuration, which causes them to function as normally closed switches. Current flows in depletion mode MOSFETs when the gate-source voltage is small but not when it is large.

MOS transistors are frequently employed in an n-channel and p-channel pair to perform logic functions. A device built with these pairs of MOS transistors is called a **complementary MOS (CMOS)** integrated circuit. The CMOS name derives from the fact that the transistors in each pair function in opposite, or complementary, fashion. Except when switching is taking place, CMOS circuits consume almost no power because the gate inputs require essentially no current. Chih-Tang Sah and Frank Wanlass of Fairchild Semiconductor developed the CMOS circuit structure in 1963.

The following diagram shows a NOT gate circuit where the NPN transistor of *Chapter 2, Digital Logic*, has been replaced by a pair of complementary MOSFETs:

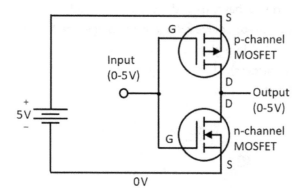

Figure 4.3: CMOS NOT gate circuit

When the *Input* signal is low (near 0 V), the lower n-channel MOSFET is switched off and the upper p-channel MOSFET is switched on. This connects the *Output* to the positive side of the voltage source, raising the *Output* signal to nearly 5 V. When *Input* is high, the upper MOSFET is switched off and the lower MOSFET is switched on, pulling the *Output* down to near 0 V. The *Output* signal is always the inverse of the *Input* signal, which represents the behavior of a NOT gate.

Today, virtually all high-density digital integrated circuits are based on CMOS technology. In addition to performing logic functions, the MOSFET is a key component of modern **random-access memory (RAM)** circuit architecture. The next section discusses the use of the MOSFET in memory circuits.

Constructing DRAM circuits with MOSFETs

A single bit in a standard **dynamic random-access memory (DRAM)** integrated circuit is composed of two circuit elements: a MOSFET and a capacitor. The following section presents a brief introduction to the electrical characteristics of capacitors.

The capacitor

A **capacitor** is a two-terminal passive circuit element capable of storing energy. Energy enters and leaves a capacitor as electrical current. The voltage across the capacitor terminals is proportional to the quantity of electrical energy contained in the capacitor.

To continue the hydraulic system analogy introduced in *Chapter 2, Digital Logic,* think of a capacitor as a balloon attached to the side of the pipe leading to a water tap. Water pressure in the pipe causes the balloon to inflate, filling it with some of the water from the pipe. Let's assume this is a strong balloon, and that as it inflates, the balloon stretches, increasing the pressure within. The balloon fills until the pressure in the balloon equals the pressure in the pipe, and it then stops filling.

If you open the tap at the end of the pipe all the way, the release of water causes the pressure in the pipe to decrease. Some of the water in the balloon will flow back into the pipe until the balloon pressure again equals the pipe pressure.

Hydraulic devices called **water hammer arrestors** function in exactly this manner to solve the problem of pipes that make banging noises when water taps are turned on and off. A water hammer arrestor uses balloon-stretching-like behavior to smooth out the abrupt changes in water pressure that result from taps opening and closing.

The quantity of electrical energy contained in a capacitor is analogous to the amount of water in the balloon. The voltage across the capacitor is analogous to the pressure inside the balloon exerted by stretching.

An electrical capacitor can be constructed from two parallel metal plates separated by an insulating material, such as air. One terminal is connected to each of the plates. The ratio of the quantity of stored electrical energy to the voltage across the capacitor is called **capacitance**. Capacitance depends on the size of the parallel plates, the distance separating them, and the type of material used as the insulator between them. The capacitance of a capacitor is analogous to the size of the balloon in the hydraulic example. A capacitor with a larger capacitance corresponds to a larger balloon. A large balloon requires more water to fill to a given pressure than a small balloon.

The schematic symbol for a capacitor is shown in the following figure:

Figure 4.4: Capacitor schematic symbol

The two horizontal lines with space between them represent the metal plate capacitor architecture described in this section. The unit of capacitance is the **farad**, named after the English scientist Michael Faraday, who, among many other achievements, invented the electric motor.

The DRAM bit cell

A DRAM bit cell is a readable and writeable storage location for a single bit of data. A DRAM module in a modern computer or phone contains billions of bit cells. A single bit in a DRAM circuit consists of a MOSFET and a capacitor, arranged as follows:

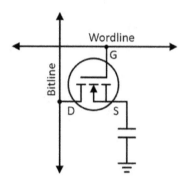

Figure 4.5: DRAM bit cell circuit

In this figure, the symbol with three horizontal lines at the bottom right is the **ground** symbol, which is the standard representation of the 0 V reference voltage we used in earlier diagrams such as *Figure 4.3*. The **wordline** and **bitline** are wires used to connect the individual bit cells into a grid.

This single-bit cell is replicated in a rectangular grid to form a complete DRAM memory bank. The following figure shows the configuration of a 16-bit DRAM bank consisting of 4 words of 4 bits each:

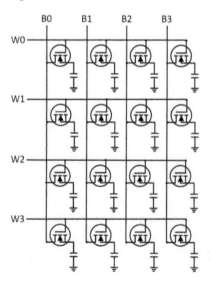

Figure 4.6: DRAM memory bank organization

Of course, real DRAM banks contain a much larger number of bits than the simple circuit shown here. Typical DRAM devices have a word size of 8 bits, rather than the 4 bits labeled B0-B3 in the figure. This means a DRAM chip can store or retrieve 8 bits in parallel.

The number of instances of bit cells along a wordline in an actual DRAM bank array is an integer multiple of the device word size. The large DRAM modules used in personal computers contain many words along each wordline. For example, a DRAM chip implemented with 8-bit words and 1,024 words per row contains 8,192 bits in a row, with all the MOSFET gate terminals along the row controlled by a single wordline signal. These devices contain additional multiplexer logic to select the specific word the processor is requesting from the many words along the row selected by the active wordline.

The vertical dimension of the DRAM bank consists of replicated copies of cell rows with one wordline controlling each row. The wordline connects all the bit cells horizontally, while the bitline signals connect the cells in all the rows vertically.

The state of each memory bit is stored in the cell's capacitor. A low voltage on the capacitor represents a 0 bit, while a high voltage represents a 1 bit. In the context of DDR5 DRAM devices, low voltage is near 0 V and high is near 1.1 V.

The wordline for each row is held low most of the time. This keeps the MOSFET turned off, maintaining the capacitor state. When it is time to read a word (actually, an entire row) of DRAM, addressing circuitry selects the appropriate wordline and drives it high while keeping all the other wordlines in the bank low. This turns on the MOSFET in each bit cell along the active wordline, allowing the cell capacitors to drive their voltages onto the connected bitlines. Bitlines for cells at the 1 (high) voltage level will have higher voltages than cells at the 0 (low) voltage level. The bitline voltages are sensed by circuitry in the DRAM device and latched into the chip's output register.

Writing to a DRAM word begins by setting the selected wordline high in the same manner as reading a word. Instead of sensing the voltage on the bitlines, the DRAM device drives each bitline with the voltage to be written to each cell, either 0 V or 1.1 V, depending on whether the data bit is a 0 or a 1. As with filling or emptying a balloon, it takes a bit of time for each capacitor to charge or discharge to the voltage presented on its bitline. After this delay has completed, the wordline is driven low to turn the MOSFETs off and lock the capacitors at their new charge levels.

DRAM circuit technology is complicated by the fact that capacitors leak. After charging a capacitor to a non-zero voltage, the charge bleeds off over time, reducing the voltage across the capacitor. Because of this, the contents of each cell must be refreshed periodically.

A refresh operation reads each cell value and writes it back to the cell. This recharges the capacitor to its "full" voltage level if it is a 1, or drives it to near 0 V if it is a 0. A typical refresh interval for modern DRAM devices is 64 milliseconds. DRAM refresh proceeds continuously, row by row, in the background during system operation and synchronizes with processor access to memory to avoid conflicts. There is a small performance penalty that occurs when processor access to memory is delayed by refresh activity.

While the need for periodic refreshing significantly complicates the design of systems using DRAM devices, the benefit of storing a bit with just one transistor and one capacitor is so immense that DRAM has supplanted all alternatives as the preferred technology for main memory in consumer, business, and scientific computer systems.

The next section will examine the architecture of the current generation of DRAM technology, DDR5.

DDR5 SDRAM

Intel produced the first commercial DRAM integrated circuit in 1970. The Intel 1103 held 1,024 bits and had a word size of 1 bit. The 1103 had to be refreshed every 2 milliseconds. By the early 1970s, MOS semiconductor DRAM overtook magnetic core memory as the preferred memory technology in computer systems. DRAM is volatile, meaning when power is removed, the charge in the bit cell capacitors dissipates and the data content is lost.

The term **double data rate** (**DDR**) refers to the transfer timing characteristics between a memory module and the processor memory controller. The original **single data rate** (**SDR**) DRAM performed one data transfer per memory clock cycle. DDR memory devices perform two transfers per clock cycle: one on the clock rising edge and one on the falling edge. The number following "DDR" identifies the generation of DDR technology. DDR5, therefore, is the fifth generation of the DDR standard. The term **synchronous DRAM** (**SDRAM**) indicates the DRAM circuitry is synchronized with the processor memory controller by a shared clock signal. The current generation of DRAM technology in widespread use is DDR4 SDRAM, while DDR5 SDRAM is beginning to roll out.

Modern personal computers and personal devices such as smartphones generally contain multiple **gigabytes** (**GB**) of RAM. One GB is 2^{30} bytes, equivalent to 1,073,741,824 (just over one billion) bytes. As the name implies, random access memory allows the processor to read or write any memory location within the RAM address space in a single operation. As of 2021, a high-end memory module available for use in laptop computers contains 32 GB of DRAM distributed across 16 integrated circuits. Each DRAM chip in this module contains 2 giga-words, (where 1 giga-word is 2^{30} words) with a word length of 8 bits.

In 2021, the leading memory module standard is DDR5 SDRAM, an evolutionary optimization of DRAM interface technologies built upon the DDR1, DDR2, DDR3, and DDR4 generations. A DDR5 memory module is packaged as a **dual inline memory module (DIMM)**. A DIMM has electrical contacts on both sides of the circuit board edge (hence the term *dual* in the name), providing connectivity to the DIMM socket in the motherboard. A standard DDR5 DIMM has 288 pins. A smaller module format called the **small outline DIMM (SODIMM)** is available for systems such as laptops where space is at a premium. A DDR5 SODIMM has 262 pins. Because of the reduced number of pins, SODIMM modules lack features that some DIMMs support, such as the ability to detect and correct bit errors in data retrieved from the device. This capability is referred to as **error correcting code (ECC)**.

DDR5 memory modules are nominally powered by 1.1 V. As a representative example, a particular DDR5 module can perform up to 4.8 billion data transfers per second, double the memory clock speed of 2,400 MHz. At 8 bytes per transfer, this DDR5 device can theoretically transfer 38.4 GB per second. DDR5 modules will be available in a variety of clock speeds, memory sizes, and price points.

Although real-world DRAM modules implement rectangular banks of single-bit cells as described in the previous section, the internal architecture of a DDR5 device is somewhat more complex. A DRAM integrated circuit generally contains multiple banks. The addressing logic selects the bank containing the desired memory location before performing a read or write operation. In DDR5 modules, banks are further arranged into **bank groups**, necessitating additional addressing logic to choose the correct group. A DDR5 device contains up to eight bank groups, each with up to four banks. The reason for partitioning the DDR5 module architecture into multiple bank groups is to maximize data transfer speed by enabling multiple simultaneous, overlapped memory access operations to proceed in parallel. This permits data transfer between processor and RAM to flow at peak speed while minimizing the need to wait for each DRAM access operation to complete.

In addition to specifying the correct address location within a DDR5 memory module, the system must provide a command via interface signals to indicate the action to be taken, specifically, whether to read from, write to, or refresh the selected row.

The DDR5 SDRAM standard, available for purchase from the **Joint Electron Device Engineering Council (JEDEC)** at https://www.jedec.org/standards-documents/docs/jesd79-5, provides the detailed definition of the DDR5 memory interface to host computer systems. This standard contains all the information needed to design memory modules compatible with any computer system supporting DDR5.

Historically, each numerical generation of the DDR SDRAM standards has been incompatible with previous generations. A motherboard built for DDR5 memory modules will only work with DDR5 modules. The slot for each DDR generation is constructed in such a way that it is not possible to insert an incorrect module. For example, a DDR4 DRAM module will not fit into a DDR5 slot.

As memory technologies evolve, the primary improvements in each new generation are increased data transfer rate and greater memory density. To assist in achieving these goals, power supply voltages have decreased in later generations, reducing system power consumption and enabling denser memory circuitry while avoiding excessive heating.

Most modern processors view system memory as a linear array of sequential addresses. In less sophisticated processor architectures, such as the 6502, the processor directly addresses RAM chips using memory addresses provided in instructions. Because of the complexity of the control signals and bank management logic in DDR5 SDRAM devices, modern computer systems provide a memory controller to translate each linear processor address into command and addressing signals selecting the appropriate DDR5 module (in a system with multiple memory modules), bank group, bank, and row/column location within the selected bank. The memory controller is a sequential logic device that manages the details of communication between the processor and DRAM memory modules. To achieve peak system performance, the memory controller must intelligently exploit the capability for overlapped operations provided by DDR5 memory modules.

Sophisticated modern processors generally integrate the memory controller function into the processor integrated circuit itself. It is also possible to design a system with a separate memory controller that sits between the processor and RAM.

A memory controller interface may contain multiple channels, where each channel is a separate communication path between the processor and one or more memory modules. The benefit of providing multiple channels in a memory architecture is that this configuration permits simultaneous memory accesses over the channels. However, a system containing multiple memory channels does not achieve an automatic increase in memory access speed. System software must actively manage the assignment of memory regions to each application or system process to balance memory usage across channels. If the operating system were to simply assign processes to physical memory regions sequentially, filling one memory module first then moving to the next, there would be no benefit from multi-channel memory because all processes would be forced to use the same memory channel.

Graphics DDR

Graphics DDR (GDDR) is a DDR memory technology optimized for use as video RAM in graphics cards. GDDR has a wider memory bus to support the high-throughput requirements of video displays. Standard DDR memory, on the other hand, is optimized to provide minimum latency access to data.

The generation numbers for GDDR and DDR are not aligned. As of 2021, GDDR6 modules have been available for a few years, while the DDR6 SDRAM standard remains in development.

Prefetching

One DRAM performance attribute that improves very little from one generation to the next is the speed of reading from or writing to an individual bit location. To achieve an increase in the average data transfer rate into and out of DRAM modules, the devices must employ other techniques to improve performance. One technique for achieving faster average data transfer speeds is **prefetching**.

The idea behind **prefetching** is to exploit the fact that whenever a particular memory location is being accessed by a processor, it is likely that addresses close to the first location will soon be accessed. Prefetching consists of reading a larger block of memory than the single address a processor instruction references and passing the entire block from the DRAM device to the processor. In the context of a DDR5 memory module, the block size is normally 64 bytes.

The DDR5 module can read 64 bytes quickly because it accesses all 512 bits of those 64 bytes simultaneously. To do this, the DDR5 module reads an integer multiple of 512 bitlines from the cells selected by a wordline. The bits of the selected row are read simultaneously, then pass through a multiplexer to extract the desired 512 bits from (perhaps) 8,192 bits in the entire row, which are then latched into an output register. The latched bits transfer from the DRAM module to the processor using DDR clocking.

With the effective use of multiple bank groups, multiple reads of memory and transfers of the resulting data can overlap in time and ensure that data moves between the memory module and the processor at the highest rate the interface can support.

Upon receiving the 64-byte block, the processor stores the data in internal cache memory and selects the specific data element (perhaps as small as 1 byte) from the block requested by the executing instruction. If a subsequent instruction accesses different data contained in the same block, the processor need only consult its local cache, resulting in much faster execution than the instruction that originally retrieved the data block from DRAM.

In addition to interacting with main memory, the processor must communicate with the outside world through input and output devices. The next section discusses the implementation of I/O interfaces in modern computer systems.

I/O subsystem

Chapter 3, Processor Elements, introduced two broad categories of I/O architecture: memory-mapped I/O and port-mapped I/O. The pros and cons of each of these approaches were significant in the early days of PCs when the number of physical address lines limited the total processor memory space to a 1 MB range. Modern processor architectures can address a far larger memory range, typically in the tens of gigabytes. A consequence of this address space expansion is the ready availability of address regions for use in I/O interfaces. Because of this, modern 32-bit and 64-bit general-purpose processors employ memory-mapped I/O for most of their interface requirements.

Sophisticated modern processors usually implement a memory controller within the processor chip, communicating directly with DDR memory modules. Most other types of I/O performed by these processors are offloaded to one or more external integrated circuits, typically referred to as a **chipset**. The term chipset is commonly used even when only one chip is needed to implement the I/O functions.

The chipset provides interfaces to a wide range of peripherals, such as disk drives, network interface, keyboard, mouse, and many others via USB. Most of these interfaces are implemented using one form or another of a serial bus. The following sections introduce the most common I/O technologies employed in modern computers.

Parallel and serial data buses

A parallel data bus communicates multiple data bits simultaneously across separate conductors between two or more communication endpoints. Early PCs used parallel buses for functions such as connecting a printer to the computer. Over time, several limitations of parallel buses became apparent:

- Depending on how many bits the bus supports, a parallel bus connection may require a lot of wires. This means cables are more expensive and there is a greater possibility of problems when cable wires break, or connectors fail to make solid electrical contact.
- As computer system developers made efforts to increase bus data rates (and thereby gain a competitive edge), another limitation of parallel buses became significant: even though the device transmitting a data word on the bus may output all the parallel bits simultaneously, the individual signals may not arrive at the destination at the same time.

This could be caused by differences in the effective path length of the conductors in the cable or across the circuit board. Because of this, there is an upper limit on the data transfer rate a parallel bus architecture can support.

Another limitation of parallel buses is they can only transfer data in one direction at a time (referred to as **half-duplex**) unless a duplicate set of connections is provided for simultaneously transferring data in the opposite direction. Parallel buses usually do not provide simultaneous bi-directional communication capability, referred to as **full-duplex** operation, for this reason.

A serial data bus transfers data between two communication endpoints a single bit at a time using a pair of conductors. Most high-speed communication paths between the processor and peripheral devices in modern computers use some form of serial bus. While, at first blush, switching from a parallel bus architecture to a serial bus seems to represent a substantial loss in throughput capability, serial buses exhibit several important advantages that make their use attractive in performance-critical applications.

High-speed serial buses in personal and business computer systems communicate over pairs of conductors using differential signaling. **Differential signaling** employs two conductors, carefully matched to be of the same length and to exhibit nearly identical electrical characteristics. When used in cables, these conductors are insulated wires twisted around each other to form **twisted pairs**.

The following figure represents a serial data bus using differential signaling:

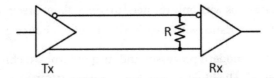

Figure 4.7: Serial bus circuit using differential signaling

The digital signal to be transmitted arrives at the transmitter (labeled **Tx**) via the input at the left side of the figure. The input is transformed into a pair of voltages on the two parallel lines crossing the center of the diagram. The small circle indicates the upper signal from the transmitter is inverted relative to the lower signal.

In a typical serial interface, a high signal level at the transmitter input will generate a voltage of 1.0 V on the top serial conductor and 1.4 V on the bottom conductor. A low signal input produces 1.4 V on the top conductor and 1.0 V on the bottom conductor.

The inputs to the receiver (which is labeled **Rx**) are high impedance, meaning the receiver draws negligible current from the circuit. The receiver measures the voltage across the resistor, which has a typical value of 100 Ω. When the Tx input is high, the upper resistor terminal is at -0.4 V relative to the lower terminal. When the Tx input is low, the upper resistor terminal is +0.4 V relative to the lower terminal.

The receiver produces its output signal by inverting one of the inputs (the top one in the figure, with the small circle) and adding the resulting voltage to the other input. In other words, the receiver only measures the difference between the voltages on the two conductors. The fundamental benefit of this approach derives from the fact that most forms of corrupting interference cause voltage variations in the conductors carrying the signal. By placing the two conductors very close together, most of the noise voltage introduced on one of the conductors will also appear on the other. The subtraction operation cancels out a large portion of the noise that would otherwise interfere with accurate signal detection by the receiver.

A serial data bus can perform several billion bit transfers per second, far more than an old PC's parallel bus. It is also possible to run several serial buses alongside each other, effectively multiplying the data transfer bandwidth by the number of buses.

A crucial difference between multiple serial buses connecting two endpoints and a parallel bus making the same connection is that, for many interface standards, the serial buses operate somewhat independently. They do not synchronize at the level of each bit transfer in the same way all the bits in a parallel bus must. This makes it easier to design interconnections capable of supporting very high data rates with only the need to worry about precisely matching conductor length and electrical characteristics within each pair of serial conductors.

The connection between a modern processor and its motherboard chipset generally consists of several serial data buses called **high-speed input output (HSIO) lanes**. Each lane is a serial connection with one data path, like *Figure 4.7*, flowing in each direction, supporting full-duplex operation.

Individual HSIO lanes can be assigned to specific types of peripheral interfaces that are implemented as serial connections, such as PCI Express, SATA, M.2, USB, and Thunderbolt. The following sections introduce each of these interface standards.

PCI Express

The original **Peripheral Component Interconnect (PCI)** bus was a 32-bit parallel bus running at 33 MHz used in PC-compatible computers from about 1995 through 2005.

The PCI slots on computer motherboards accepted a wide variety of expansion cards performing functions such as networking, video display, and audio output. By the early 2000s, the limitations of the parallel bus architecture had become constraining, and development began on a serial bus replacement for PCI named PCI Express.

PCI Express, abbreviated as **PCIe**, is a bi-directional differential signaling serial bus used primarily to connect communication endpoints on computer motherboards. PCIe performance is expressed in billions of transfers (giga-transfers) per second, or GT/s. One "transfer" is a single bit propagated across the bus from transmitter to receiver. PCIe inserts additional redundant bits in each multi-bit communication to ensure data integrity. Different generations of PCIe have varying numbers of these overhead bits, influencing the effective data transfer rate. The following table shows the major generations of PCIe, the year each was introduced, the single-lane transfer rate in GT/s, and the effective data transfer rate in MB/s:

PCIe generation	Year introduced	Transfer rate	Effective one-way data rate
1.0a	2003	2.5 GT/s	250 MB/s
2.0	2007	5 GT/s	500 MB/s
3.0	2010	8 GT/s	985 MB/s
4.0	2017	16 GT/s	1,969 MB/s
5.0	2019	32 GT/s	3,938 MB/s
6.0 (proposed)	2021	64 GT/s	7,877 MB/s

Table 4.1: PCI Express generations

The effective data rate presented here is for one-way communication. PCIe supports full-speed data transfer in both directions simultaneously.

The PCIe standards support multi-lane connections indicated by the notations x1, x2, x4, x8, x16, and x32. Most modern motherboards implement, as a minimum, PCIe x1 and x16 slots. PCI x1 slots are compatible with a board edge connector length of 25 mm, while x16 slots expect a board edge connector length of 89 mm. A PCIe card will operate correctly in any slot in which it can physically fit. For example, a PCIe x1 card can be plugged into an x16 slot and will use just one of the 16 available lanes.

A primary application for PCIe x16 slots is to interface between the processor and a graphics card, with the goal of providing peak performance for graphics-intensive applications such as gaming. A PCIe 5.0 x16 interface is capable of unidirectional data transfer at 63 GB/s.

In modern computer architectures, the processor chip usually provides a direct PCIe 16-lane connection to a graphics board installed in a PCIe x16 slot. This means you don't need to pass the graphics card PCIe signals through the chipset.

Other than the graphics display and the DDR SDRAM interface, most I/O in modern computer systems is managed by the chipset. The processor and chipset communicate through a collection of HSIO lanes. The chipset provides interfaces to peripheral devices such as disk drives, network interfaces, keyboard, and mouse. The interfaces to these devices commonly use the SATA, M.2, and USB serial interface standards, discussed next.

SATA

The **Serial AT Attachment (SATA)** is a bi-directional serial interface standard for connecting computer motherboards to storage devices. The "AT" in SATA refers to the IBM PC AT. Like a single PCIe lane, SATA contains two differential signaling pairs of conductors, where one pair carries data in each direction. Unlike PCIe, SATA is intended for operation over cables rather than over metal signal traces on motherboards. In addition to electrical and data format requirements, the SATA standard contains detailed specifications for compatible cables and connectors.

A SATA cable contains one bi-directional lane supporting communication between a processor and a storage device such as a magnetic disk drive, optical disk drive, or solid-state drive. The following table shows the major revisions of the SATA standard, the year each was introduced, and performance parameters:

SATA generation	Year introduced	Transfer rate	Effective one-way data rate
1.0	2003	1.5 GT/s	150 MB/s
2.0	2004	3 GT/s	300 MB/s
3.0	2009	6 GT/s	600 MB/s

Table 4.2: SATA generations

The data transfer rate in this table describes one-way communication, though, like PCIe, SATA supports full-duplex data transfer.

The SATA standard continues to undergo incremental improvements, but as of 2021 there has been no announcement of an upcoming SATA generation with a faster data transfer rate.

M.2

Modern **solid-state drives (SSDs)** employ flash memory to store data rather than the rotating magnetic platters in traditional hard drives. Because of the radically different technology of SSDs, the SATA interface that works well enough in most cases for rotating disks has proven to be a significant performance barrier for SSDs.

To access an arbitrary block of data (called a **sector**) on a magnetic disk drive, the drive head must physically move to the track containing the sector, then wait for the beginning of the sector to rotate to the head position before the drive can begin reading data. In contrast, an SSD can directly address any data sector in a manner very similar to the way a processor accesses a DRAM location.

The M.2 specification was developed to provide a small form factor and high-performance interface for flash memory storage in small, portable devices. The performance limitations of the SATA interface are removed, and it is possible to transfer data at rates several times faster than SATA can support.

In addition to mass storage devices, M.2 supports other interfaces including PCIe, USB, Bluetooth, and Wi-Fi. Modern motherboards include M.2 slots, which, in addition to the higher performance, consume much less space in the computer case than traditional disk drive bays.

USB

The **USB** interface provides a simple (from the user's viewpoint) interface for connecting a wide variety of peripheral devices to a computer system. In addition to communication protocols, the USB standard revisions define requirements for cables, connectors, and for supplying power to connected devices via the USB cable.

USB cables have easily identifiable connector types and USB-enabled devices support hot-plugging (plugging devices together while powered on). USB devices are self-configuring, and, in most cases, users need not concern themselves with installing drivers when attaching a new device to a computer with a USB cable.

Early USB data cables (through generation 2.0) contained a single differential signaling pair that could pass data in only one direction at a time. Later versions of the USB standard (USB 3.2 Gen 1 onward) support simultaneous bi-directional data transfer. In addition, USB generations 3.2 and USB4 provide up to two lanes, doubling the data transfer rate.

The following table shows the major revisions of the USB standard, the date each was introduced, the maximum number of lanes supported, and peak data transfer performance:

USB generation	Year introduced	Number of lanes	Transfer rate	Effective one-way data rate
1.1	1998	1	0.012 GT/s	1.5 MB/s
2.0	2000	1	0.48 GT/s	60 MB/s
3.2 Gen 1	2008	1	5 GT/s	500 MB/s
3.2 Gen 2	2013	1	10 GT/s	1,200 MB/s
3.2 Gen 2x2	2017	2	20 GT/s	2,400 MB/s
USB4	2019	2	40 GT/s	4,800 MB/s

Table 4.3: USB generations

In USB generations through 2.0, communication takes place entirely under the control of the host. The host initiates each communication interaction by sending packets addressed to a particular device and performs data transfers to or from the device. Beginning with USB 3.2 Gen 1, devices can initiate communication with the host, effectively providing an interrupt capability for connected peripherals.

Thunderbolt

Thunderbolt is a collection of high-speed serial interface standards introduced in 2011. The original Thunderbolt interface combined PCIe and DisplayPort signal transmission using two serial Thunderbolt lanes.

Thunderbolt 4 is the latest generation of the Thunderbolt standard, adding USB4 compatibility while supporting connectivity to PCIe devices and to multiple high-resolution displays from a single computer port. Thunderbolt 4 uses the same connector as USB 3.2 and later generations (the USB-C connector) and supports the 40 Gbit/s USB4 data rate. USB devices operate properly when connected to a computer's Thunderbolt 4 port. Thunderbolt 4 peripheral devices are not compatible with non-Thunderbolt 4 USB-C ports, however.

The next section provides an overview of the most popular graphics display interface standards.

Graphics displays

In the domains of gaming, video editing, graphic design, and animation, video processing performance is critical. Generating and displaying high-resolution graphics requires an enormous number of mathematical calculations.

While general-purpose processors can perform the necessary computations, such processors lack the performance users of these applications have come to expect.

High-performance graphics cards, called **graphics processing units (GPUs)**, are essentially miniature supercomputers, heavily optimized to perform graphical computational tasks such as 3D scene rendering. Because the computations involved in scene rendering are highly repetitive, substantial performance gains can be achieved with the use of hardware parallelization. Graphics processors contain a large number of relatively simple computational units, each performing a small portion of the overall task.

A GPU may contain thousands of individual processing units that each function like an ALU. While the original driving force that led to the development of high-performance GPUs was 3-D scene generation, later generations of this technology have found broad use in fields such as big data analytics and machine learning. Any numerically intensive computational task that can be broken into a collection of parallel operations is suitable for acceleration with a GPU architecture.

Of course, not all users require extreme video performance. To accommodate users with modest graphics needs and more limited budgets, modern processors often integrate a moderately capable GPU into the processor chip. In many applications, this approach provides more than adequate graphical performance. This configuration is referred to as **integrated graphics**, meaning the GPU function is integrated into the processor die and shares system memory with the processor. Computer systems with integrated graphics are lower in cost while providing adequate graphics performance for basic computing tasks such as email, web browsing, and watching videos.

Many desktop computer systems, as well as some laptops, provide integrated graphics while offering the option of installing a high-performance graphics card. This allows users to tailor the computer system to their price and performance needs.

Several different video standards are in current use for connecting displays to computers. Because the output generated by a computer graphics interface must be compatible with the connected display, it is common for computers to provide more than one type of video connector. Computer monitors and high-definition televisions usually provide a selection of video connection types as well.

Chapter 6, Specialized Computing Domains, will cover the processing architecture of GPUs in more detail. The following sections describe video interface standards used in computer applications past and present.

VGA

The **Video Graphics Array (VGA)** video standard for personal computers was introduced by IBM in 1987. VGA is an analog interface that remains in widespread use today, though most modern computers do not provide a VGA connector. It is not uncommon to find older computers with VGA outputs using a converter cable to present a display on a monitor supporting DVI or HDMI video input.

Modern versions of the VGA standard support display resolutions up to 1,920 pixels wide by 1,200 pixels high, refreshed at 60 Hz. Because the VGA video signal is analog, some loss of signal quality occurs during transmission to the display. This effect is most noticeable at high screen resolutions.

DVI

The **Digital Visual Interface (DVI)** video standard was developed to improve the visual quality of computer displays by transferring the video signal digitally from the computer to the monitor. To maintain backward compatibility with older computers and monitors, DVI cables can carry VGA analog signals as well.

Like the high-speed serial interfaces discussed earlier in this chapter, DVI uses differential serial signaling to transfer video data. A DVI connector contains four serial lanes. Individual lanes carry red, green, and blue color information, and the fourth lane carries the common clock signal.

Three DVI variants are defined, depending on the combination of digital and analog video signal types supported:

- **DVI-A** supports only the analog video signal. This option is intended to provide backward compatibility with VGA computers and monitors. The DVI-A connector has a different pin layout than traditional VGA connectors, so an adapter cable is required to connect to legacy VGA devices.
- **DVI-D** is a digital-only interface supporting single-link and dual-link options. The dual-link option provides additional serial data lanes to increase the video bandwidth for higher-resolution displays. Dual-link does not mean the cable supports dual monitors.
- **DVI-I** is an integrated interface supporting both the analog interface of DVI-A and the digital modes of DVI-D. A DVI-I digital interface may be single- or *dual*-link.

DVI interfaces are used primarily in computer display applications. The effective data rate of a single-link DVI-D connection is 3.96 gigabits per second (Gbit/s). Dual-link DVI-D transfers video data at twice the single-link rate, 7.92 Gbit/s.

HDMI

High-Definition Media Interface (HDMI) is supported by most modern computers and monitors, and by virtually all modern televisions and related video entertainment devices such as DVD players. HDMI supports digital video only (there is no analog capability) and uses the same differential serial bus as DVI-D. In addition to video data, HDMI cables also transport multi-channel digital audio.

The HDMI standard has undergone several revisions since its introduction in 2002. Each successive revision has maintained backward compatibility while adding new capabilities. Later versions of the standard have increased video bandwidth, increased the range of supported screen resolutions, added high-definition audio capabilities, added support for Ethernet communication over the HDMI cable, and added features to support gaming. Although each HDMI version is backward compatible, newer features are only available in configurations where the signal source device, the display device, and the connecting cable are all compatible with the newer standard.

HDMI version 2.1 was released in 2017. This standard supports an effective data rate of 42.6 Gbit/s over four differential serial lanes.

DisplayPort

DisplayPort, introduced in 2006, is a digital interface standard supporting digital video, audio, and USB connections. While HDMI is targeted at consumer electronics such as televisions and home theater systems, DisplayPort is oriented toward computer applications. DisplayPort transfers data in packets with clocking information embedded in each packet, eliminating the need for a separate clock channel.

A single computer DisplayPort output can drive multiple monitors connected in a daisy chain, where one cable connects the computer to the first monitor, a second cable connects the first and second monitors, and so on. The monitors must provide support for this capability. The maximum number of displays that can be combined in this manner is limited only by the capabilities of the graphics card, the maximum cable data throughput, and the resolution and refresh rate of the displays.

DisplayPort 2.0 was released in 2019, with an effective data rate of up to 77.4 Gbit/s over four differential serial lanes.

Network interface

A computer network is a collection of digital devices interacting over a shared communication medium. A **local area network (LAN)** consists of a limited number of computers that might reside in a single physical location such as a home or office building. The connected computers, phones, and other digital devices in your home represent a LAN. Device connections within a LAN environment may use a wired interface, usually Ethernet, or a wireless interface, typically Wi-Fi.

Geographically separated computers and LANs communicate using a **wide area network (WAN)**. WAN services are often provided by a telecommunication company such as a cable television provider or telephone company. Your home LAN most likely connects to the Internet via WAN services provided by your telephone or cable company.

Home and business network interface devices (called **routers**) provided by WAN service providers usually offer Ethernet and Wi-Fi options for connecting local devices to the WAN. The following sections introduce the technologies of Ethernet and Wi-Fi.

Ethernet

Ethernet is a set of networking standards for connecting computers using cables in a LAN environment. The original version of Ethernet was developed by Robert Metcalfe at Xerox Palo Alto Research Center in 1974. Ethernet was released commercially in 1980 as a 10 Mbit/s communication technology for groups of computers connected with coaxial cabling. The name of the technology was derived from the historical term *luminiferous aether*, a hypothesized medium filling all of space and enabling the propagation of electromagnetic waves. The Ethernet cable serves as a conceptually similar communication medium.

The **Institute of Electrical and Electronic Engineers (IEEE)** began developing standards for LAN technologies, including Ethernet, in 1980. The IEEE 802.3 Ethernet standard was published in 1985. Since then, the standard has undergone several revisions supporting increased data rates and different network topologies. The most obvious difference in modern computer networks from the original Ethernet standard is the use of point-to-point twisted pair cables in place of the original shared coaxial cable.

Modern computers commonly use Gigabit Ethernet interfaces to communicate over **unshielded twisted-pair (UTP)** cabling. Gigabit Ethernet is formally defined in the IEEE 802.3ab standard and supports 1.0 Gbit/s with an effective data transfer rate up to 99% of the raw bit rate, though the amount of overhead varies considerably depending on the communication protocol in use.

Ethernet communication is composed of variable-size data units called **frames** containing up to 1,518 bytes. The header of each frame contains addressing information identifying the source and destination Ethernet interfaces. Because modern twisted-pair connections are point-to-point, the most common structure for connecting a group of computers is to run a cable from each computer to a **switch**. A **switch** is a device that receives frames transmitted by the connected computers and, based on the destination address contained in each frame, immediately forwards it to the correct recipient. Ethernet cables are limited to a maximum recommended length of 100 meters, constraining the physical size of an Ethernet LAN to an area such as a single office building or home.

Modern motherboards usually contain a built-in Ethernet interface, which eliminates the need to consume a PCIe slot with an Ethernet card. An Ethernet interface, whether built into the motherboard or installed in a PCIe expansion slot, consumes one HSIO lane connecting from the processor through the chipset to the Ethernet interface.

Wi-Fi

The IEEE released the first version of the 802.11 wireless communication standard in 1997 supporting a raw data rate of 2 Mbit/s in the 2.4 GHz radio frequency band. The 802.11b standard, released in 1999 with an 11 Mbit/s raw data rate, proved to be commercially popular. The technology was named **Wi-Fi** in 1999 as a reference to the term "hi-fi," referring to high-fidelity sound reproduction.

The 802.11g standard, released in 2003, has a raw bit rate of 54 Mbit/s. 802.11n, released in 2009, supports **multiple-input-multiple-output** (**MIMO**) antennas and optional operation in the 5 GHz band. The 802.11ac standard, published in 2013, supports a bit rate in the 5 GHz band of over 500 Mbit/s using MIMO antenna configurations. The 802.11ax standard, approved in 2021, provides a 39% throughput increase relative to 802.11ac for a single client and offers enhanced support for use in dense radio signal environments. 802.11ax is marketed as Wi-Fi 6 for operation in the 2.4 and 5 GHz bands and as Wi-Fi 6E for operation in the 6 GHz band.

Wi-Fi devices can suffer from interference produced by household appliances such as cordless phones, microwave ovens, and other Wi-Fi networks in the area. Wi-Fi signal propagation is affected by factors such as walls and other obstacles between transmitter and receiver, **multipath** (the destructive interference between a direct-path signal and a reflected copy of the signal), and is limited by the maximum power a Wi-Fi transmitter is authorized to emit. The use of multiple antennas in 802.11n, 802.11ac, and 802.11ax configurations significantly mitigates multipath-related performance issues.

Modern WAN routers supplied by telecommunication service providers usually contain a combination of Ethernet and Wi-Fi communication interfaces. A primary benefit of Wi-Fi in comparison to Ethernet for these applications is the reduction in the amount of cabling required.

One drawback of Wi-Fi is the potential for security issues because the radio frequency signal can propagate far outside the building containing the communicating systems. The Wi-Fi standards provide substantial support for secure communication using protocols such as **Wi-Fi Protected Access 2 (WPA2)**, but system administrators and users must ensure the appropriate security features are enabled and that secret information such as the network password is sufficiently complex and is stored securely.

Support for Wi-Fi is ubiquitous in portable digital devices such as laptops, smartphones, and tablets, and is directly built into many motherboards.

The next section presents the computer interfaces with the lowest bandwidth requirements: the keyboard and mouse.

Keyboard and mouse

Compared to the high-speed interfaces discussed earlier in this chapter, the bandwidth requirements for a keyboard and mouse are quite modest. These devices are the sole input methods used by the human operator in most computer configurations, and thus are only required to operate at the speed of human actions. Even the fastest typist can only press one or two dozen keys per second.

Keyboard

A mechanical computer keyboard consists of a collection of keys, each of which activates an electrical momentary switch. A standard full-size keyboard typically contains 104 keys including the arrow keys, control keys (Home, Scroll Lock, and so on), and the numeric keypad. Modern keyboards commonly use a USB cable for connection to the computer or connect wirelessly.

Because the bandwidth requirements for human interaction are so low, some computer motherboards provide a slower USB 2.0 port for keyboard connection while offering higher-performance USB 3.2 or faster interfaces for high-speed peripherals. This results in a small cost reduction for the motherboard components.

Because the keyboard reports the press and release of each key separately, the computer can process combinations of keys pressed simultaneously. For example, holding the *Shift* key down while pressing the *A* key produces a capital *A*.

Some computers and digital devices such as tablets and phones provide a touchscreen interface. When text input is required on these devices, the system displays a keyboard pattern on the screen and the user touches letter locations to produce keypresses.

Mechanical keyboards tend to provide more accurate input and are favored by users entering significant quantities of text. Because the surface of a touchscreen is completely flat, there is no feedback to the user's fingers indicating alignment with the keys. This results in more frequent input errors when using touchscreen keyboards. Of course, when using a touchscreen keyboard there is no need to provide a mechanical keyboard, which represents a substantial benefit for portable devices. In addition, touchscreen keyboard input does not suffer from mechanical failures that can affect the components of mechanical keyboards. For users wearing gloves, text input is more difficult on both touchscreens and keyboards.

Mouse

A computer mouse is a hand-held device that moves a pointer horizontally and vertically across a computer screen. The user initiates actions based on the pointer location by pressing buttons on the mouse. Modern mice often provide a small wheel capable of rolling in either direction, used to perform tasks such as scrolling through a document.

As with the keyboard, the mouse commonly connects to the computer via USB over a wired or wireless connection. The mouse has low bandwidth requirements and can be supported on a USB 2.0 port.

The operation of a mouse requires a flat horizontal surface, typically a table top, upon which the user moves the mouse. Modern mice most commonly use optical emitters and sensors to detect motion across the surface. Many mice have difficulty operating on highly reflective surfaces such as glass tabletops.

A trackball is similar in concept to a mouse, except that rather than moving a mouse across a surface, a ball is retained at a fixed location but is allowed to rotate in any direction using hand motion. By rolling the ball forward, backward, left, and right, the user can move the pointer on the computer display.

A trackball does not require the amount of surface space a mouse needs, and the trackball can be fixed at a stationary location. The ability to firmly attach a trackball to a structure makes the trackball the preferred pointing device for computer stations installed in ground vehicles, ships, and aircraft.

As with the keyboard, the computer detects the press and release of each mouse button as separate events. Users can exploit this capability to perform operations such as dragging an icon across the screen by following these steps:

1. Place the pointer over the icon
2. Press and hold the left mouse button
3. Move the pointer (with the icon now attached) to the new location
4. Release the mouse button

Together, the keyboard and mouse provide all the input capability most computer users need to perform their interactive tasks.

The next section brings together the subsystem descriptions in this chapter to examine the specifications of a modern computer motherboard.

Modern computer system specifications

With the information in this chapter, you should be able to interpret most of the technical features of a modern computer motherboard, processor, and chipset. This section provides an example of the specifications for a current (2021) motherboard with some explanation of the individual features.

The designers of a computer motherboard must make several decisions such as the number of PCIe expansion ports, number of DIMM slots, number of USB ports, and the number of SATA ports to be provided in a particular motherboard model. These decisions are guided by the target customer demographic, whether it be gamers, business users, or cost-conscious home users.

The example motherboard presented here is the ASUS Prime X570-Pro, a higher-performance board intended for gaming applications that supports gaming-related features such as overclocking. **Overclocking** refers to increasing the clock frequencies for the processor and other system components with the goal of improving performance.

Overclocking increases heat generation and may lead to unstable performance if a component is driven at an excessive frequency.

Feature	Specification	Notes
Processor	AMD AM4 socket compatible with third-generation AMD Ryzen™ processors	The socket contains 1,331 contacts. The processor interfaces directly to DDR4 system memory. There is a PCI 4.0 x16 interface direct from the processor to the GPU.
Chipset	AMD X570 16 x PCIe 4.0 lanes 6 SATA 6 Gbit/s ports	The processor-to-chipset interface is PCIe 4.0 x4.
Graphics card	Up to 3 x PCIe 4.0 x16 slots	Multiple GPUs can run in parallel using Nvidia **Scalable Link Interface (SLI)** or AMD Crossfire technology.
Expansion slots	3 x PCIe 4.0 x16 slots 3 x PCIe 4.0 x1 slots	The x16 slots can be configured as a single x16 slot, as two x8 slots, or as two x8 slots plus an x4 slot.
System memory	4 x dual-channel DDR4 3200 MHz DIMM slots containing up to 128 GB total	Up to four DDR4 modules can be installed with up to 32 GB each. Overclocking up to 5,100 MHz is possible.
Disk interface	2 x M.2 6 x SATA 6 Gbit/s	The M.2 slots use PCI 4.0 for high-performance SSDs. The SATA ports support traditional disks.
Ethernet	1 x Gigabit Ethernet	High-speed Ethernet networking.
USB	2 x USB 2.0 headers 1 x USB 3.2 Gen 1 (5 Gbit/s) headers 1 x USB 3.2 Gen 2 (10 Gbit/s) headers	Each header supports multiple USB ports. Use USB 2.0 for keyboard and mouse. Use USB 3.2 for fast peripherals such as external drives.

Table 4.4: Example motherboard specifications

This example is intended to provide some perspective on the specifications of higher-end consumer-grade computer capabilities as of 2021.

 If you are looking to purchase a computer, you can use the information in this chapter to make yourself a more informed consumer.

Summary

This chapter began with an introduction to the computer memory subsystem, the MOSFET, and the capacitor. We examined the circuitry that implements the DRAM bit cell. We reviewed the architecture of DDR5 memory modules and the operation of multichannel memory controllers. Other types of I/O devices were introduced, with a focus on high-speed differential serial connections and their ubiquitous use in technologies such as PCIe, SATA, USB, and video interfaces.

Popular video standards were presented, including VGA, DVI, HDMI, and DisplayPort. We looked at the Ethernet and Wi-Fi networking technologies as well. We continued with a discussion of standard computer peripheral interfaces including the keyboard and mouse. The chapter concluded with a description of an example modern motherboard, highlighting some of its interesting features.

With the information presented in this chapter, you should have a solid understanding of modern computer components from the level of technical specifications down to the technologies used in implementing the circuitry.

In the next chapter, we will explore the high-level services computer systems must implement, such as disk I/O, network communications, and interactions with users. We'll examine the software layers that implement these features, starting at the level of the processor instruction set and registers. Several key aspects of operating systems will be covered including booting, multithreading, and multiprocessing.

Exercises

1. Create a circuit implementation of a NAND gate using two CMOS transistor pairs. Unlike NPN transistor gate circuits, no resistors are required for this circuit.
2. A 16-gigabit DRAM integrated circuit has two bank group selection inputs, two bank selection inputs, and 17 row address inputs. How many bits are in each row of a bank in this device?

Join our community Discord space

Join the book's Discord workspace for a monthly *Ask me Anything* session with the author:
`https://discord.gg/7h8aNRhRuY`

5

Hardware-Software Interface

Most computer software is not written at the processor instruction level in assembly language. Almost all the applications we work with daily are written in one high-level programming language or another, relying on pre-existing libraries of capabilities that the application programmers built upon during the software development process. Practical programming environments, consisting of high-level languages and their associated libraries, offer many services, including disk **input/output (I/O)**, network communication, and interactions with users, all easily accessible from program code.

This chapter describes the software layers that implement these features, beginning with processor instructions within device drivers. Several key aspects of operating systems will be covered in this chapter, including booting, process scheduling, multithreading, and multiprocessing.

After completing this chapter, you will understand the services provided by operating systems and the functions of **Basic Input/Output System (BIOS)** and **Unified Extensible Firmware Interface (UEFI)** firmware. You will have learned how threads of execution function at the processor level and how multiple processor cores coordinate within a computer system. You will also have a broad understanding of the process of booting into an operating system securely, beginning with the first instruction executed.

We will cover the following topics:

- Device drivers
- BIOS
- The boot process
- Operating systems

- Processes and threads
- Multiprocessing

Technical requirements

Files for this chapter, including answers to the exercises, are available at `https://github.com/PacktPublishing/Modern-Computer-Architecture-and-Organization-Second-Edition`.

Device drivers

A device driver provides a standardized interface for software applications to interact with a category of peripheral devices such as disk storage. This avoids the need for the application developer to understand and implement all of the technical details required for the proper operation of each device type. Device drivers also manage the coordination needed when applications written by different developers attempt to access the same device at the same time. Most device drivers allow multiple simultaneously executing applications to interact with multiple instances of associated peripherals in a secure and efficient manner.

At the lowest level, the device driver code provides software instructions that manage communication interactions with the peripheral, including handling interrupts generated by device service requests. A device driver controls the operation of hardware resources in the processor, in the peripheral device, and in other system components such as the processor chipset.

In computer systems supporting privileged execution modes, device drivers usually operate at an elevated privilege level. Code running with elevated privilege, which is typically restricted to the operating system and device drivers, has full access to the capabilities of the processor and peripheral devices. This grants access to I/O resources that are inaccessible to less privileged code. Only trusted code is permitted to interact with these interfaces directly. If unprivileged application code were able to access a peripheral's hardware interface, a programming error that caused the device to behave improperly would immediately affect all applications that attempt to use the device. The steps involved in transitioning the flow of instruction execution and data between unprivileged user code and privileged driver code will be discussed in *Chapter 9, Specialized Processor Extensions*.

Drivers that execute with elevated privilege are referred to as **kernel-mode drivers**. The **kernel** is the central core of the operating system, serving as the interface between computer hardware and higher-level operating system functions such as the scheduler.

As we learned in *Chapter 3, Processor Elements,* the two principal mechanisms for accessing I/O devices are port-mapped I/O and memory-mapped I/O. Although memory-mapped I/O is predominant in modern computers, some architectures, such as x86, continue to support and use port-mapped I/O. In an x86 system, many modern peripheral devices provide a system interface that combines port-mapped I/O and memory-mapped I/O.

Programming tools for modern operating systems, such as those available for Linux and Windows, provide resources for developing device drivers capable of interacting with peripheral devices using port-mapped and memory-mapped I/O techniques. Installing a device driver in these operating systems requires elevated privilege, but users of the driver do not require such privilege.

Although device drivers for sophisticated peripheral devices can be quite complex and difficult to understand for those not intimately familiar with the device's hardware and internal firmware, some legacy devices are fairly simple. One example of this is the parallel printer port that was introduced in early PCs and remained a standard component of personal computers for many years. Even though modern computers rarely include these interfaces, inexpensive parallel port expansion cards remain readily available and modern operating systems provide driver support for these interfaces. Electronics hobbyists often use a parallel port as a simple interface for interacting with external circuits using **Transistor-Transistor Logic** (**TTL**) 5V digital signals on PCs running Linux or Windows.

The next section will examine the device-driver-level details of the parallel port interface.

The parallel port

The programming interface for the PC parallel printer port consists of three 8-bit registers, originally located at sequential I/O port numbers beginning at hexadecimal 378. This collection of ports provides the interface for system printer number 1, identified as LPT1 in PC-compatible computers running MS-DOS and Windows. Modern PCs may map the parallel port to a different range of I/O ports during **Peripheral Component Interconnect** (**PCI**) device initialization, but operation of the interface is otherwise unchanged from early PCs.

Device drivers for the parallel port in modern computers perform the same functions using the same instructions as in early PCs. In this section, we'll assume the printer port is mapped to the legacy I/O port range in a 64-bit version of Linux.

To interact with parallel port hardware, the x86 processor executes in and out instructions to read from and write to I/O ports.

If we assume the parallel port driver has been installed and initialized, a user application can call a driver function to read from, and write to, the parallel port data lines.

The following driver instructions read the digital voltage levels present on the eight data lines of the parallel port and store the resulting 8-bit value in the processor's al register:

```
mov edx,0x378
in al,dx
```

In x86 assembly language, instructions with two operands are written in the form opcode destination, source. This example uses the al, edx, and dx processor registers. The al register is the lowest 8 bits of the 32-bit eax register, and dx is the lowest 16 bits of the 32-bit edx register. This sequence of instructions loads the immediate value 0x378 into the edx register, and then reads the 8-bit data value from the port number contained in dx into al.

The C language source code that generated the preceding assembly instructions is:

```
char input_byte;
input_byte = inb(0x378);
```

The inb function is provided by the Linux operating system to perform 8-bit input from an I/O port. This code will only function properly if it is running at elevated privilege under the operating system. An application running at the user privilege level will fail if it attempts to execute these instructions because it is not authorized to perform port I/O directly.

The instructions for writing a byte to the parallel port data register, and thus setting the state of the eight digital output signals, are shown in the following code segment:

```
mov edx,0x378
movzx eax,BYTE PTR [rsp+0x7]
out dx,al
```

These instructions set the edx register to the port number, and then load eax from a variable on the stack. rsp is the 64-bit stack pointer. rsp is 64 bits because this driver is running on a 64-bit version of Linux. movzx stands for "move with zero extension," which means moving the 8-bit data value (designated by BYTE PTR) at the address given as rsp+0x7 into the lower 8 bits of the 32-bit eax register, and filling the 24 remaining bits in eax with zeros. The final instruction writes the byte in al to the port number in dx.

The C source code that produces these instructions is:

```
char output_byte = 0xA5;
outb(output_byte,0x378);
```

Similar to `inb`, the `outb` function is provided by Linux to enable device drivers to write an 8-bit value to the given I/O port.

This example demonstrates how interaction between software executing on the processor and peripheral device hardware registers happens at the lowest level of device driver operation. Drivers for more complex devices on x86 systems usually combine port-mapped I/O, as shown in the preceding examples, with memory-mapped I/O, which accesses a device interface using reads and writes that are, in terms of processor instructions, identical to memory accesses.

These examples presented hardware access methods used by drivers on the original parallel PCI bus architecture. The next section discusses features that allow unmodified legacy PCI drivers to operate properly on modern PCIe-based computers, taking full advantage of PCIe's high-speed serial communication technology.

PCIe device drivers

As we saw in the previous chapter, PCIe uses high-speed serial connections as communication paths between the processor and PCIe peripheral devices. You may be wondering about the steps a device driver must perform to interact with this impressive hardware. The simple answer is that drivers do not need to do anything special to take full advantage of the high-performance capabilities of PCIe. PCIe was expressly designed to be software-compatible with the parallel PCI bus used in PCs of the 1990s. Device drivers written for PCI continue to work properly in computers using the serial PCIe bus. The task of translating between processor I/O instructions such as `in` and `out` and the sequential serial data transfers necessary to communicate with PCIe devices is handled transparently by the PCIe subsystems in the processor, chipset, and PCIe devices.

PCI and PCIe devices perform an automated configuration operation during system startup and when a device is hot-plugged in a running system. **Hot plugging** is the installation of hardware in a system that is powered on.

Once the configuration is complete, the device interface is known to the operating system. The interface between a PCI or PCIe peripheral and the processor may include any combination of the following communication mechanisms:

- One or more I/O port ranges
- One or more memory address regions supporting memory-mapped I/O
- Connection to a processor interrupt handler

The interface configuration procedure applies to both PCI and PCIe drivers, enabling legacy PCI drivers to work properly in PCIe systems. Of course, the physical card interface differs greatly between parallel PCI and serial PCIe devices, so the cards themselves are not interchangeable across bus technologies. The bus slots for PCIe are intentionally different from PCI slots to prevent the accidental insertion of PCI devices into PCIe slots, and vice versa.

Bulk data transfer to and from peripheral devices generally relies on **Direct Memory Access (DMA)** in both PCI and PCIe systems. In PCIe systems, DMA operations take full advantage of the high data rates possible with multi-lane serial connections, blasting data across the interface at close to the theoretical maximum bit rate that each single- or multi-lane link can support. The technological evolution that supplanted legacy parallel bus PCI technology with the vastly higher-performing multi-lane serial technology of PCIe, all while retaining seamless device driver compatibility, has been quite remarkable.

Device driver structure

A device driver is a software module that implements a set of predefined functions that allow the operating system to associate the driver with compatible peripheral devices and perform controlled access to those devices. This allows system processes and user applications to perform I/O operations on shared devices.

This section provides a brief overview of some of the functions a Linux device driver must implement for use by application developers. This example prefixes the function names with the fictitious device name mydevice and is written in the C programming language.

The following functions perform initialization and termination of the driver itself:

```
int mydevice_init(void);
void mydevice_exit(void);
```

The operating system calls mydevice_init to configure the device driver at system startup or at a later time if the device is connected by hot plugging. The mydevice_init function returns an integer code indicating if the initialization was successful or, if unsuccessful, the error that occurred. Successful driver initialization is indicated by a return code of zero.

When the driver is no longer needed, such as during system shutdown or when the device is removed while the system is running, the system calls mydevice_exit to cleanly end access to the device and release any system resources allocated by the driver.

The next two functions allow system processes and user applications to initiate and terminate communication sessions with the device:

```
int mydevice_open(struct inode *inode, struct file *filp);
int mydevice_release(struct inode *inode, struct file *filp);
```

mydevice_open attempts to initiate access to the device and reports any errors that may occur while doing so. The inode parameter is a pointer to a data structure containing the information required to access a specific file or another device. The filp parameter is a pointer to a data structure containing information about the open file. In Linux I/O, all device types are consistently represented as files, even if the device itself is not inherently file-based. The name filp is short for *file pointer*. All functions operating on the file receive a pointer to this structure as an input. Among other details, the filp structure indicates whether the file is opened for reading, writing, or both.

The mydevice_release function closes the device or file and deallocates any resources allocated in the call to mydevice_open.

Following a successful call to mydevice_open, application code can begin to read from and write to the device. The functions performing these operations are:

```
ssize_t mydevice_read(struct file *filp, char *buf,
    size_t count, loff_t *f_pos);
ssize_t mydevice_write(struct file *filp, const char *buf,
    size_t count, loff_t *f_pos);
```

The mydevice_read function reads from the device or file and transfers the resulting data to a buffer in application memory space. The count parameter indicates the requested amount of data and f_pos indicates the offset from the start of the file at which to begin reading. The buf parameter is the address of the destination for the data. The number of bytes actually read (which may be less than the number requested) is provided as the function return value, with a data type of ssize_t.

The mydevice_write function has most of the same parameters as mydevice_read, except the buf parameter is declared const (constant) because mydevice_write reads starting at the memory address indicated by buf and writes the data to the file or device.

One point of interest in the implementation of these functions is that the privileged driver code cannot (or at least should not, if the operating system permits it) access user memory directly. This is to prevent driver code from accidentally or intentionally reading from or writing to inappropriate memory locations such as kernel space.

To avoid this potential problem, special functions named copy_to_user and copy_from_user are provided by the operating system for use by drivers to access user memory. These functions take the necessary steps to validate the user space addresses provided in the function call before copying data.

This section provided a brief introduction to the hardware interface operations performed by device drivers and introduced the top-level structure of a device driver.

During system power-up, before the operating system boots and initializes its drivers, firmware must execute to perform low-level self-testing and system configuration. The next section presents an introduction to the code that first executes when the computer receives power: the BIOS.

BIOS

A computer's BIOS contains code that first executes at system startup. In the early days of personal computers, the BIOS provided a set of programming interfaces that abstracted the details of peripheral interfaces such as keyboards and video displays.

In modern PCs, the BIOS performs system testing and peripheral device configuration during startup. After that process has been completed, the processor (under software control) interacts with peripheral devices directly without further intervention by the BIOS.

Early PCs stored the BIOS code in a **read-only memory (ROM)** chip on the motherboard. This code was permanently programmed and could not be altered. Modern motherboards generally store the motherboard BIOS in a reprogrammable flash memory device. This allows BIOS updates to be installed to add new features or to fix problems found in earlier firmware versions. The process of updating the BIOS is commonly known as *flashing the BIOS*.

One downside of BIOS reprogrammability is that this capability makes it possible for malicious code to be introduced into a system by writing to the BIOS flash memory. When this type of attack is successful, it enables the malicious code to execute every time the computer starts up. Fortunately, successful BIOS firmware attacks have proven to be quite rare.

As the BIOS takes control during system startup, one of the first things it does is run a **Power-On Self-Test (POST)** of key system components. During POST, the BIOS attempts to interact with system components, including the keyboard, video display, and the boot device, typically a disk drive. Although the computer may contain a high-performance graphics processor, the video interface used by the BIOS during startup may be a primitive video mode, supporting text display only.

The BIOS uses the video and keyboard interfaces to display any errors detected during system testing and to allow the user to change stored settings. The keyboard and video interfaces provided by the BIOS enable the initial setup and configuration of a computer that does not yet contain a boot device.

When the video interface is not working properly, the BIOS will be unable to display information related to the error. In this situation, the BIOS attempts to use the PC speaker, if present, to indicate the error using a pattern of beeps. Motherboard documentation provides information about the type of error indicated by each beep pattern. Some motherboards provide a numerical digit display for reporting POST errors and other status information.

Depending on the system configuration, either the BIOS or the operating system manages the initialization of PCIe devices during system startup. Upon completion of a successful configuration process, all PCIe devices have been assigned compatible I/O port ranges, memory-mapped I/O ranges, and interrupt numbers.

As startup proceeds, the operating system identifies the appropriate driver to associate with each peripheral based on the manufacturer and device identification information provided by the peripheral over PCIe. Following successful initialization, the driver interacts directly with each peripheral to perform I/O operations upon request. System processes and user applications call a set of standardized driver functions, introduced in the previous section, to initiate access to the device, perform read and write operations, and close the device.

One common BIOS-related procedure that users perform is selecting the boot order among the available storage devices. For example, this feature lets you configure the system to first attempt to boot from an optical disk containing a valid operating system image, if such a disk is in the drive. If no bootable optical disk is found, the system might then attempt to boot from the main disk drive. Several mass storage devices can be configured to attempt booting an operating system in priority order.

BIOS configuration mode is sometimes accessed by pressing a specific key, such as *Esc* or the *F2* function key, during the early stage of the boot process. The appropriate key to press is usually indicated on screen shortly after power-on. Upon entering BIOS configuration mode, settings are displayed in a menu format. You can select among different screens to modify parameters associated with features such as boot priority order. After making parameter changes, an option is provided to save the changes to **nonvolatile memory** (**NVM**) and resume the boot process. Be careful when doing this because making inappropriate changes to the BIOS settings can leave the computer unbootable.

The capabilities of BIOS implementations have grown substantially over the years since the introduction of the IBM PC. As PC architectures grew to support 32-bit and then 64-bit operating systems, the legacy BIOS architecture failed to keep pace with the needs of the newer, more capable systems. Major industry participants undertook an initiative to define a system firmware architecture that left behind the limitations of the BIOS. The result of this effort was the UEFI standard, which has replaced the traditional BIOS capabilities in modern motherboards.

UEFI

The **Unified Extensible Firmware Interface (UEFI)** is a 2007 standard defining an architecture for firmware that implements the functions provided by the legacy BIOS and adds several significant enhancements. As with BIOS, UEFI contains code executed immediately upon system startup.

UEFI supports a number of design goals, including enabling support for boot disk devices larger than 2 **terabytes (TB)**, faster startup times, and improved security during the boot process. UEFI provides several features that, when enabled and used properly, substantially reduce the possibility of accidental or malicious corruption of firmware stored in UEFI flash memory.

In addition to the capabilities provided by legacy BIOS implementations described previously, the UEFI supports these features:

- **UEFI applications** are executable code modules stored in UEFI flash memory. UEFI applications provide extensions to capabilities available in the motherboard pre-boot environment and, in some cases, provide services for use by operating systems during runtime. One example of a UEFI application is the UEFI shell, which presents a command-line interface for interacting with the processor and peripheral devices. The UEFI shell supports device data queries and permits the modification of nonvolatile configuration parameters.

- The GNU **GRand Unified Bootloader (GRUB)** is another example of a UEFI application. GRUB supports *multi-boot* configurations by presenting a menu from which the user selects one of multiple available operating system images to boot during system startup.

- **Architecture-independent device drivers** provide processor-independent implementations of device drivers for use by UEFI. This enables a single implementation of UEFI firmware to be used on architectures as diverse as x86 and **Advanced RISC Machine (ARM)** processors. Architecture-independent UEFI drivers are stored in a byte-code format that is interpreted by processor-specific firmware. These drivers enable UEFI interaction with peripherals such as graphics cards and network interfaces during the boot process.

- **Secure Boot** employs cryptographic certificates to ensure that only legitimate device drivers and operating system loaders are executed during system startup. This feature validates the digital signature of each firmware component before allowing it to execute. This validation process protects against many classes of malicious firmware-based attacks.
- **Faster booting** is achieved by performing operations in parallel that took place sequentially under BIOS. In fact, booting is so much faster that many UEFI implementations do not offer the user an option to press a key during booting because waiting for a response would delay system startup. Instead, operating systems such as Windows enable entry to UEFI settings by allowing the user to request access while the operating system is running, followed by a reboot to enter the UEFI configuration screen.

UEFI does not simply replace the functions of the old BIOS. It is a miniature operating system that supports advanced capabilities, such as allowing a remotely located technician to use a network connection to troubleshoot a PC that refuses to boot.

Following POST and the low-level configuration of system devices, and having identified the appropriate boot device based on boot priority order, the system begins the operating system boot process.

The boot process

The procedure for booting a system image varies depending on the partition style of the mass storage device containing the image and the security features enforced during boot. The goal of the boot process is to bring up the system following power application and initialize the operating system, leaving the computer in a known state and ready to perform useful work.

Beginning in the early 1980s, the standard disk partition format was called the **master boot record (MBR)**. An MBR partition has a boot sector located at the logical beginning of its storage space. The MBR boot sector contains information describing the device's logical partitions. Each partition contains a filesystem organized as a tree structure of directories and the files within them.

Due to the fixed format of MBR data structures, an MBR storage device can contain a maximum of four logical partitions and can be no larger than 2 TB in size, equal to 2^{32} 512-byte data sectors. These limits have become increasingly constraining as commercially available disk sizes grew beyond 2 TB. To resolve these issues, and in tandem with the development of UEFI, a new partition format called **GUID partition table (GPT)**, where **GUID** stands for **globally unique identifier**, was developed to eliminate restrictions on disk size and the number of partitions, while providing some additional enhancements.

A GPT-formatted disk has a maximum size of 2^{64} 512-byte sectors, accommodating over 8 billion TB of data. As normally configured, GPT supports up to 128 partitions per drive. The type of each partition is indicated by a 128-bit GUID, allowing an effectively unlimited number of new partition types to be defined in the future. Most users do not need very many partitions on a single disk, so the most obvious user benefit of GPT is its support for larger drives.

The boot process takes place with some differences between BIOS and UEFI motherboards, as described in the following sections.

BIOS boot

In a BIOS motherboard, following POST and PCIe device configuration, BIOS begins the boot process. BIOS attempts to boot from the first device in the configured priority sequence. If a valid device is present, the firmware reads a small piece of executable code called the *boot loader* from the MBR boot sector and transfers control to it. At that point, the BIOS firmware has completed execution and is no longer active for the duration of system operation. The boot loader initiates the process of loading and starting the operating system.

If a *boot manager* is used with a BIOS motherboard, the MBR boot sector code must start the manager rather than directly loading an operating system. The boot manager (such as GRUB) displays a list from which the user selects the desired operating system image. The BIOS firmware itself has no knowledge of multi-booting, and the boot manager operating system selection process takes place without BIOS involvement.

MULTI-BOOTING VERSUS BOOT PRIORITY ORDER

Multi-booting within a boot manager allows the user to select the desired operating system from a menu of available choices. This differs from the boot priority list maintained by the BIOS, which empowers the BIOS itself to select the first available operating system image.

UEFI boot

In a UEFI motherboard, after the POST and device configuration stages have been completed (in a manner very similar to the corresponding BIOS steps), UEFI begins the boot process. A boot manager may be displayed as part of the startup procedure. A UEFI boot manager, which is part of the UEFI firmware, presents a menu from which the user can select the desired operating system image.

If the user does not select an operating system from the boot manager within a few seconds (or if no boot manager menu is displayed), the UEFI attempts to boot from the first device in the configured priority sequence.

The UEFI firmware reads the boot manager executable code (which is separate from the UEFI boot manager) and boot loader files from configured locations on the system disk and executes these files during the startup process.

The following screenshot shows portions of the system **boot configuration data (BCD)** information stored on a Windows 10 system. To display this information on your computer, you must run the `bcdedit` command from Command Prompt with Administrator privileges:

```
C:\>bcdedit
Windows Boot Manager
--------------------
identifier              {bootmgr}
device                  partition=\Device\HarddiskVolume1
path                    \EFI\MICROSOFT\BOOT\BOOTMGFW.EFI
…
Windows Boot Loader
-------------------
identifier              {current}
device                  partition=C:
path                    \WINDOWS\system32\winload.efi
…
```

In this example, the **Windows Boot Manager** is located at `\EFI\MICROSOFT\BOOT\BOOTMGFW.EFI`. This file is normally stored on a hidden disk partition and is not readily available for display in directory listings.

The **Windows boot loader** is identified as `\WINDOWS\system32\winload.efi` and is located at `C:\Windows\System32\winload.efi`.

The possibility of malicious modification of the UEFI firmware or the on-disk software components executed during system boot necessitates additional protection, provided by the trusted boot process.

Trusted boot

The goal of the trusted boot process is to perform the entire boot sequence while ensuring that all of the software executed is properly authorized and has not been modified. The boot software verifies the authenticity of firmware and software files by computing the cryptographic hash of the bytes comprising each software component and ensuring the resulting hash value is valid using a digital signature.

A **cryptographic hash function** generates a "fingerprint" for a block of data of arbitrary length. The output value produced by a given cryptographic hash function is always the same length. For example, the 256-bit **Secure Hashing Algorithm (SHA-256)** always produces a 256-bit output regardless of the length of its input data.

The feature that makes a cryptographic hash function secure is that any attempt to modify the data used as input to the function will cause the hash output to also be different. It is effectively impossible to identify a set of modifications to an input data block that will produce the same hash output as the original data block. While it is theoretically possible to come up with a modified data set with the same SHA-256 hash as the original block, the computations required to do so would take longer than the Earth's remaining lifetime on the fastest supercomputer available today. We will discuss hash functions in more detail in *Chapter 15, Blockchain and Bitcoin Mining Architectures*.

To create a digital signature, a software publisher must compute the hash value for the firmware or software component memory image. The next step is to encrypt the hash using a private signing key. This private encryption key is associated with a publicly available decryption key. The encrypted hash value forms the digital signature that is stored along with the firmware or software file.

The public key is made available to users of the firmware or software. This key is used to decrypt the encrypted hash value, which can then be compared to the hash value computed from the file's data bytes during boot. The two hash values must match exactly to confirm the file is unmodified since it was digitally signed. As with the cryptographic hash algorithm, it is effectively impossible to alter the digital signature in a manner that would enable undetected modifications to the protected data file.

While the public key is not intended to be kept secret, the users of the key must ensure that the key itself is from the trusted source and has not been replaced by a malicious actor. To ensure that only trusted public keys are used during the boot process, these keys are stored in a hardware location that is inaccessible to unauthorized users and software. The standard mechanism for this in PCs is called the **Trusted Platform Module (TPM)**. A TPM is, in effect, a microcontroller carefully designed to protect cryptographic keys.

In newer computers, TPM keys are configured by the system manufacturer to ensure the system boots in a secure manner from the moment of its initial power-on.

Unlike personal computers, most embedded devices use a much simpler boot process that does not involve the BIOS or UEFI. The next section discusses the boot process in embedded devices.

Embedded devices

Most embedded systems, such as smartphones, do not generally have separate boot firmware such as the BIOS or UEFI in a PC. As we saw with the 6502, these devices perform a processor hardware reset when power is applied and begin code execution at a specified address. All code in these devices is typically located in a nonvolatile storage region such as flash memory.

During startup, embedded devices perform a sequence of events similar to the PC boot process. Peripheral devices are tested for proper operation and initialized prior to first use. The boot loader in such devices may need to select among multiple memory partitions to identify an appropriate system image. As with UEFI, embedded devices often incorporate security features during the boot process to ensure that the boot loader and operating system image are authentic before allowing the boot process to proceed.

In both PC and embedded systems, execution of the boot loader is the first step in bringing up the operating system. We'll look at operating system startup next.

Operating systems

An operating system is a multilayer suite of software that provides an environment in which applications perform useful functions such as word processing, placing telephone calls, or managing the operation of a car engine. Applications running under control of the operating system execute algorithms implemented as processor instruction sequences and perform I/O interactions with peripheral devices as required to complete their tasks.

The operating system provides standardized programming interfaces that application developers use to access system resources such as processor execution threads, disk files, input from a keyboard or other peripherals, and output to devices such as a computer screen or instruments on an automotive dashboard.

Operating systems can be broadly categorized as real-time or non-real-time:

- A **real-time operating system (RTOS)** provides features to ensure that responses to inputs occur within a defined time limit.

Processors performing tasks such as managing the operation of a car engine or a kitchen appliance typically run an RTOS to ensure that the electrical and mechanical components they control receive responses to any change in inputs within a bounded time.

- Non-real-time operating systems do not attempt to ensure that responses are generated within any particular time limit. Instead, these systems attempt to perform processing as quickly as possible, even if it sometimes takes a long time to complete the work.

REAL-TIME VERSUS NON-REAL-TIME OPERATING SYSTEMS

RTOSes are not necessarily faster than non-real-time operating systems. A non-real-time operating system may be faster on average compared to an RTOS, but the non-real-time system may occasionally exceed the time limits specified for applications executing on the RTOS. The goal of the RTOS is to never exceed the response time limit.

For the most part, general-purpose operating systems such as Linux and Windows are non-real-time operating systems. They try to get assigned tasks—such as reading a file into a word processor or computing a spreadsheet—finished as quickly as possible, although the time to complete an operation may vary widely, depending on other tasks the system may be performing.

Some aspects of general-purpose operating systems, particularly audio and video output, have specific real-time requirements. We've all seen poor video replay at one time or another, in which the video stutters and appears jerky. This behavior is the result of failing to meet the real-time performance demands of a video display. Cell phones have similar real-time requirements for supporting two-way audio during telephone calls.

For both real-time and non-real-time operating systems, in standard PCs as well as in embedded devices, operating system startup tends to follow a similar sequence of steps. The boot loader either loads the operating system kernel into memory or simply jumps to an address in nonvolatile memory to begin executing the operating system code.

Once started, an operating system kernel performs the following steps, though not necessarily in this order:

- The processor and other system devices are configured. This includes setting up any required registers internal to the processor and any associated I/O management devices, such as a chipset.
- In systems using paged virtual memory (introduced in *Chapter 7, Processor and Memory Architectures*), the kernel configures the memory management unit.

- Base-level system processes, including the **scheduler** and the **idle process**, are started. The scheduler manages the sequence of execution for process threads. The idle process contains the code that executes when there are no other threads ready for the scheduler to run.
- Device drivers are enumerated and associated with each peripheral in the system. Initialization code is executed for each driver, as discussed earlier in this chapter.
- Interrupts are configured and enabled. Once interrupts have been enabled, the system begins to perform I/O interactions with peripheral devices.
- System services are started. These processes support non-operating system activities (such as networking) and persistent, installed capabilities (for example, a web server).
- For PC-type computers, a user interface process is started, which presents a login screen. This screen allows the user to initiate an interactive session with the computer. In embedded devices, the real-time application begins execution. The basic operational sequence for a simple embedded application is to read inputs from I/O devices, execute a computation algorithm to generate outputs, and write the outputs to I/O devices, repeating this procedure at fixed time intervals.

This section uses the term *process* to indicate a program running on a processor. The term *thread* indicates a flow of execution within a process, of which there may be more than one. The next section examines these topics in more detail.

Processes and threads

Many, but not all, operating systems support the concept of multithreaded execution. A **thread** is a sequence of program instructions that logically executes in isolation from other threads. An operating system running on a single-core processor creates the illusion of multiple simultaneously running threads by performing **time-slicing**.

In time-slicing, an operating system scheduler grants each ready-to-run thread a period of time in which to execute. As a thread's execution interval ends, the scheduler interrupts the running thread and continues executing the next thread in its queue. In this manner, the scheduler gives each thread a bit of time to run before going back to the beginning of the list and starting over again.

In operating systems capable of supporting multiple runnable programs simultaneously, the term *process* refers to a running instance of a computer program. The system allocates resources, such as memory and membership in the scheduler's queue of runnable threads, to each process.

When a process first begins execution, it consists of a single thread. The process may create more threads as it executes.

Programmers create multithread applications for various reasons, including the following:

- One thread can perform I/O while a separate thread executes the main algorithm of the program. For example, a primary thread can periodically update a user display with received information while a separate thread waits in a blocked state for user input from the keyboard.
- Applications with substantial computational requirements can take advantage of multi-processor and multi-core computer architectures by splitting large computational jobs into groups of smaller tasks capable of execution in parallel. By running each of these smaller tasks as a separate thread, programs enable the scheduler to assign different threads to execute on multiple cores simultaneously.

A process passes through a series of states during its life cycle. Some process states assigned by operating systems are as follows:

- **Initializing**: When a process is first started, perhaps as the result of a user double-clicking an icon on the desktop, the operating system begins loading the program code into memory and assigning system resources for its use.
- **Waiting**: After process initialization has been completed, it is ready to run. At this point, its thread is assigned to the scheduler's queue of runnable threads. The process remains in the Waiting state until the scheduler permits it to start running.
- **Running**: The thread is executing the program instructions contained in its code section.
- **Blocked**: The thread enters this state when it requests I/O from a device that causes execution to pause. For example, reading data from a file normally causes blocking. In this state, the thread waits in the Blocked state for the device driver to finish processing the request. As soon as a running thread becomes blocked, the scheduler is free to switch to another runnable thread while the first thread's I/O operation is in progress. When the operation completes, the blocked thread returns to the Waiting state in the scheduler queue and eventually returns to the Running state, where it processes the results of the I/O operation.

Ready-to-run processes rely on the scheduler to receive the execution time. The scheduler process is responsible for granting execution time to all system and user threads.

The scheduler is an interrupt-driven routine that executes at periodic time intervals, as well as in response to actions taken by threads, such as the initiation of I/O operations. During operating system initialization, a periodic timer is attached to the scheduler interrupt handler and the scheduler timer is started.

While each process is in the Initializing state, the kernel adds a data structure called a **process control block (PCB)** to its list of running processes. The PCB contains information the system requires to maintain and interact with the process over its lifetime, including its memory allocations and details regarding the file containing its executable code. In most operating systems, a process is identified by an integer that remains unique during its lifetime.

In Windows, the **Resource Monitor** tool (you can start this tool by typing Resource Monitor into the Windows search box and clicking on the result identified as **Resource Monitor**) displays information about running processes, including the **process identifier (PID)** associated with each process. In Linux, the top command displays the processes consuming the most system resources, identifying each by its PID.

The scheduler maintains information associated with each thread in a **thread control block (TCB)**. Each process has a list of associated TCBs, with a minimum of one entry in the list. The TCB contains information related to the thread, such as processor context. The **processor context** is the collection of information the kernel uses to resume execution of a waiting or blocked thread, consisting of these items:

- The saved processor registers
- The stack pointer
- The flags register
- The instruction pointer

Similar to the PID, each thread has an integer **thread identifier (TID)** that remains unique during its lifetime.

The scheduler uses one or more scheduling algorithms to ensure execution time is allocated equitably among system and user processes. Two main categories of thread scheduling algorithms have been widely used since the early days of computing—non-preemptive and preemptive:

- **Non-preemptive scheduling** grants a thread complete execution control, allowing it to run until it terminates or until it voluntarily releases control to the scheduler so that other threads can have a chance to run.
- In **preemptive scheduling**, the scheduler has the authority to stop a running thread and hand execution control to another thread without approval from the first thread.

When a preemptive scheduler switches execution from one thread to another, it performs the following steps:

1. Either a timer interrupt occurs that causes the scheduler to begin execution, or a running thread performs an action that causes blocking, such as initiating an I/O operation.
2. The scheduler copies the departing thread's processor registers into the context fields of the thread's TCB.
3. The scheduler consults its list of runnable threads and determines which thread to place in the Running state.
4. The scheduler loads the context of the incoming thread into the processor registers.
5. The scheduler resumes execution of the incoming thread by jumping to the instruction pointed to by the thread's program counter.

Thread scheduling occurs at a high frequency, which implies the code involved in scheduler activity has to be as efficient as possible. In particular, storing and retrieving processor context takes some time, so operating system designers make every effort to optimize the performance of the scheduler's context switching code.

Because there may be numerous processes competing for execution time at a given moment, the scheduler must ensure that critical system processes can execute at their required intervals. At the same time, from the user's point of view, applications must remain responsive to user inputs while providing an acceptable level of performance during lengthy computations.

Various algorithms have been developed over the years to efficiently manage these competing demands. A key feature of most thread scheduling algorithms is the use of process priorities. The next section introduces several priority-based thread scheduling algorithms.

Scheduling algorithms and process priority

Operating systems supporting multiple processes generally provide a prioritization mechanism to ensure that the most important system functions receive adequate processing time, even when the system is under heavy load, while continuing to provide adequate time for the execution of lower-priority user processes. Several algorithms have been developed to meet various performance goals for different types of operating systems. Some scheduling algorithms that have been popular over the years, beginning with the simplest, are as follows:

- **First come, first served (FCFS)**: This non-preemptive approach was common in legacy batch processing operating systems. In an FCFS scheduling algorithm, each process is granted execution control and retains control until execution is completed. There is no prioritization of processes, and the time to complete any process is dependent on the execution time of processes preceding it in the input queue.

- **Cooperative multithreading**: Early versions of Windows and macOS used a non-preemptive multithreading architecture that relied on each thread to voluntarily relinquish control to the operating system at frequent intervals. This required significant effort by application developers to ensure a single application did not starve other applications of opportunities to execute by failing to release control at appropriate intervals. Each time the operating system received control, it selected the next thread to execute from a prioritized list of runnable threads.

- **Round-robin scheduling**: A preemptive round-robin scheduler maintains a list of runnable threads and grants each an execution interval in turn, starting over at the beginning of the list when reaching the end. This approach effectively sets all process priorities as equal, giving each an opportunity to execute for defined periods of time, at time intervals dependent on the number of processes in the scheduler's list.

- **Fixed-priority preemptive scheduling**: In this algorithm, each thread is assigned a priority indicating the importance of its receiving execution control when it is in the Waiting state. When a thread enters the Waiting state, if it has a higher priority than the currently running thread, the scheduler immediately stops the running thread and turns control over to the incoming thread. The scheduler maintains the list of Waiting processes in priority order, with the highest priority threads at the head of the line. This algorithm can result in the failure of lower-priority threads to get any execution time if higher-priority threads monopolize the available execution time.

- **Rate-monotonic scheduling (RMS)**: This is a fixed-priority preemptive scheduling algorithm commonly used in real-time systems with hard deadlines (a hard deadline is one that cannot be missed). Under RMS, threads that execute more frequently are assigned higher priorities. As long as a few criteria are satisfied (the thread execution interval equals the deadline, there can be no delay-inducing interactions between threads, and the context switch time is negligible), if the maximum possible execution time of each thread is below a mathematically derived limit, deadlines are guaranteed to be met.

- **Fair scheduling**: Fair scheduling attempts to maximize the utilization of processor time while ensuring each user is granted an equal amount of execution time. Rather than assigning numeric priorities to threads, the effective priority of each thread is determined by the amount of execution time it has consumed. As a thread uses more and more processor time, its priority declines, giving other threads more opportunities to run. This approach has the benefit that, for interactive users who do not consume much execution time, the responsiveness of the system is improved. The Linux kernel uses a fair scheduling algorithm as its default scheduler.

- **Multilevel feedback queue**: This algorithm uses multiple queues, each with a different priority level. New threads are added at the tail of the highest-priority queue. At each scheduling interval, the scheduler grants execution to the thread at the head of the high-priority queue and removes the thread from the queue, which moves the remaining threads closer to execution. Eventually, the newly created thread receives an opportunity to execute. If the thread consumes all the execution time granted to it, it is preempted at the completion of its interval and added to the tail of the next lower-priority queue. The Windows Scheduler is a multilevel feedback queue.

The system idle process contains the thread that executes when there is no user- or system-assigned thread in the Waiting state. An idle process may be as simple as a single processor instruction that forms an infinite loop, jumping to itself. Some operating systems place the system in a power-saving mode during idle periods rather than executing an idle loop.

The percentage of processor time consumed by running processes is computed by determining the fraction of time the system was executing a non-idle thread over a measurement period. The following screenshot is a Windows Resource Monitor view of running processes consuming the highest average share of processor time:

Chapter 5

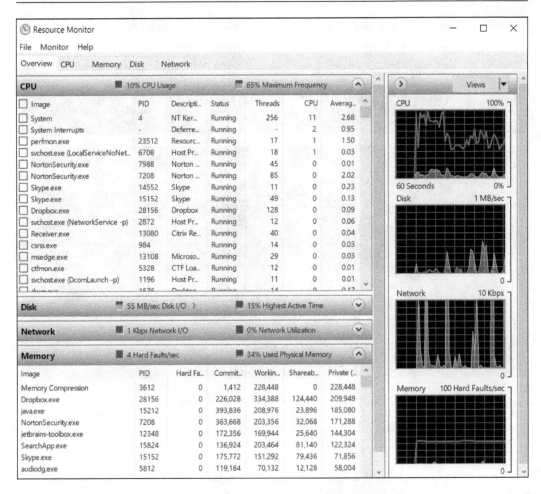

Figure 5.1: Windows Resource Monitor process display

In this figure, the **PID** column displays the numeric process identifier and the **Threads** column shows the number of threads in the process. The process state is **Running** for all of the processes visible in this display.

The following screenshot shows the result of running the top command on a Linux system:

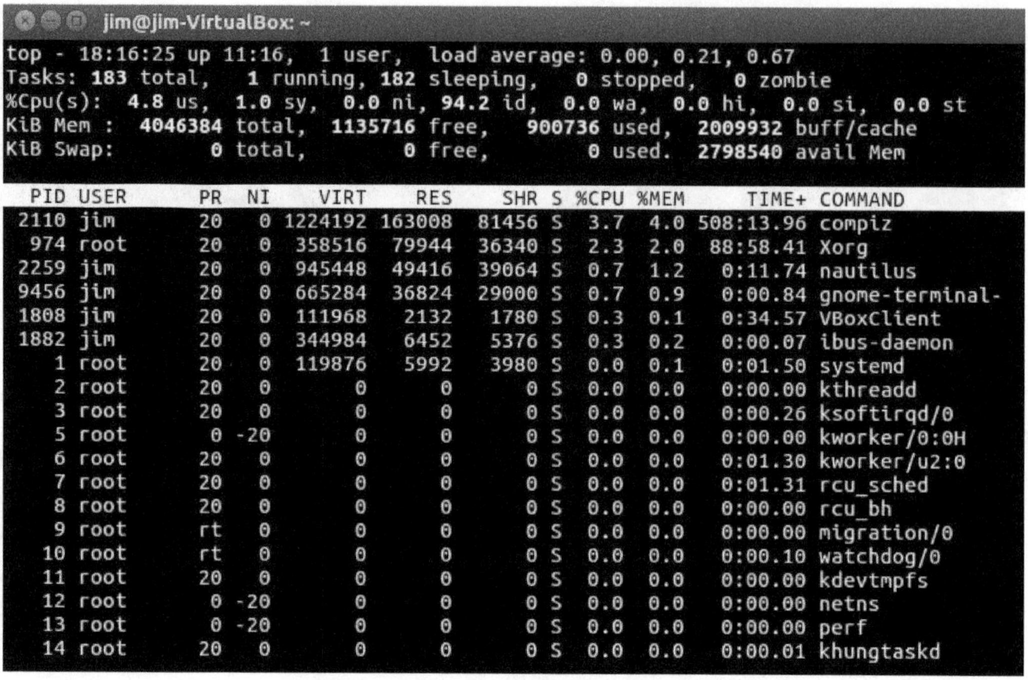

```
top - 18:16:25 up 11:16,  1 user,  load average: 0.00, 0.21, 0.67
Tasks: 183 total,   1 running, 182 sleeping,   0 stopped,   0 zombie
%Cpu(s):  4.8 us,  1.0 sy,  0.0 ni, 94.2 id,  0.0 wa,  0.0 hi,  0.0 si,  0.0 st
KiB Mem :  4046384 total,  1135716 free,   900736 used,  2009932 buff/cache
KiB Swap:        0 total,        0 free,        0 used.  2798540 avail Mem

  PID USER      PR  NI    VIRT    RES    SHR S %CPU %MEM     TIME+ COMMAND
 2110 jim       20   0 1224192 163008  81456 S  3.7  4.0 508:13.96 compiz
  974 root      20   0  358516  79944  36340 S  2.3  2.0  88:58.41 Xorg
 2259 jim       20   0  945448  49416  39064 S  0.7  1.2   0:11.74 nautilus
 9456 jim       20   0  665284  36824  29000 S  0.7  0.9   0:00.84 gnome-terminal-
 1808 jim       20   0  111968   2132   1780 S  0.3  0.1   0:34.57 VBoxClient
 1882 jim       20   0  344984   6452   5376 S  0.3  0.2   0:00.07 ibus-daemon
    1 root      20   0  119876   5992   3980 S  0.0  0.1   0:01.50 systemd
    2 root      20   0       0      0      0 S  0.0  0.0   0:00.00 kthreadd
    3 root      20   0       0      0      0 S  0.0  0.0   0:00.26 ksoftirqd/0
    5 root       0 -20       0      0      0 S  0.0  0.0   0:00.00 kworker/0:0H
    6 root      20   0       0      0      0 S  0.0  0.0   0:01.30 kworker/u2:0
    7 root      20   0       0      0      0 S  0.0  0.0   0:01.31 rcu_sched
    8 root      20   0       0      0      0 S  0.0  0.0   0:00.00 rcu_bh
    9 root      rt   0       0      0      0 S  0.0  0.0   0:00.00 migration/0
   10 root      rt   0       0      0      0 S  0.0  0.0   0:00.10 watchdog/0
   11 root      20   0       0      0      0 S  0.0  0.0   0:00.00 kdevtmpfs
   12 root       0 -20       0      0      0 S  0.0  0.0   0:00.00 netns
   13 root       0 -20       0      0      0 S  0.0  0.0   0:00.00 perf
   14 root      20   0       0      0      0 S  0.0  0.0   0:00.01 khungtaskd
```

Figure 5.2: Linux top command process display

The upper part of the display contains summary information, including the number of processes (referred to as Tasks here) in each state.

Each row in the lower part of the display presents information about one running process. As in Windows, the PID column indicates the PID. The state of each process is shown in the S column, with these possible values:

- **R**: Runnable, meaning either running or in the queue of ready-to-run threads
- **S**: Sleeping: Paused while blocked; waiting for an event to complete
- **T**: Stopped in response to a job control command (pressing *CTRL+Z* will do this)
- **Z**: Zombie, which occurs when a child process belonging to another process terminates, but the child process information continues to be maintained by the system until the parent process ends

The PR column displays the scheduling priority of the process. Smaller numbers represent higher priorities.

Up to this point, we have referred to the computer processor as a singular entity. In most modern PCs, the processor integrated circuit contains two or more processor cores, each implementing the features of a complete, independent processor, including a control unit, register set, and ALU. The next section discusses the attributes of systems containing multiple processing units.

Multiprocessing

A **multiprocessing** computer contains two or more processors that simultaneously execute sequences of instructions. The processors in such a system typically share access to system resources, such as main memory and peripheral devices. The processors in a multiprocessing system may be of the same architecture, or individual processors may be of different architectures to support unique system requirements. Systems in which all processors are treated as equal are referred to as **symmetric multiprocessing** systems. Devices that contain multiple processors within a single integrated circuit package are called **multi-core processors**.

At the level of the operating system scheduler, a symmetric multiprocessing environment simply provides more processors for use in thread scheduling. In such systems, the scheduler treats additional processors as resources when assigning threads for execution.

In a well-designed symmetric multiprocessing system, throughput can approach the ideal scenario of scaling linearly with the number of available processor cores, assuming contention for shared resources is minimal. If multiple threads on separate cores attempt to perform heavy simultaneous access to main memory, for example, there will be inevitable performance degradation as the system arbitrates access to the resource and shares it among competing threads. A multichannel interface to DRAM can improve overall system performance in this scenario.

A symmetric multiprocessing system is an example of a **multiple instruction, multiple data (MIMD)** architecture. MIMD is a parallel processing configuration in which each processor core executes an independent sequence of instructions on its own set of data. A **single instruction, multiple data (SIMD)** parallel processing configuration, in comparison, executes the same instruction operation on multiple data elements simultaneously.

Modern processors implement SIMD instructions to perform parallel processing on large datasets such as graphical images and audio data sequences. In current-generation PCs, the use of multi-core processors enables MIMD execution parallelism, while specialized instructions within the processors provide a degree of SIMD execution parallelism. SIMD processing will be discussed further in *Chapter 8, Performance-Enhancing Techniques*.

Processor clock speeds have grown from the 4.77 MHz of the original PC to over 4 GHz in modern processors, nearly a thousand-fold increase. Future clock speed increases are likely to be more limited as fundamental physical limits present looming obstacles. To compensate for limited future performance gains from increases in clock speed, the processor industry has turned to emphasizing various forms of execution parallelism in personal computer systems and smart devices. Future trends are likely to continue the growth in parallelism as systems integrate dozens, then hundreds, and eventually thousands of processor cores executing in parallel in PCs, smartphones, and other digital devices.

Summary

This chapter began with an overview of device drivers, including details on the instruction sequences used by driver code to read from and write to a simple I/O device: the PC parallel port. We continued with a discussion of the legacy BIOS and the newer UEFI, which provide the code that first executes on PC power-up, performs device testing and initialization, and initiates loading of the operating system. We saw how the trusted boot process can ensure that only authorized and unmodified code can be permitted to execute during system startup.

We continued with a description of some of the fundamental elements of operating systems, including processes, threads, and the scheduler. Various scheduling algorithms used in past computers and the systems of today were introduced. We examined the output of tools available in Linux and Windows that present information about running processes.

The chapter concluded with a discussion of multiprocessing and its performance impact on the computer systems of today, as well as the implications of MIMD and SIMD parallel processing for the future of computing.

The next chapter will introduce specialized computing domains and their unique processing requirements in the areas of real-time computing, digital signal processing, and **graphics processing unit (GPU)** processing.

Exercises

1. Restart your computer and enter the BIOS or UEFI settings. Examine each of the menus available in this environment. Does your computer have BIOS or does it use UEFI? Does your motherboard support overclocking? When you are finished, be sure to select the option to quit without saving changes unless you are absolutely certain you want to make changes.

2. Run the appropriate command on your computer to display the currently running processes. What is the PID of the process you are using to run this command?

Join our community Discord space

Join the book's Discord workspace for a monthly *Ask me Anything* session with the author: `https://discord.gg/7h8aNRhRuY`

6
Specialized Computing Domains

Most computer users are at least superficially familiar with key performance-related attributes of personal computers and smart digital devices such as processor speed and **random-access memory (RAM)** size. This chapter explores the performance requirements of computing domains that tend to be less directly visible to users, including real-time systems, digital signal processing, and **graphics processing unit (GPU)** processing.

We will examine the unique computing features associated with each of these domains and review some examples of modern devices implementing these concepts.

After completing this chapter, you will be able to identify application areas that require real-time computing and will understand the underlying concepts of digital signal processing, with an emphasis on its widespread use in wireless communication. You will also understand the basic architecture of modern GPUs and will be familiar with some modern implementations of components in the computing domains discussed in this chapter.

This chapter covers the following topics:

- Real-time computing
- Digital signal processing
- GPU processing
- Examples of specialized architectures

Technical requirements

The files for this chapter, including the answers to the exercises, are available at https://github.com/PacktPublishing/Modern-Computer-Architecture-and-Organization-Second-Edition.

Real-time computing

The last chapter provided a brief introduction to some of the requirements of real-time computing in terms of a system's responsiveness to changes in its inputs. These requirements are specified in the form of timing deadlines that limit how long the system can take to produce an output in response to a change in its input. This section looks at these timing specifications in more detail and presents some of the features real-time computing systems implement to ensure timing requirements are met.

Real-time computing systems can be categorized as providing soft or hard guarantees of responsiveness. A **soft real-time system** is considered to perform acceptably if it meets its desired response time most, but not necessarily all, of the time. An example of a soft real-time application is the clock display on a cell phone. When opening the clock display, some implementations momentarily present the time that was shown the last time the clock display was used before quickly updating to the correct, current time. Of course, users would like the clock to show the correct time whenever it is displayed, but momentary glitches such as this are not typically seen as significant problems.

A hard real-time system, on the other hand, is considered to have failed if it ever misses any of its response-time deadlines. Safety-critical systems such as airbag controllers in automobiles and flight control systems for commercial aircraft have hard real-time requirements. Designers of these systems take timing requirements very seriously and devote substantial effort to ensuring the real-time processor satisfies its timing requirements under all possible operating conditions.

The control flow of a simple real-time system is shown in the following figure:

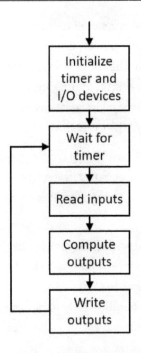

Figure 6.1: Real-time system control flow

Figure 6.1 represents a real-time computing system using a hardware interval timer to control the time sequencing of its operation. A down-counting interval timer performs a repetitive cycle of the following steps:

1. Load the counter register with a predefined numeric value
2. Decrement the counter at a fixed clock rate
3. When the count reaches zero, generate an event such as setting a bit in a register or triggering a processor interrupt
4. Go back to *step 1*

An interval timer generates a periodic sequence of events with timing accuracy that depends on the characteristics of the system clock, which is often driven by a quartz crystal. By waiting for the timer event at the top of each loop, the system in *Figure 6.1* begins each execution pass at fixed, nearly equal time intervals.

To satisfy the demands of hard real-time operation, the execution time of the code inside the loop (the code contained in the **Read inputs**, **Compute outputs**, and **Write outputs** blocks in *Figure 6.1*) must always be less than the timer interval. Prudent system developers ensure that no path through the code results in execution time anywhere close to the hard real-time limit. A conservative system design rule might insist that the longest execution path for code inside the loop consumes no more than 50% of the timer interval.

Practical real-time systems constructed in this configuration might be based on an 8-, 16-, or 32-bit processor running at a clock frequency in the tens to hundreds of MHz range. The timer employed in the main loop of such systems generates events at a developer-selected frequency, often in the 10 to 1,000 Hz range.

In less complex system designs, the code represented by the blocks in *Figure 6.1* often runs directly on the processor hardware with no intervening software layers. This configuration contains no operating system of the type described in *Chapter 5, Hardware-Software Interface*. A sophisticated real-time application, on the other hand, is likely to have more extensive needs than can be met by this simplistic architecture, which makes the use of a real-time operating system attractive.

Real-time operating systems

A **real-time operating system (RTOS)** contains several features superficially similar to the general-purpose operating systems discussed in *Chapter 5, Hardware-Software Interface*. An RTOS design differs significantly from general-purpose operating systems, however, in that all RTOS aspects—from kernel internals to device drivers, to system services— are focused on meeting hard real-time requirements.

Most RTOS designs employ preemptive multithreading, often referred to as **multitasking** in RTOS literature. The terms *task* and *thread* are somewhat synonymous in the RTOS context, so for consistency we will continue to use the term *thread* to indicate an RTOS task.

RTOS designs at the lower end of the sophistication scale typically support multithreading within the context of a single application process. These simpler RTOSes support thread prioritization but often lack memory protection features.

More sophisticated RTOS architectures provide operating system features such as memory protection in addition to prioritized preemptive multithreading. These RTOSes allow multiple processes to be in the Running state simultaneously, each potentially containing several threads. In protected memory systems, kernel memory access by application threads is prohibited and applications cannot reach into each other's memory regions.

RTOS environments, from lower to higher levels of sophistication, provide several data structures and communication techniques geared toward efficient data transfer between threads, and to support controlled access to shared resources. These capabilities are often available in non-RTOSes as well. Some examples of these features are:

- **Mutex:** A mutex (short for *mutual exclusion*) is a mechanism for a thread to claim access to a shared resource without blocking the execution of other threads. In its simplest form, a mutex is a variable accessible to all threads that holds the value 0 when the resource is free and 1 when the resource is in use. A thread that wants to use the resource reads the current value of the mutex variable and, if it is 0, sets it to 1 and performs the operation using the resource. After completing the operation, the thread sets the mutex back to 0. There are, however, some potential problems with mutexes:

    - **Thread preemption:** Let's say a thread reads the mutex variable and sees that it is 0. Because the scheduler can interrupt an executing thread at any time, that thread might be interrupted before it has a chance to set the mutex to 1. A different thread then resumes execution and takes control of the same resource because it sees the mutex is still 0. When the original thread resumes, it finishes setting the mutex to 1 (even though, by now, it has already been set to 1). At this point, both threads incorrectly believe they have exclusive access to the resource, which is likely to lead to serious problems when both threads try to use the resource.

        To prevent this scenario, many processors implement some form of a **test-and-set instruction**. A test-and-set instruction reads a value from a memory address and sets that location to 1 in a single, uninterruptable (also referred to as **atomic**) action. In the x86 architecture, the BTS (bit test and set) instruction performs this operation. In processor architectures that lack a test-and-set instruction (such as the 6502), the risk of preemption can be eliminated by disabling interrupts before checking the state of the mutex variable, and then re-enabling interrupts after setting the mutex to 1. This approach has the disadvantage of reducing real-time responsiveness while interrupts are disabled.

    - **Priority inversion:** Priority inversion occurs when a higher-priority thread attempts to gain access to a resource while the corresponding mutex is held by a lower-priority thread. In this situation, RTOS implementations generally place the higher-priority thread in a blocked state, allowing the lower-priority thread to complete its operation and release the mutex.

The priority inversion problem occurs when a thread with a priority between the upper and lower levels of the two threads begins execution. While this mid-priority thread is running, it prevents the lower-priority thread from executing and releasing the mutex. The higher-priority thread must now wait until the mid-priority thread finishes execution, effectively disrupting the entire thread prioritization scheme. This can lead to a failure of the high-priority thread to meet its deadline.

One method to prevent priority inversion is **priority inheritance**. In an RTOS implementing priority inheritance, whenever a higher-priority thread (`hi_thread`) requests a mutex held by a lower-priority thread (`lo_thread`), `lo_thread` is temporarily raised in priority to the level of `hi_thread`. This eliminates any possibility of a mid-priority thread delaying the completion of the (originally) lower-priority `lo_thread`. When `lo_thread` releases the mutex, the RTOS restores the thread's original priority.

- **Deadlock**: Deadlock can occur when multiple threads attempt to lock multiple mutexes. If `thread1` and `thread2` both require `mutex1` and `mutex2`, a situation may arise in which `thread1` locks `mutex1` and attempts to lock `mutex2`, while at the same time `thread2` has already locked `mutex2` and attempts to lock `mutex1`. Neither task can proceed from this state, hence the term *deadlock*. Some RTOS implementations check the ownership of mutexes during lock attempts and report an error in a deadlock situation. In simpler RTOS designs, it is up to the system developer to ensure deadlock cannot occur.

- **Semaphore**: A semaphore is a generalization of the mutex. Semaphores can be of two types: binary and counting. A **binary semaphore** is like a mutex except that rather than controlling access to a resource, the binary semaphore is intended to be used by one task to send a signal to another task. If `thread1` attempts to *take* semaphore1 while it is unavailable, `thread1` will block until another thread or interrupt service routine *gives* semaphore1.

  - A **counting semaphore** contains a counter with an upper limit. Counting semaphores are used to control access to multiple interchangeable resources. When a thread takes the counting semaphore, the counter increments and the task proceeds. When the counter reaches its limit, a thread attempting to take the semaphore blocks until another thread gives the semaphore, decrementing the counter.

    Consider the example of a system that supports a limited number of simultaneously open files.

A counting semaphore can be used to manage file open and close operations. If the system supports up to 10 open files and a thread attempts to open an 11th file, a counting semaphore with a limit of 10 will block the file open operation until another file is closed and its descriptor becomes available.

- **Queue**: A **queue** (also referred to as a **message queue**) is a unidirectional communication path between processes or threads. The sending thread places data items into the queue and the receiving thread retrieves those items in the same order they were sent. The RTOS synchronizes access between the sender and receiver so the receiver only retrieves complete data items. Queues are commonly implemented with a fixed-size storage buffer. The buffer will eventually fill and block further insertions if a sending thread adds data items faster than the receiving thread retrieves them.

  RTOS message queues provide a programming interface for the receiving thread to check if the queue contains data. Many queue implementations also support the use of a semaphore to signal a blocked receiving thread when data becomes available.

- **Critical section**: It is common for multiple threads to require access to a shared data structure. When using shared data, it is vital that read and write operations from different threads do not overlap in time. If such an overlap occurs, the reading thread may receive inconsistent information if it accesses the data structure while another thread is performing an update. The mutex and semaphore mechanisms provide options for controlling access to shared data structures. The use of a critical section is an alternate approach that isolates the code accessing the shared data structure and allows only one thread to execute that sequence at a time.

- A simple method to implement a critical section is to disable interrupts just before entering a critical section and re-enable interrupts after completing the critical section. This prevents the scheduler from running and ensures the thread accessing the data structure has sole control until it exits the critical section. However, this method has the disadvantage of impairing real-time responsiveness by preventing responses to interrupts, including thread scheduling, while interrupts are disabled.

Some RTOS implementations provide a more sophisticated implementation of the critical section technique involving the use of critical section data objects. Critical section objects typically provide options to allow a thread to either enter a blocked state until the critical section becomes available or test if the critical section is in use without blocking. The option for testing critical section availability allows the thread to perform other work while waiting for the critical section to become free.

There are far more real-time computing systems in operation today than there are PCs we think of as *computers*. General-purpose computers represent less than 1% of the digital processors produced each year. Devices ranging from children's toys to digital thermometers, to televisions, to automobiles, to spacecraft contain at least one, and often dozens of, embedded processors, each running some type of RTOS.

This section provided a brief introduction to some of the communication and resource management capabilities common in RTOS implementations. The next section introduces processing architectures used in the processing of digital samples of analog signals.

Digital signal processing

A **digital signal processor** (DSP) is optimized to perform computations on digitized representations of analog signals. Real-world signals such as audio, video, cell phone **radio frequency** (RF) transmissions, and radar are **analog** in nature, meaning the information being processed is the response of an electrical sensor to a continuously varying input voltage. Before a digital processor can begin to work with an analog signal, the signal voltage must be converted to a digital representation by an **analog-to-digital converter** (ADC). The following section describes the operation of ADCs and **digital-to-analog converters** (DACs).

ADCs and DACs

An ADC measures an analog input voltage and produces a digital output word representing the input voltage. A DAC performs the reverse operation of an ADC, converting a digital word to an analog voltage. ADCs often use a DAC internally during the conversion process.

A variety of circuit architectures are used in DAC applications, generally with the goal of achieving a combination of low cost, high speed, and high precision. One of the simplest DAC designs is the **R-2R ladder**, shown here in a 4-bit input configuration:

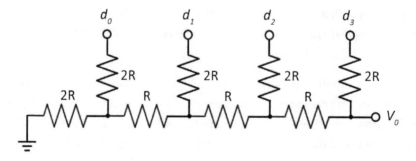

Figure 6.2: R-2R ladder DAC

This DAC uses a 4-bit data word on the inputs $d_0 - d_3$, where d_3 is the most significant bit, to generate an analog voltage, V_0 If we assume each bit of the 4-bit word d is driven at either 0 V (for a 0 bit) or 5 V (for a 1 bit), the output V_0 equals $(d / 2^4) * 5$ V, where d is a data value in the range 0 to 15. An input word of 0 has an output of 0 V, and an input word of 15 has an output of $(15/16) * 5$ V $= 4.6875$ V. Intermediate values of d produce equally spaced output voltages at intervals of $(1/16) * 5$ V $= 0.3125$ V.

An ADC can use an internal DAC like this (though usually with a larger number of bits, with a correspondingly finer voltage resolution) together with a sample-and-hold circuit to determine the digital equivalent of an analog input voltage. Because the analog input signal can vary continuously over time, ADC circuits generally use a **sample-and-hold** circuit to maintain a constant analog input voltage during the conversion process. A sample-and-hold circuit is an analog device with a digital **hold** input signal. When the hold input is inactive, the sample-and-hold output tracks the input voltage. When the hold input is asserted, the sample-and-hold circuit freezes its output voltage at the input voltage that was present at the moment the hold signal became active.

With the sample-and-hold output held constant, the ADC uses its DAC to determine the digital equivalent of the input voltage. To make this determination, the ADC uses a **comparator**, which is a circuit that compares two analog voltages and produces a digital output signal indicating which is the higher voltage. The ADC feeds the sample-and-hold output voltage into one input of the comparator and the DAC output into the other input, as shown in the following diagram, in which the DAC input word size is n bits:

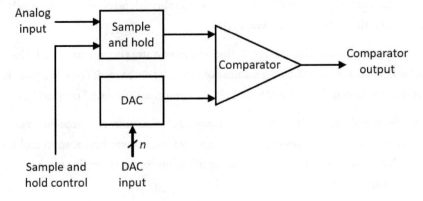

Figure 6.3: ADC architecture

The job of the ADC is to determine the DAC input word that causes the comparator to change state. A simple way to do this is to count upward from zero, writing each numeric value to the DAC inputs and observing the comparator output to see if it changed state. The DAC output that first causes the comparator to change state is the smallest DAC output voltage that is greater than the sample-and-hold output voltage. The actual sampled analog voltage is between this DAC output and the DAC output when driven by a data word one count smaller. This ADC configuration is called a **counter type ADC**.

While simple in concept, the counter type ADC can be quite slow, especially if the word size is large. A faster method is to sequentially compare each bit in the DAC data word, beginning with the most significant bit (d_3). Starting with a data word of 1000b in our 4-bit example, the first comparator reading indicates if the analog input voltage is above or below the DAC voltage midpoint. This determines if bit d_3 of the ADC reading is 0 or 1. Using the now-known value of d_3, d_2 is set to 1 to indicate which quarter of the full-scale range the input voltage lies within. This procedure is repeated to sequentially determine each of the remaining bits, ending with the least significant bit.

This ADC conversion technique is referred to as **successive approximation**. A successive approximation ADC is much faster than a counter type ADC. In our example, the maximum possible number of comparisons drops from 16 to 4. In a 12-bit successive approximation ADC, the maximum possible number of comparisons drops from 4,096 to 12. In general, using successive approximation rather than counting reduces the maximum number of steps for an n-bit ADC from 2^n to n. Successive approximation ADCs are available with resolutions from 8 to 18 bits, with maximum conversion rates up to several MHz.

ADCs and DACs are characterized by resolution and maximum conversion speed. The resolution of an ADC or DAC is determined by the number of bits in its data word. The maximum conversion speed determines how quickly the ADC or DAC can produce sequential outputs.

To process real-time data, an ADC produces a sequence of measurements at periodic time intervals for use as input to further processing. Requirements for data word resolution and sample rate vary widely depending on the particular DSP application. Some examples of standard digitized analog data formats are:

- Compact disk digital audio is sampled at 44.1 kHz with 16 bits per sample in two channels, corresponding to the left and right speakers.

- Video cameras measure the analog light intensity received at each pixel in a two-dimensional array and convert the reading to a digital word, usually 8 bits wide. Separate closely spaced sensors with color filters produce red, green, and blue measurements for each pixel in the image. The complete dataset for a single pixel consists of 24 bits, composed of three 8-bit color values. A single image can contain tens of millions of pixels, and video cameras typically output 30 to 60 frames per second. Because digital video produces such an enormous quantity of data, compression algorithms are generally used to reduce storage and transmission requirements.
- A mobile phone contains an RF transceiver that down-converts the received RF signal to a frequency range suitable for input to an ADC. Typical parameters for a mobile phone ADC are 12 bits of resolution and a sample rate of 50 MHz.
- An automotive radar system samples RF energy reflected from nearby obstacles with a resolution of 16 bits at a rate of 5 MHz.

In the next section, we will examine signal processing requirements for data sequences captured by ADCs.

DSP hardware features

DSPs are optimized to execute signal processing algorithms on digitized samples of analog information. The **dot product** is a fundamental operation used in many algorithms performed by DSPs. If A and B are two equal-length vectors (a **vector** is a one-dimensional array of numeric values), the dot product of A and B is formed by multiplying each element of A by the corresponding element of B, and summing the resulting products. Mathematically, if the length of each vector is n (indexed from 0 to n-1), the dot product of the vectors is:

$$A \cdot B = \sum_{i=0}^{n-1} A_i B_i = A_0 B_0 + A_1 B_1 + A_2 B_2 + \cdots + A_{n-1} B_{n-1}$$

The repetitive nature of the dot product calculation provides a natural path for performance optimization in digital systems. The basic operation performed in the dot product computation is called **multiply-accumulate (MAC)**.

A single MAC operation consists of multiplying two numbers together and adding the result to an accumulator, which was initialized to zero at the beginning of the dot product calculation. The mathematical performance of DSP chips is commonly measured in terms of MACs per second. Many DSP architectures are capable of performing one MAC per instruction clock cycle.

To perform a MAC operation on every clock cycle, a DSP cannot dedicate separate clock cycles to read a MAC instruction from program memory, read each of the vector elements to be multiplied from data memory, compute the product, and add it to the accumulator. All of these things must happen in one step.

The **von Neumann architecture**, introduced in *Chapter 1, Introducing Computer Architecture*, uses a single memory region for program instructions and data. This configuration results in a limitation known as the **von Neumann bottleneck**, resulting from the need to pass both program instructions and data values across a single processor-to-memory interface.

This effect can be mitigated with an architecture that separates program instructions and data storage into two separate memory regions, each with its own processor interface. This configuration, called the **Harvard architecture**, allows program instruction and data memory access to occur in parallel, enabling instructions to execute in a smaller number of clock cycles. The Harvard architecture will be discussed in more detail in *Chapter 7, Processor and Memory Architectures*.

A DSP with a Harvard architecture must perform two data memory accesses to retrieve the elements of the *A* and *B* vectors to be multiplied in a MAC operation. This normally requires two clock cycles, failing to meet the performance goal of one MAC per clock cycle. A **modified Harvard architecture** supports the use of program memory to store data values in addition to instructions. In many DSP applications, the values of one of the vectors (let's say the *A* vector) involved in the dot product are constant values known at the time the application is compiled. In a modified Harvard architecture, the elements of the *A* vector can be stored in program memory and the elements of the *B* vector, representing input data read from an ADC, are stored in data memory.

To perform each MAC operation in this architecture, one element of the *A* vector is read from program memory, one element of the *B* vector is read from data memory, and the accumulated product is stored in an internal processor register. If the DSP contains cache memory for program instructions, the MAC instruction performing each step of the dot product will be retrieved from the cache once the first MAC operation reads it from program memory, avoiding further memory access cycles to bring in the instruction. This configuration (a modified Harvard architecture with program instruction caching) enables single-cycle MAC operations for all iterations of the dot product once the first MAC operation is complete. Since the vectors involved in real-world dot product computations commonly contain hundreds or even thousands of elements, the overall performance of the dot product operation can closely approach the ideal of one MAC operation per DSP clock cycle.

A DSP can be categorized as having a fixed-point or a floating-point architecture. Fixed-point DSPs use signed or unsigned integers to perform mathematical operations such as MAC. Fixed-point DSPs are generally less costly than floating-point DSPs. However, fixed-point mathematics has the potential for numeric issues such as overflow, which can manifest by exceeding the range of the dot product accumulator.

To reduce the possibility of overflow, DSPs often implement an extended range accumulator, sometimes 40 bits wide in a 32-bit architecture, to support dot products on lengthy vectors. Due to concerns regarding overflow and related numerical issues, programming fixed-point DSPs requires extra effort to ensure these effects don't result in unacceptable performance degradation.

Floating-point DSPs often use a 32-bit wide numeric format for internal calculations. Once an integer ADC reading has been received by the DSP, all further processing is performed using floating-point operations. By taking advantage of floating-point operations, the potential for issues such as overflow is drastically reduced, resulting in quicker software development cycles.

The use of floating-point processing also improves the fidelity of computation results, realized in terms of improved **signal-to-noise ratio (SNR)** in comparison to an equivalent fixed-point implementation. Fixed-point calculations quantize the result of each mathematical operation at the level of the integer's least significant bit. Floating-point operations generally maintain accurate results from each operation that are within a small fraction of the corresponding fixed-point least significant bit.

Signal processing algorithms

Building upon our understanding of DSP hardware and the operations it supports, we will next look at some examples of digital signal processing algorithms in real-world applications.

Convolution

Convolution is a formal mathematical operation on a par with addition and multiplication. Unlike addition and multiplication, which operate on pairs of numbers, convolution operates on pairs of signal vectors. In the DSP context, a signal is a series of digitized samples of a time-varying input measured at equally spaced time intervals. Convolution is the most fundamental operation in the field of DSP.

In many practical applications, one of the two signals involved in a convolution operation is a fixed vector of numbers stored in memory. The other signal is a sequence of samples originating from ADC measurements.

To implement the convolution operation, as each ADC measurement is received, the DSP computes an updated output, which is simply the dot product of the fixed data vector (let's say the length of this vector is n) and the most recent n input samples received from the ADC. To compute the convolution of these vectors, the DSP must perform n MAC operations each time it receives an ADC sample.

The fixed vector in this example, referred to as h, is called the **impulse response**. A digital impulse is defined as a theoretically infinite sequence of samples in which one sample is 1 and all the preceding and following samples are 0. Using this vector as the input to a convolution with the vector h produces an output identical to the sequence h, surrounded by preceding and following zeros. The single 1 value in the impulse sequence multiplies each element of h on successive iterations, while all other elements of h are multiplied by 0.

The particular values contained in the h vector determine the effects of the convolution operation on the input data sequence. Digital filtering is one common application of convolution.

Digital filtering

A **frequency selective filter** is a circuit or algorithm that receives an input signal and attempts to pass desired frequency ranges to the output without distortion while eliminating, or at least reducing to an acceptable level, frequency ranges outside the desired ranges.

We are all familiar with the bass and treble controls in audio entertainment systems. These are examples of frequency selective filters. The bass function implements a variable gain **lowpass filter**, meaning the audio signal is filtered to select the lower frequency portion of the signal, and this filtered signal is fed to an amplifier that varies its output power in response to the position of the bass control. The treble section is implemented similarly, using a **highpass filter** to select the higher frequencies in the audio signal. The outputs of these amplifiers are combined to produce the signal sent to the speakers.

Frequency selective filters can be implemented with analog technology or with digital signal processing techniques. Simple analog filters are cheap and only require a few circuit components. However, the performance of these simple filters leaves much to be desired.

Some key parameters of a frequency selective filter are **stopband suppression** and the width of the transition band. Stopband suppression indicates how good a job the filter does of eliminating undesired frequencies in its output. In general, a filter does not entirely eliminate undesired frequencies, but for practical purposes, these frequencies can be reduced to a level that is so small they are irrelevant.

The transition band of a filter describes the frequency span between the passband and the stopband. The **passband** is the range of frequencies to be passed through the filter, and the **stopband** is the range of frequencies to be blocked by the filter. It is not possible to have a perfectly sharp edge between the passband and stopband. Some separation between the passband and the stopband is required, and trying to make the transition from passband to stopband as narrow as possible requires a more complex filter than one with a wider transition band.

A digital frequency selective filter is implemented with a convolution operation using a carefully selected set of values for the h vector. With the proper selection of elements in h, it is possible to design highpass, lowpass, bandpass, and bandstop filters. As discussed in the preceding paragraphs, highpass and lowpass filters attempt to pass the high and low frequency ranges, respectively, while blocking other frequencies. A **bandpass filter** attempts to pass only the frequencies within a specified range and block all other frequencies outside that range. A **bandstop filter** attempts to pass all frequencies except those within a specified range.

The goals of a high-performance frequency selective filter are to impart minimal distortion of the signal in the passband, provide effective blocking of frequencies in the stopband, and have as narrow a transition band as possible.

An analog filter implementing high-performance requirements may require a complex circuit design involving costly precision components. A high-performance digital filter, on the other hand, is still just a convolution operation. A digital circuit implementing a high-performance lowpass filter with minimal passband distortion and a narrow transition band may require a lengthy h vector, possibly containing hundreds—or even thousands—of elements. The design decision to implement such a filter digitally depends on the availability of cost-effective DSP resources capable of performing MAC operations at the rate required by the filter design.

Fast Fourier transform (FFT)

The **Fourier transform**, named after the French mathematician Jean-Baptiste Joseph Fourier, decomposes a time-domain signal into a collection of sine and cosine waves of differing frequencies and amplitudes. The original signal can be reconstructed by summing these waves together through a process called the **inverse Fourier transform**.

DSPs operate on time-domain signals sampled at fixed intervals. Because of this sampling, the DSP implementation of the Fourier transform is called the **discrete Fourier transform** (DFT). In general, a DFT converts a sequence of n equally spaced time samples of a function into a sequence of n DFT samples, equally spaced in frequency.

Each DFT sample is a complex number, composed of a real number and an imaginary number. An **imaginary number**, when squared, produces a negative result.

We won't delve into the mathematics of imaginary numbers here. An alternative way to view the complex number representing a DFT frequency component (called a **frequency bin**) is to consider the real part of the complex number to be a multiplier for a cosine wave at the bin frequency and the imaginary part to be a multiplier for a sine wave at the same frequency. Summing these wave components produces the time-domain representation of that DFT frequency bin.

The simplest implementation of the DFT algorithm for a sequence of length n is a double-nested loop in which each loop iterates n times. If an increase in the length of the DFT is desired, the number of mathematical operations increases as the square of n. For example, to compute the DFT for a signal with a length of 1,000 samples, at least a million operations are required.

In 1965, James Cooley of IBM and John Tukey of Princeton University published a paper describing the computer implementation of a more efficient DFT algorithm, which came to be known as the **Fast Fourier Transform, (FFT)**. The algorithm they described was originally invented by the German mathematician Carl Friedrich Gauss around 1805.

The FFT algorithm breaks a DFT into smaller DFTs, where the lengths of the smaller DFTs can be multiplied together to form the number of samples in the original DFT. The efficiency improvement provided by the FFT algorithm is greatest when the DFT length is a power of 2, enabling recursive decomposition through each factor of 2 in the DFT length. A 1,024-point FFT requires only a few thousand operations compared to over a million for the double-nested loop DFT implementation.

It is important to understand that the FFT operates on the same input data sequence as the DFT and produces the same output as the DFT; the FFT just does it *much* faster for longer sequences.

The FFT is used in many practical applications in signal processing. Some examples are:

- **Spectral analysis**: The output from a DFT on a time-domain sample sequence is a set of complex numbers representing the amplitude of sine and cosine waves over a frequency range representing the signal. These amplitudes directly indicate which frequency components are present at significant levels in the signal and which frequencies contribute smaller or negligible content.

    Spectral analysis is used in applications such as audio signal processing, image processing, and radar signal processing. Laboratory instruments called **spectrum analyzers** are commonly used for testing and monitoring RF systems such as radio transmitters and receivers.

A spectrum analyzer displays a periodically updated image representing the frequency content of its input signal, derived from an FFT computed over samples of that signal.

- **Filter banks:** A filter bank is a series of individual frequency-selective filters, each processing a separate frequency band. The complete set of filters in the bank covers the entire frequency range of the input signal. A **graphic equalizer**, used in high-fidelity audio applications, is an example of a filter bank.

  An FFT-based filter bank is useful for decomposing multiple frequency-separated data channels transmitted as a single combined signal. At the receiver, the FFT separates the received signal into multiple bands, each of which contains an independent data channel. The signal contained in each of these bands is further processed to extract its data content.

  The use of FFT-based filter banks is common in radio receivers for wideband digital data communication services such as digital television and 5G mobile communications.

- **Data compression:** A signal can be compressed to a smaller size by performing an FFT and discarding frequency components considered unimportant. The remaining frequency components form a smaller dataset that can be further compressed using standardized coding techniques.

  This approach is referred to as **lossy compression** because some of the information in the input signal is lost. Lossy compression will generally produce a greater degree of signal compression compared to **lossless compression**. Lossless compression algorithms are used in situations where any data loss is unacceptable, such as when compressing computer data files.

- **Discrete cosine transform (DCT):** The DCT is similar in concept to the DFT except that rather than decomposing the input signal into a set of sine and cosine functions as in the DFT, the DCT decomposes the input signal into only cosine functions, each multiplied by a real number. Computation of the DCT can be accelerated using the same technique the FFT employs to accelerate the computation of the DFT.

  The DCT has the valuable property that, in many data compression applications, most of the signal information is represented in a smaller number of DCT coefficients in comparison to alternative algorithms such as the DFT. This allows a larger number of the less significant frequency components to be discarded, thereby increasing data compression effectiveness.

DCT-based data compression is employed in many application areas that computer users and consumers of audio and video entertainment interact with daily, including MP3 audio, **Joint Photographic Experts Group (JPEG)** images, and **Moving Picture Experts Group (MPEG)** video.

DSPs are often used in applications involving one- and two-dimensional data sources. Some examples of one-dimensional data are audio signals and the RF signal received by a mobile phone radio transceiver. One-dimensional signal data consists of one sample value at each instant of time for each of possibly several input channels.

A photographic image is an example of two-dimensional data. A two-dimensional image is described in terms of the width and height of the image in pixels, and the number of bits representing each pixel. Each pixel in the image is separated from the surrounding pixels by horizontal and vertical spatial offsets. Every pixel in the image is (theoretically) sampled at the same point in time.

Motion video represents three-dimensional information. One way to define a video segment is as a sequence of two-dimensional images presented sequentially at regular time intervals. While traditional DSPs are optimized to work with the two-dimensional data of a single image, they are not necessarily ideal for processing sequential images at a high update rate.

The next section introduces GPUs, which are processors dedicated to handling the computing requirements of video synthesis and display.

GPU processing

A GPU is a digital processor optimized to perform the mathematical operations associated with generating and rendering graphical images for display on a computer screen. The primary applications for GPUs are playing video recordings and creating synthetic images of three-dimensional scenes.

The performance of a GPU is measured in terms of screen resolution (the pixel width and height of the image) and the image update rate in frames per second. Video playback and scene generation are hard real-time processes in which any deviation from smooth, regularly time-spaced image updates is likely to be perceived by users as unacceptable graphical stuttering.

As with the video cameras described earlier in this chapter, GPUs generally represent each pixel using three 8-bit color values, which indicate the intensities of red, green, and blue. Any perceptible color can be produced by combining appropriate values for each of these three colors. Within each color channel, the value 0 indicates the color is absent, and 255 is maximum intensity. Black is represented by the color triple (red, green, blue) = (0, 0, 0), and white is (255, 255, 255). With 24 bits of color data, over 16 million unique colors can be displayed. The granularity between adjacent 24-bit color values is, in general, finer than the human eye can distinguish.

In modern personal computers, GPU functionality is available in a variety of configurations:

- A GPU card can be installed in a PCIe slot
- A system can provide a GPU as one or more discrete integrated circuits on the main processor board
- GPU functionality can be built into the central processor's integrated circuit

The most powerful consumer-class GPUs are implemented as PCIe expansion cards. These high-end GPUs contain dedicated graphics memory and feature a fast communication path with the main system processor (typically using a PCIe x16 slot) for receiving commands and data representing the scene to be displayed. Some GPU designs support the use of multiple identical cards in a single system to generate scenes for a single graphical display. This technology features a separate high-speed communication bus linking the GPUs to each other. The use of multiple GPUs in a system effectively increases the parallelization of graphics processing.

GPUs exploit the concept of **data parallelism** to perform identical computations simultaneously on a vector of data items, producing a corresponding vector of outputs. Modern GPUs support thousands of simultaneously executing threads, providing the capability to render complex, three-dimensional images containing millions of pixels at 60 or more frames per second.

The architecture of a typical GPU consists of one or more multi-core processors, each supporting multithreaded execution of data-parallel algorithms. The interface between the GPU processors and graphics memory is optimized to provide maximum average data throughput, rather than attempting to minimize access latency (which is the design goal for main system memory). GPUs can afford to sacrifice a degree of latency performance to achieve peak streaming rate between the GPU and its dedicated memory because maximizing throughput results in the highest possible frame update rate.

In computer systems with less extreme graphical performance demands, such as business applications, a lower-performance GPU integrated within the same circuit die as the main processor is a lower-cost and often perfectly acceptable configuration. Integrated GPUs can play streaming video and provide a more limited level of three-dimensional scene-rendering capabilities compared to the higher-end GPUs.

Rather than relying on dedicated graphics memory, integrated GPUs use a portion of system memory for graphics rendering. Although the use of system memory rather than specialized graphics memory results in a performance hit, such systems provide sufficient graphics performance for most home and office uses.

Smart devices such as portable phones and tablets contain GPUs as well, providing the same video playback and three-dimensional scene-rendering capabilities as larger personal computer systems. The constraints of small physical size and reduced power consumption necessarily limit the performance of portable device GPUs. Nevertheless, modern smartphones and tablets are fully capable of playing high-definition streaming video and rendering sophisticated gaming graphics.

GPUs as data processors

For many years, GPU architectures were designed very specifically to support the computational needs of real-time three-dimensional scene rendering. In recent years, users and GPU vendors have increasingly recognized that these devices are in fact small-scale supercomputers suitable for use across a much broader range of applications. Modern GPUs provide floating-point execution speed measured in trillions of floating-point operations per second (**teraflops**). As of 2021, a high-end GPU provides floating-point performance measured in the dozens of teraflops and can execute data-parallel mathematical algorithms hundreds of times faster than a standard desktop computer.

Taking advantage of the immense parallel computing power available in high-end GPUs, vendors of these devices provide programming interfaces and expanded hardware capabilities to enable the implementation of more generalized algorithms. Of course, GPUs, even with enhancements to support general computing needs, are only truly effective at speeding up algorithms that exploit data parallelization.

Some application areas that have proven to be suitable for GPU acceleration are as follows.

Big data

In fields as diverse as climate modeling, genetic mapping, business analytics, and seismic data analysis, problem domains share the need to analyze enormous quantities of data, often measured in **terabytes (TB)** or **petabytes (PB)** (1 PB is 1,024 TB) in as efficient a manner as possible. In many cases, these analysis algorithms iterate over large datasets searching for trends, correlations, and more sophisticated connections among what may seem at first to be disparate, unrelated masses of samples.

Until recently, analysis of these datasets at the appropriate level of granularity has often been dismissed as infeasible due to the extensive execution time required for such processing. Today, however, many big data applications produce results in a reasonable length of time by combining the use of GPU processing, often on machines containing multiple interconnected GPUs, and splitting the problem across multiple computer systems in a cloud environment. The use of multiple computers, each containing multiple GPUs, to execute highly parallel algorithms across an enormous dataset can be accomplished at a surprisingly low cost these days in comparison to the historical costs associated with supercomputing systems.

Deep learning

Deep learning is a category of **artificial intelligence** (**AI**) that uses multilayer networks of artificial neurons to model the fundamental operations of human brain cells. A biological **neuron** is a type of nerve cell that processes information. Neurons are interconnected via **synapses** and use electrochemical impulses to pass information among themselves. During learning, the human brain adjusts the connections among neurons to encode the information being learned for later retrieval. The human brain contains tens of billions of neurons.

Artificial neural networks (**ANNs**) employ a software model of neuron behavior to mimic the learning and retrieval processes of the human brain. Each artificial neuron receives input from potentially many other neurons and computes a single numeric output. Some neurons are driven directly by the input data to be processed and others produce outputs that are retrieved as the result of the ANN computation. Each communication path between neurons has a weighting factor associated with it, which is simply a number that multiplies the strength of the signal traveling along that path. The numeric input to a neuron is the sum of the input signals it receives, each multiplied by the weight of the associated path.

The neuron computes its output using a formula called the **activation function**. The activation function determines the degree to which each neuron is "triggered" by its inputs.

The next diagram represents an example of a single neuron that sums the inputs from three other neurons ($N_1 - N_3$), each multiplied by a weighting factor ($w_1 - w_3$) The sum passes to the activation function, $F(x)$, which produces the neuron's output signal.

The use of three inputs in this example is arbitrary; in actual applications, each neuron can receive input from any number of other neurons.

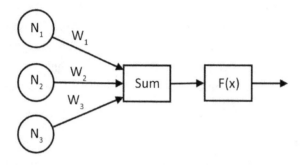

Figure 6.4: A neuron receiving inputs from three neurons

ANNs are organized in layers, where the first layer, called the **input layer**, is followed by one or more internal layers (called **hidden layers**), which are followed by an **output layer**. Some ANN configurations are arranged in a data flow sequence from input to output, called a **feedforward network**, while other configurations provide feedback from some neurons to neurons in preceding layers. This configuration is called a **recurrent network**.

The next figure shows an example of a simple feedforward network with three input neurons, a hidden layer consisting of four neurons, and two output neurons. This network is fully connected, meaning each neuron in the input and hidden layers connects to all neurons in the following layer. The connection weights are not shown in this diagram.

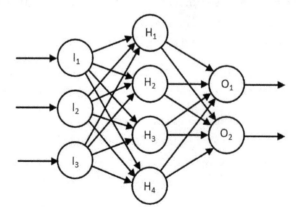

Figure 6.5: A three-layer feedforward network

Training an ANN consists of adjusting the weighted connections between neurons so that, when presented with a particular set of input data, the desired output is produced.

Using an appropriate learning algorithm, an ANN can be trained with a dataset composed of known correct outputs for a wide variety of inputs.

Training a large, sophisticated ANN to perform a complex task such as driving a car or playing chess requires a tremendous number of training iterations drawn from a very large dataset. During training, each iteration makes small adjustments to the weighting factors within the network, slowly driving the network to a state of convergence. Once fully converged, the network is considered trained and can be used to produce outputs when presented with novel input. In other words, the network generalizes the information it learned during training and applies that knowledge to new situations.

The feature that makes ANNs particularly suitable for GPU processing is their parallel nature. The human brain is effectively a massively parallel computer with billions of independent processing units. This form of parallelism is exploited during the ANN training phase to accelerate network convergence by performing the computations associated with multiple artificial neurons in parallel.

The next section will present some examples of computer system types in terms of the architectural concepts presented in this chapter.

Examples of specialized architectures

This section examines some application-focused computing system configurations and highlights the specialized architectural requirements addressed in each. The configurations we will look at are:

- **Cloud compute server**: Several vendors offer access to computing platforms accessible to customers via the internet. These servers allow users to load software applications onto the cloud server and perform any type of computation they desire. In general, these services bill their customers based on the type and quantity of computing resources allocated and the length of time they are in use. The advantage for the customer is that these services cost nothing when they are not in use.

  At the higher end of performance, servers containing multiple interconnected GPU cards can be harnessed to perform large-scale, floating-point intensive computations on huge datasets. In the cloud context, it is straightforward and often cost-effective to break a computation into smaller parts suitable for parallel execution across multiple GPU-enabled servers.

This makes it feasible for organizations—and even individuals with limited funding—to harness computing capabilities that, until just a few years ago, were the exclusive province of government, big business, and research universities possessing the wherewithal to implement supercomputing facilities.

- **Business desktop computer**: Business information technology managers strive to provide employees with the computing capability they need to do their jobs at the lowest cost. Most office workers do not require exceptional graphics or computing performance, though their computer systems must support modest video presentation requirements for such purposes as displaying employee training videos.

  For business users, the GPU integrated into modern processors is usually more than sufficient. For a reasonable price, business buyers can purchase computer systems with processors in the midrange of performance that include integrated graphics. These systems provide full support for modern operating systems and standard office applications such as word processing, email, and spreadsheets. Should the need arise to expand a system's capabilities with higher-performance graphics, the installation of a GPU in an expansion slot is a straightforward upgrade.

- **High-performance gaming computer**: Computer gaming enthusiasts running the latest 3D games demand an extreme level of GPU performance to display detailed, high-resolution scenes at the highest achievable frame rate. These users are willing to invest in a powerful, power-hungry, and costly GPU (or even multiple GPUs) to achieve the best possible graphics performance.

  Almost as important as the graphics performance, a high-performance gaming computer must have a fast system processor. The processor and GPU work together over the high-speed interface connecting them (typically PCIe x16) to compute the position and viewing direction of the scene observer as well as the type, location, visual characteristics, and orientation of all objects in the scene. The system processor passes this geometric information to the GPU, which performs the mathematical operations necessary to render a lifelike image for display. This process repeats at a rate sufficient to deliver a smooth presentation of complex, rapidly changing scenes.

- **High-end smartphone**: Today's smartphones combine high-performance computational and graphical display capabilities with strict limits on power consumption and heat generation. Users insist on fast, smooth, vibrant graphics for gaming and video display, but will not tolerate short battery life or a device that becomes hot to the touch.

Modern phone displays contain millions of full-color pixels, up to 12 GB of RAM, and support up to 1 TB of flash storage. These devices generally come with two high-resolution cameras (one on the front and one on the back), capable of capturing still images and recording video. High-end phones contain a 64-bit multi-core processor with an integrated GPU, as well as a variety of features providing an optimal combination of energy efficiency and high performance.

Smartphone architectures contain DSPs to perform tasks such as encoding and decoding voice audio during telephone calls and processing the received and transmitted RF signals flowing through the phone's various radio transceivers. A typical phone supports digital cellular service, Wi-Fi, Bluetooth, and **near-field communication (NFC)**. Modern smartphones are powerful, well-connected computing platforms optimized for operation under battery power.

This section discussed computer system architectures representing just a tiny slice of current and future applications of computing technology. Whether a computer system sits on an office desk, resides in a smartphone, or is flying a passenger aircraft, a common set of general architectural principles applies during the process of system design and implementation.

Summary

This chapter examined several specialized domains of computing, including real-time systems, digital signal processing, and GPU processing. After completing this chapter, you should have greater familiarity with the features of modern computers related to real-time operation, the processing of analog signals, and graphics processing in application areas including gaming, voice communication, video display, and the supercomputer-like applications of GPUs. These capabilities are important extensions to the core computing tasks performed by the central processor, whether in a cloud server, a desktop computer, or a smartphone.

The next chapter will take a deeper look at modern processor architectures, specifically the von Neumann, Harvard, and modified Harvard variants. The chapter will also cover the use of paged virtual memory and the features and functions of a memory management unit.

Exercises

1. **Rate monotonic scheduling (RMS)** is an algorithm for assigning thread priorities in preemptive, hard, real-time applications in which threads execute periodically.

RMS assigns the highest priority to the thread with the shortest execution period, the next-highest priority to the thread with the next-shortest execution period, and so on. An RMS system is schedulable, meaning all tasks are guaranteed to meet their deadlines (assuming no inter-thread interactions or other activities such as interrupts cause processing delays) if the following condition is met:

$$\sum_{k=1}^{n} \frac{C_i}{T_i} \leq n \left(2^{1/n} - 1\right)$$

This formula represents the maximum fraction of available processing time that can be consumed by n threads. In this formula, C_i is the maximum execution time required for thread i, and T_i is the execution period of thread i.

Is the following system composed of three threads schedulable?

Thread	Execution Time (C_i), ms	Execution Period (T_i), ms
Thread 1	50	100
Thread 2	100	500
Thread 3	120	1000

2. A commonly used form of the one-dimensional discrete cosine transform is:

$$X_k = \sum_{n=0}^{N-1} x_n \cos\left[\frac{\pi}{N}\left(n + \frac{1}{2}\right)k\right]$$

In this formula, k, the index of the DCT coefficient, runs from 0 to N-1.

Write a program to compute the DCT of the following sequence:

$$x = \{0.5, 0.2, 0.7, -0.6, 0.4, -0.2, 1.0, -0.3\}$$

The cosine terms in the formula depend only on the indexes n and k, and do not depend on the input data sequence x. This means the cosine terms can be computed once and stored as constants for later use. Using this as a preparatory step, the computation of each DCT coefficient reduces to a sequence of MAC operations.

This formula represents the unoptimized form of the DCT computation, requiring N^2 iterations of the MAC operation to compute all N DCT coefficients.

3. The hyperbolic tangent is often used as an activation function in ANNs. The hyperbolic tangent function is defined as follows:

$$\tanh(x) = \frac{e^x - e^{-x}}{e^x + e^{-x}}$$

Given a neuron with inputs from three preceding neurons as depicted in *Figure 6.4*, compute the neuron's output with the hyperbolic tangent as the activation function F(x) using the following preceding neuron outputs and path weights:

Neuron	Neuron Output	Weight
N_1	0.6	0.4
N_2	-0.3	0.8
N_3	0.5	-0.2

Join our community Discord space

Join the book's Discord workspace for a monthly *Ask me Anything* session with the author:
`https://discord.gg/7h8aNRhRuY`

7

Processor and Memory Architectures

This chapter takes a deeper look at modern processor architectures, specifically the von Neumann, Harvard, and modified Harvard variants, as well as the computing domains in which each architecture tends to be applied. The concepts and benefits of paged virtual memory, employed extensively in consumer and business computing and in portable smart devices, are also introduced. We will examine the practical details of memory management in the real-world context of Windows NT and later Windows versions. The chapter concludes with a discussion of the features and functions of the memory management unit.

After completing this chapter, you will have learned the key features of modern processor architectures and the use of physical and virtual memory. You will also understand the benefits of memory paging and the functions of the memory management unit.

This chapter covers the following topics:

- The von Neumann, Harvard, and modified Harvard architectures
- Physical and virtual memory
- Paged virtual memory
- Memory management unit

Technical requirements

The files for this chapter, including the answers to the exercises, are available at https://github.com/PacktPublishing/Modern-Computer-Architecture-and-Organization-Second-Edition.

The von Neumann, Harvard, and modified Harvard architectures

In earlier chapters, we touched briefly on the history and modern applications of the von Neumann, Harvard, and modified Harvard processor architectures. In this section, we'll examine each of these configurations in greater detail and look at the computing applications in which each of these architectures tends to be applied.

The von Neumann architecture

The von Neumann architecture was introduced by John von Neumann in 1945. This processor configuration consists of a control unit, an arithmetic logic unit, a register set, and a memory region containing program instructions and data. The key feature distinguishing the von Neumann architecture from the Harvard architecture is the use of a single area of memory for program instructions and the data acted upon by those instructions. It is conceptually straightforward for programmers, and relatively easier for circuit designers, to locate all the code and data a program requires in a single memory region.

This diagram shows the elements of the von Neumann architecture:

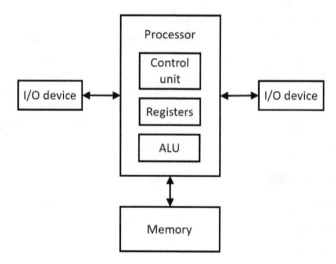

Figure 7.1: The von Neumann architecture

Although the single-memory architectural approach simplified the design and construction of early generations of processors and computers, the use of shared program and data memory has presented some challenges related to system performance and, in recent years, security.

Some of the more significant issues are as follows:

- **The von Neumann bottleneck**: Using a single interface between the processor and the main memory for instruction and data access often requires multiple memory cycles to retrieve a processor instruction and access the data it requires. In the case of an immediate value stored next to its instruction opcode, there might be little or no bottleneck penalty because, at least in some cases, the immediate value gets loaded along with the opcode in a single memory access. Most programs, however, spend much of their time working with data stored in memory allocated separately from the program instructions. In this situation, multiple memory access operations are required to retrieve the opcode and any required data items.

  The use of cache memories for program instructions and data, discussed in detail in *Chapter 8, Performance-Enhancing Techniques*, can significantly mitigate this limitation. However, when working with code sequences and data objects that exceed the size of cache memory, the benefit of caching is reduced, possibly by a substantial amount. There is no avoiding the fact that placing code and data in the same memory region with a shared communication path to the processor will, at times, act as a limitation on system performance.

- **von Neumann security considerations**: The use of a single memory area for code and data opens possibilities for creative programmers to store sequences of instructions in memory as "data," and then direct the processor to execute those instructions. Programs that write code into memory and then execute it are implementing **self-modifying code**. Besides being difficult to troubleshoot (because many software debugging tools expect the program in memory to contain the instructions that were originally compiled into it), this capability has been exploited for years by hackers with more sinister motives.

  Buffer overflow is a distressingly common flaw in widely used software tools such as operating systems, web servers, and databases. Buffer overflow occurs when a program requests input and stores that input in a fixed-length data buffer. If the code is not careful to check the length of the input provided by the user, it is possible for the user to enter an input sequence longer than the available storage space. When this happens, the additional data overwrites memory intended for other purposes.

  If the buffer being overwritten is stored on the program's stack, it is possible for a creative user to provide a lengthy input sequence that overwrites the return address of the currently executing function, which happens to be stored on the same stack.

By carefully crafting the contents of the input data sequence, the attacker can seize control of the executing application and direct it to execute any desired sequence of instructions. To do this, the hacker must prepare an input sequence that overflows the input buffer, overwrites the function's return address with a different, carefully chosen, address, and writes a sequence of instructions into memory that begins execution at this address. The sequence of instructions inserted by the attacker begins execution when the function that originally requested user input returns, transferring control to the hacker's code. This allows the hacker to "own" the computer.

Various attempts to resolve the buffer overflow problem have occupied an enormous amount of computer security researchers' time over the years since the first widespread occurrence of this type of attack in 1988. Processor vendors and operating system developers have implemented a variety of features to combat buffer overflow attacks, with names such as **data execution prevention** (**DEP**) and **address space layout randomization** (**ASLR**). While these fixes have been effective to some degree, the fundamental processor feature that enables this type of exploitation is the use of the same memory region for program instructions and data in the von Neumann architecture.

The Harvard architecture

The Harvard architecture was originally implemented in the Harvard Mark I computer in 1944. A strict Harvard architecture uses one address space and memory bus for program instructions and a separate address space and memory bus for data. This configuration has the immediate benefit of enabling simultaneous access to instructions and data, thereby implementing a form of parallelism. Of course, this enhancement comes at the expense of essentially duplicating the number of address lines, data lines, and control signals that must be implemented by the processor to access both memory regions.

The following diagram shows the layout of a processor implementing the Harvard architecture:

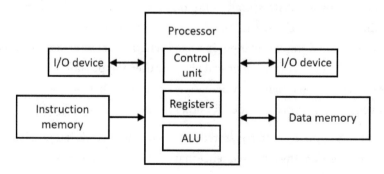

Figure 7.2: Harvard architecture

The Harvard architecture potentially provides a higher performance level by parallelizing accesses to instructions and data. This architecture also removes the entire class of security issues associated with maliciously executing data as program instructions, provided the instruction memory cannot be modified by program instructions. This assumes the program memory is loaded with instructions in a trustworthy manner.

In hindsight, with knowledge of the proliferation of von Neumann architecture-enabled security threats, there is reason to wonder if the entire information technology industry would not have been vastly better off had there been early agreement to embrace the Harvard architecture and its complete separation of code and data memory regions, despite the costs involved.

In practice, a strict Harvard architecture is rarely used in modern computers. Several variants of the Harvard architecture are commonly employed, collectively called **modified Harvard architectures**. These architectures are the topic of the next section.

The modified Harvard architecture

Computers designed with a **modified Harvard architecture** have, in general, some degree of separation between program instructions and data. This reduces the effects of the von Neumann bottleneck and mitigates the security issues we've discussed. The separation between instructions and data is rarely absolute, however. While systems with modified Harvard architectures contain separate program instruction and data memory regions, these processors typically support some means of storing data in program memory and, in some cases, storing instructions in data memory.

The following diagram shows a modified Harvard architecture representing many real-world computer systems:

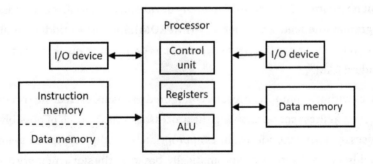

Figure 7.3: Modified Harvard architecture

As we saw in the previous chapter, **digital signal processors** (**DSPs**) achieve substantial benefits from the use of a Harvard-like architecture.

By storing one numeric vector in instruction memory and a second vector in data memory, a DSP can execute one **multiply-accumulate (MAC)** operation per processor clock cycle. In these systems, instruction memory and the data elements it contains are typically read-only memory regions. This is indicated by the unidirectional arrow connecting the instruction memory to the processor in *Figure 7.3*. Consequently, only constant data values are suitable for storage in the instruction memory region.

Besides DSPs, most modern general-purpose processors contain separate instruction and data caches, thereby implementing significant aspects of the Harvard architecture. Processor architectures such as x86 and ARM support parallel and independent access to instructions and data when the requested items happen to reside in the first level of cache memory. When on-chip cache lookups are unsuccessful, the processor must access the main memory over the von Neumann-style shared bus, which takes significantly longer.

As a practical matter, the implementation details of a particular processor in terms of von Neumann versus Harvard architectural features seldom matter to software developers, other than in terms of performance considerations. Programmers generally develop programs in their high-level language of choice and the compiler or interpreter handles the details related to allocating data and instructions to the appropriate memory regions.

The next section discusses the benefits of memory virtualization.

Physical and virtual memory

Memory devices in computers can be categorized as **random-access memory (RAM)**, which can be read from and written to at will, and **read-only memory (ROM)**, which, as the name indicates, can be read but not written. Some types of memory devices, such as flash memory and **electrically erasable programmable read-only memory (EEPROM)**, inhabit a middle ground, where the data content of the devices can be changed, just not as easily, or as quickly, or updated as many times, as standard RAM.

Memory devices within a computer must be configured to ensure each device occupies a unique span of the system address space, enabling the processor to access each of possibly several RAM and ROM devices by setting the address lines appropriately. Modern computer systems generally perform this address space allocation automatically, based on the slot a memory device occupies.

Software running on early computer systems, and on the less-sophisticated computers and embedded processors of today (such as 6502-based systems), uses addresses within RAM and ROM devices in program instructions to perform reads and writes.

For example, a 6502 instruction such as `JMP $1000` instructs the processor to load its instruction pointer with the hexadecimal value $1000 and execute the instruction at that memory location. In processing this instruction, the 6502 control unit places the value $1000 on the 6502's 16 address lines and reads the byte from that memory address. This byte is interpreted as the opcode of the next instruction to be executed. Similarly, loading a byte from memory with an instruction such as `LDA $0200` places the value $0200 on the address lines and copies the byte at that address into the A register.

In systems that use physical addressing, the memory addresses in instructions are the actual addresses of the referenced instruction or data item. This means the memory address contained in an instruction is the same address used to electrically access the appropriate location in a memory device.

This architectural approach is conceptually straightforward to implement in processor designs, but in a scenario involving multiple simultaneously executing programs (referred to as **multiprogramming**), the burden of software development can quickly become excessive. If each one of multiple programs is developed in isolation from the others (in a scenario involving multiple independent developers, for example), there must be some way to allocate the available RAM and ROM address spaces to individual programs to ensure multiple programs can be in the running state simultaneously (perhaps in the context of an RTOS) without interfering with each other's use of memory.

One well-known early effort to support the execution of multiple programs in a single address space on PCs is the MS-DOS **terminate and stay resident** (**TSR**) program concept. TSR programs allocate memory and load their code into it, and then return control to the operating system. Users can continue to work with the system normally, loading and using other applications (one at a time, of course), but they can also access the TSR as needed, typically by typing a special key combination. It is possible to load multiple TSR programs into memory simultaneously, each accessible via its own key combination. After activating a TSR program, the user interacts with it as needed, then executes a TSR command to return to the currently running main application.

While limited in many ways (including consuming a portion of the maximum of 640 KB of RAM available in early PCs), TSR programs effectively enabled the execution of multiple programs in a single RAM address space.

Developing TSR applications was a challenging task. The more advanced TSR programs available in the 1980s and 1990s took advantage of undocumented MS-DOS features to provide the maximum utility to their users.

As a result of this complexity, TSR programs developed a reputation for causing system instability. A different approach for supporting multiprogramming was clearly needed.

The use of **virtual memory** overcomes the biggest challenges that prohibited the widespread use of multiprogramming in the original PC design. Virtual memory is a method of memory management that enables each application to operate in its own memory space, seemingly independent of any other applications that may be in the running state simultaneously on the same system. In a computer with virtual memory management, the operating system is responsible for the allocation of physical memory to system processes and to user applications. The memory management hardware and software translate memory requests originating in the application's virtual memory context to physical memory addresses.

Apart from easing the process of developing and running concurrent applications, virtual memory also enables the allocation of a larger amount of memory than exists in the computer. This is possible using secondary storage (typically a disk file) to temporarily hold copies of sections of memory removed from physical memory to enable a different program (or a different part of the same program) to run in the now-free memory.

In modern general-purpose computers, memory sections are usually allocated and moved in multiples of a fixed-size chunk, called a **page**. Memory pages are typically 4 KB or larger. Moving memory pages to and from secondary storage in virtual memory systems is called **page swapping**. The file containing the swapped-out pages is the **swap file**.

In a virtual memory system, neither application developers nor the code itself need to be concerned with how many other applications are running on the system or how full the physical memory may be getting. As the application allocates memory for data arrays and places calls to library routines (which requires the code for those routines to be loaded into memory), the operating system manages the allocation of physical memory and takes the steps necessary to ensure each application receives memory upon request. Only in the unusual case of completely filling the available physical memory while also filling the swap file to its limit is the system forced to return a failure code in response to a memory allocation request.

Virtual memory provides several notable benefits in addition to making things easier for programmers:

- Not only are applications able to ignore each other's presence, they are prevented from interfering with each other, accidentally or intentionally. The virtual memory management hardware is responsible for ensuring each application can only access memory pages that have been assigned to it.

- Attempts to access another process's memory, or any other address outside its assigned memory space, result in an **access violation** exception.
- Each memory page has a collection of attributes that restrict the types of operations supported within it. A page may be marked read-only, causing any attempts to write data to the page to fail. A page may be marked executable, meaning it contains code that can be executed as processor instructions. A page may be marked read-write, indicating the application is free to modify the page at will. By setting these attributes appropriately, operating systems can improve system stability by ensuring processor instructions can't be modified and that the execution of data as instructions cannot occur, whether such an attempt is the result of an accident or malicious intent.
- Memory pages can be marked with a minimum required privilege level, allowing pages to be restricted for access only by code running with kernel privilege. This restriction ensures the operating system continues operating properly even in the presence of misbehaving applications. This allows system memory to be mapped into each process's address space while prohibiting application code from interacting directly with that memory. Applications can only access system memory indirectly, via a programming interface consisting of system calls.
- Memory pages can be marked as shareable among applications, meaning a page can be explicitly authorized as accessible from more than one process. This enables efficient interprocess communication.

Early versions of Microsoft Windows implemented some features of memory virtualization using 80286 and 80386 processor memory segmentation capabilities. In the Windows context, the use of virtual memory came into its own with the introduction of Windows NT 3.1 in 1993. The Windows NT system architecture was based on the Digital Equipment Corporation **Virtual Address Extension (VAX)** architecture, developed in the 1970s. The VAX architecture implemented a 32-bit virtual memory environment with a 4 GB virtual address space available to each of potentially many applications running in a multiprogramming context. One of the key architects of the VAX operating system, **Virtual Memory System (VMS)**, was David Cutler, who later led the development of Microsoft Windows NT.

Windows NT has a flat 32-bit memory organization, meaning any address in the entire 32-bit space is directly accessible using a 32-bit address. No additional programmer effort is required to manipulate segment registers. By default, the Windows NT virtual address space is divided into two equal-sized chunks: a 2 GB user address space in the lower half of the range, and a 2 GB kernel space in the upper half.

The next section delves into the implementation of paged virtual memory in 32-bit Windows NT on Intel processors. While Windows NT is not entirely representative of virtual memory implementations in other operating systems, similar principles apply even though other environments differ in the details. This introduction provides background on the concepts of virtual memory while deferring additional details related to more modern architectures, such as 64-bit processors and operating systems, until later chapters.

Paged virtual memory

In 32-bit Windows NT on Intel processors, memory pages are 4 KB in size. This implies that addressing a location within a particular page requires 12 address bits (because 2^{12}=4,096). The remaining 20 bits of a 32-bit virtual address are used in the virtual-to-physical translation process.

In Windows NT, all memory addresses in a program (both those referenced in the source code and in compiled executable code) are virtual addresses. They are not associated with physical addresses until the program runs under the control of the memory management unit.

A contiguous 4 KB section of Windows NT physical memory is called a **page frame**. The page frame is the smallest unit of memory managed by the Windows virtual memory system. Each page frame starts on a 4 KB boundary, meaning the lower 12 address bits are all zero at the base of any page frame. The system tracks information related to page frames in page tables.

A Windows NT **page table** is sized to occupy a single 4 KB page. Each 4-byte entry in a page table enables the translation of a 32-bit address from the virtual address space used by program instructions to a physical address required to access a location in RAM or ROM. A 4 KB page table contains 1,024 page address translations. A single page table manages access to 4 MB of address space: each page table contains 1,024 page frames multiplied by 4 KB per page. A process may have several associated page tables, all of which are managed by a page table directory.

A **page table directory** is a 4 KB page containing a series of 4-byte references to page tables. A page table directory can contain 1,024 page table references. A single page table directory covers the entire 4 GB address space (4 MB per page table multiplied by 1,024 page table references) of 32-bit Windows NT.

Each Windows NT process has a page table directory, set of page tables, and collection of page frames allocated for its use. The process page tables apply to all threads within the process because all the process's threads share the same address space and memory allocations.

When the system scheduler switches from one process to another, the virtual memory context of the incoming process replaces the context of the outgoing process.

Intel x86 processors maintain the address of the current process page table directory in the CR3 register, also known as the **Page Directory Base Register (PDBR)**. This is the entry point to the page table directory and the page tables, enabling the processor to translate any valid virtual address to the corresponding physical address.

In accessing an arbitrary (valid) location in memory, and assuming information that would expedite the access is not already stored in the cache of recent virtual-to-physical address translations, the processor first looks up the physical address of the relevant page table in the page table directory using the upper 10 bits of the virtual address. It then accesses the page table and uses the next most significant 10 address bits to select the physical page containing the requested data. The lower 12 bits of the address then specify the memory location in the page frame requested by the executing instruction.

Page frames do not represent actual divisions in physical memory.

Physical memory is not actually divided into page frames. The page structure is merely a method the system uses to keep track of the information required to translate virtual addresses to physical memory locations.

To meet users' performance expectations, each memory access must be as fast as possible. At least one virtual-to-physical translation takes place during the execution of every instruction to fetch instruction opcodes and data. Due to the high-frequency repetition of this process, processor designers expend great effort ensuring virtual address translation takes place as efficiently as possible.

In modern processors, a translation cache retains the results of recent virtual memory translation lookups. This approach enables a very high percentage of virtual memory translations to occur internally in the processor, without any of the clock cycles that would be required if the processor needed to look up a page table address in the page table directory and then access the page table to determine the requested physical address.

The data structures used in virtual-to-physical address translations are not accessible to applications running at user privilege level. All the activity related to address translation takes place in processor hardware and in kernel mode software processes.

To help clarify the operation of virtual address translation, the following diagram presents an example of how Windows translates a 32-bit virtual address to a physical address:

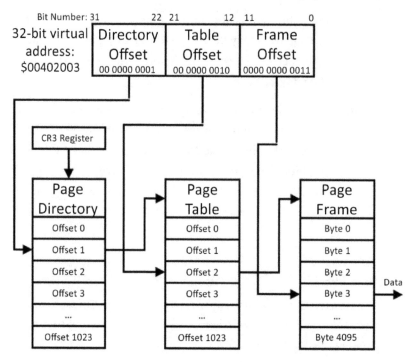

Figure 7.4: Virtual to physical address translation

We'll go through the translation process in *Figure 7.4* step by step. Assume the processor is requesting the 8-bit data value stored at virtual address $00402003 with an instruction such as mov al, [ebx], where ebx has previously been loaded with the value $00402003. We will assume the translation for this address is not already stored in the processor's cache of recent virtual-to-physical address translations, and we'll also assume the page is resident in the main memory. The following procedure describes the translation process:

1. The processor attempts to execute the mov al, [ebx] instruction, but it cannot complete it because it does not have immediate access to the information needed to perform the translation of the virtual address in ebx to the corresponding physical address. This generates a **page fault**, which transfers control to the operating system so it can resolve the address translation. The use of the term *fault* here does not imply that an error of some kind occurred. Page faults are a routine part of application execution.

2. The requested virtual address is shifted right by 22 bit positions, leaving the 10-bit directory offset, which has the value 1 in this example.

3. The directory offset is shifted left by 2 bit positions (because each entry in the page directory is 4 bytes) and is added to the content of processor register CR3 (the PDBR). The result is the address of the page table directory entry containing the address of the relevant page table.

4. The requested virtual address is shifted right by 12 bit positions and masked to leave only the 10-bit table offset, which has the value 2 in this example.

5. The table offset is shifted left by 2 bit positions (because each entry in this table is also 4 bytes) and added to the page table address identified in *Step 3*. The 32-bit address read from this location is the physical address of the page frame containing the requested data.

6. The processor stores the translation, which is a conversion from the upper 20 bits of a virtual address to the corresponding upper 20 bits of a page frame address, in its translation cache.

7. The processor restarts the mov al, [ebx] instruction, which will now succeed in moving the requested data byte into the al register using the cached virtual-to-physical translation. The lower 12 bits of the virtual address (the frame offset), which contain the value 3 in this example, are added to the page frame address computed in *Step 5* to access the requested byte.

Once these steps are complete, the translation for the requested page frame remains available in processor translation cache memory for some time. As long as the translation remains in the cache, subsequent requests for the same virtual address or for other locations in the same page frame will execute without delay until the cache entry for this page frame is overwritten by subsequent code execution.

The page fault procedure described in the preceding steps is called a **soft fault**. A soft fault sets up the virtual-to-physical translation for a page that is already accessible by the processor but is not in the translation cache.

A **hard fault** occurs when accessing a page that has been swapped to secondary storage. Processing a hard fault requires several additional steps, including allocating a page frame to receive the requested frame, requesting the page from secondary storage, and updating the page table with the physical address of the page. Because hard faults involve a disk transfer, this type of fault has a much greater impact on application performance than a soft fault.

The translation process converts the upper 20 bits of a virtual address to the corresponding 20 bits of the physical address. This leaves 12 bits in each page table entry available to contain status and configuration information for the page frame. The use of these bits is described in the following section.

Page status bits

The following table describes each of the 12 status bits in a 32-bit Windows NT page table entry:

Bit	Name	Description
0	Valid	1 indicates that this page table entry is usable for translation. If this bit is 0, the remaining bits may have different meanings, as defined by the operating system. The following bit descriptions assume that the valid bit is 1.
1	Write	1 indicates that the page is writeable. 0 means that the page is read-only.
2	Owner	1 indicates that the page is user mode. 0 indicates that the page is kernel mode.
3	Write through	1 indicates that changes to the page are to be flushed to disk immediately. 0 indicates that page changes will be maintained in RAM.
4	Cache disabled	1 indicates that caching is disabled for the page. 0 means that caching is enabled.
5	Accessed	1 indicates that the page has been read or written. 0 indicates the page has not been accessed in any way.
6	Dirty	1 indicates that the page has been written. 0 indicates that no writes have occurred.
7	Reserved	Unused.
8	Global	1 indicates that this translation applies to all processes. 0 means that this entry applies to only one process.
9	Reserved	Unused.
10	Reserved	Unused.
11	Reserved	Unused.

Table 7.1: Page status bits

The processor uses the page status bits to maintain information about the page content and to control access to each page by system and user processes. The `Owner` bit identifies a page as owned by the kernel or by a user. User processes cannot read or write any pages owned by the kernel. Any attempt to write to a page marked read-only (where the page's `Write` bit is 0) results in an access violation exception.

The system uses the page status bits to manage memory as efficiently as possible. If the `Accessed` bit is clear, the page has been allocated but never used. When the system needs to free physical memory, pages that have never been accessed are prime candidates for removal because there is no need to save their contents when removing them from memory. Similarly, if a page's `Dirty` bit is clear, the page's data has not been modified since it was brought into memory. The memory manager can release pages that do not have the `Dirty` bit set with the knowledge that when the page is needed again, it can be reloaded from its source location (typically a disk file) to restore it accurately.

Pages with the `Dirty` bit set must be stored in the swap file when they are removed from memory. When a page is moved to the swap file, the page table entry is updated using a different format from *Table 7.1* to indicate it is not valid for translation (the `Valid` bit is clear) and to store its location within the swap file.

The format of page table entries is defined by the processor architecture, which, in this case, is the Intel x86 family.

The processor hardware accesses the page table entries to perform virtual-to-physical address translation and to enforce page protections as the processor runs at full speed.

In addition to managing the memory used by each process, the system must keep track of all the RAM and ROM page frames in the computer, whether in use by a process or not. The system maintains this information in lists called **memory pools**, described next.

Memory pools

Windows NT categorizes memory pools into two types, non-paged and paged:

- **Non-paged pool:** The non-paged pool contains all page frames that are guaranteed to always remain resident in memory. Code for interrupt service routines, device drivers, and the memory manager itself must always remain immediately accessible by the processor for system performance reasons and, in the case of the memory manager itself, for the system to function at all. Non-paged virtual addresses reside in the system portion of a process's virtual address space.

- **Paged pool:** The paged pool contains virtual memory pages that can be temporarily swapped out of physical memory whenever needed.

The system tracks the status of every page frame in physical memory in a structure called the **page frame number (PFN)** database. The PFN is the upper 20 bits of the physical base address of a page frame. Each page frame can be in one of several states depending on its current and previous usage. These are some of the key page frame states:

- **Active:** The page frame is part of a system or user process working set. A **working set** is the portion of a process's virtual address space currently present in physical memory.
- **Standby:** Standby pages are pages that have been removed from process working sets and have not been modified.
- **Modified:** Modified pages have been removed from process working sets and have been modified. These pages must be written to disk before their page frames are reused.
- **Free:** Free pages are unused but still contain data from their last membership in a working set. For security reasons, these pages cannot be made available to a user process until their data content has been overwritten with zeros.
- **Zeroed:** Zeroed pages are free and have been overwritten with zeros. These pages are available for allocation by user processes.
- **Bad:** Bad pages have generated hardware errors during processor accesses. Bad pages are tracked in the PFN database and are not used by the operating system.

As system services and applications start up, run, and shut down, page frames transition between states under system control. In Windows NT, a system task runs during idle periods and converts free pages to zeroed pages by overwriting those pages with zeros.

This discussion has focused on the implementation of virtual memory in the x86 processor architecture under Windows NT. Other processor architectures and operating systems implement virtual memory using similar concepts.

The processor component in control of the memory allocation, address translation, and protection functions is called a **memory management unit**. The next section examines the memory management unit as a generalized computer system component.

Memory management unit

Processor architectures supporting paged virtual memory either implement the **memory management unit (MMU)** functionality within the processor itself or, sometimes, particularly in the case of older designs, as a separate integrated circuit. Within the MMU, the processor's virtual address space is divided into page-sized allocation units.

Pages may be of a fixed size, as in the Windows NT example, or an MMU may support multiple sizes. Modern processors, including later generation x86 processors, often support two page sizes, one small and one large. Small pages are typically a few KB while a large page may be a few MB. Large page support avoids the inefficiencies associated with allocating numerous smaller pages when working with large data objects.

As discussed earlier, the MMU generally contains a cache to improve the speed of memory access by avoiding the need to traverse the page table directory and perform a page table lookup for each memory access. Although the use of caching for performance enhancement is a topic of *Chapter 8, Performance-Enhancing Techniques*, we introduce the virtual-to-physical address translation cache here because this capability is a core feature of most MMU designs.

The caching component within the MMU that stores previously used virtual-to-physical translations is called a **translation lookaside buffer** (**TLB**). To avoid looking up a page table in the page table directory and then looking up the page frame in the referenced page table on every memory access, the TLB stores translations between virtual addresses and page frames resulting from those lookups in a hardware structure called **associative memory**.

Each time the processor needs to access physical memory, which may occur multiple times during the execution of a single instruction, it first checks the TLB's associative memory to determine whether the translation information is resident in the TLB. If it is, the instruction immediately uses the information stored in the TLB to access physical memory. If the TLB does not contain the requested translation, a page fault occurs and the processor must traverse the page table directory and a page table to determine the translated address, assuming the referenced page is resident in memory.

The following diagram represents the operation of the TLB:

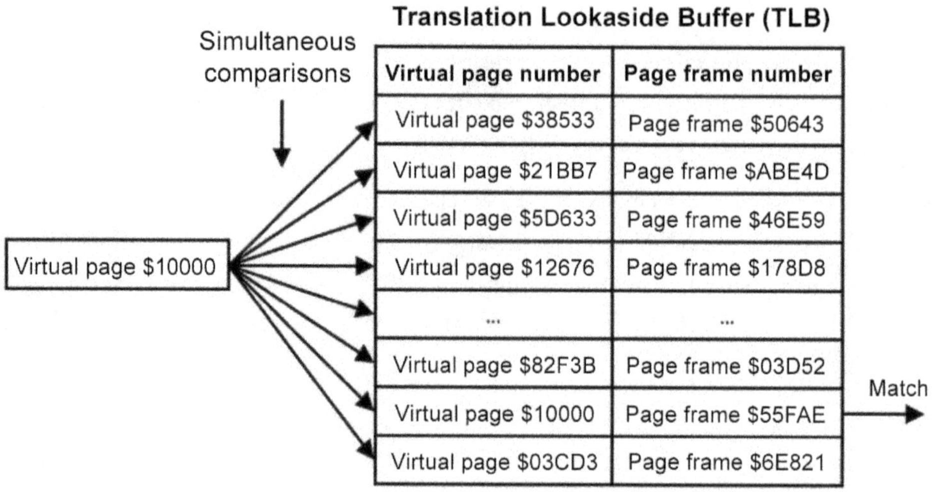

Translation Lookaside Buffer (TLB)

Virtual page number	Page frame number
Virtual page $38533	Page frame $50643
Virtual page $21BB7	Page frame $ABE4D
Virtual page $5D633	Page frame $46E59
Virtual page $12676	Page frame $178D8
...	...
Virtual page $82F3B	Page frame $03D52
Virtual page $10000	Page frame $55FAE
Virtual page $03CD3	Page frame $6E821

Simultaneous comparisons from Virtual page $10000; Match at Virtual page $10000 → Page frame $55FAE.

Figure 7.5: Translation lookaside buffer operation

On each memory access, the processor extracts the upper 20 bits of the virtual address to identify the virtual page number. This page number, $10000 in this example, is used to search the TLB for a matching entry. The TLB hardware simultaneously compares the requested virtual page number with all the virtual page numbers resident in the TLB. If a match is found, the corresponding page frame number is immediately provided for use in accessing physical memory.

The TLB contains a limited number of entries, typically 64 or fewer. The processor must manage which TLB entries are retained and which are discarded as address requests for disparate memory locations are processed. When the TLB is full, the MMU decides which of the existing TLB entries to overwrite with new information. The MMU may choose a TLB entry at random for replacement, or a more sophisticated implementation may use age information to replace the TLB entry that has gone the longest without being used.

In addition to performing virtual-to-physical address translation, MMUs generally perform the following functions:

- **Separation of virtual memory into kernel space and user space**: Kernel memory is reserved for use by the operating system and related components such as device drivers. User space is available for applications and for other actions initiated by users, such as processing commands typed into a command prompt window. User-level code cannot access system memory directly. Instead, user code must call system functions to request services such as the allocation of memory and I/O operations.

- **Isolation of process memory:** Each process has its own address space, which is the only memory it is allowed to access. Unless a system-authorized memory sharing region is set up between processes, each process is prohibited from accessing memory in use by another process. One process cannot erroneously or intentionally modify memory that is private to another process.

- **Page-level access restrictions:** In addition to protecting system pages from user access and protecting process-private pages from access by other processes, a process can set protections on individual pages it owns. Pages can be marked read-only, which prohibits the modification of the contents. Pages marked no-execute cannot be used to provide instructions for processor execution. In some architectures, pages can be marked no-access, which prohibits both reading and writing. Pages containing executable code may be marked read-only to prevent the accidental or intentional modification of instructions in memory.

- **Detection of software problems:** In some programming languages, particularly the C language, it is unfortunately common to attempt to use a pointer (a **pointer** is a variable that contains the address of another variable) containing an invalid address. The most common invalid address encountered in this situation is 0, because variables are often initialized to zero. This problem is so common that the system's response to it has its own name: the **null pointer exception**. When a C program attempts to access a memory location that is not in the program's valid virtual address range, such as the address $00000000, the MMU triggers an exception, which, unless handled by the program, typically results in a program crash with an error message printed to the console window. In systems without virtual memory, accesses to erroneous locations may simply read or write the memory at the referenced address without any indication of an error, leading to the incorrect operation of the application or the entire system. Such bugs in systems without an MMU can be extremely difficult to fix if the problem does not become apparent immediately.

Modern processors running Linux, Windows, and most smart-device operating systems generally require their host systems to use virtual memory management and provide the page protection mechanisms described in this chapter.

Real-time embedded processors performing safety-critical tasks such as operating aircraft flight controls or managing automotive airbag operation may or may not support the full feature set of an MMU. One drawback related to the use of virtual memory in hard real-time systems is the variable time delay resulting from the need to process soft faults and, if page swapping is implemented, hard faults. Because execution timing must be strictly controlled in many real-time systems, their designers often avoid the use of virtual memory. Such systems do not contain an MMU, but they often implement many of the other features an MMU provides, such as hardware protection of system memory and access control for RAM regions.

Summary

This chapter examined the principal modern processor architectural categories, including the von Neumann, Harvard, and modified Harvard variants, and their use in different computing domains. The concepts of paged virtual memory were examined, including some details pertaining to the implementation of paged virtual memory in Windows NT on the x86 processor.

The general structure of MMUs was discussed, with emphasis on the use of the TLB as a virtual-to-physical translation performance optimization technique.

The next chapter will expand beyond the performance enhancement provided by the TLB to look in depth at widely used processor acceleration methods including caching, instruction pipelining, and instruction parallelism.

Exercises

1. A 16-bit embedded processor has separate memory regions for code and data. Code is stored in flash memory and modifiable data is stored in RAM. Some data values, such as constants and initial values for RAM data items, are stored in the same flash memory region as the program instructions. RAM and ROM reside in the same address space. Which of the processor architectures discussed in this chapter best describes this processor?

2. The processor described in *Exercise 1* has memory security features that prevent code under execution from modifying program instruction memory. The processor uses physical addresses to access instructions and data. Does this processor contain an MMU?

3. The order of accessing sequential elements in a large data structure can have a measurable impact on processing speed due to factors such as the reuse of TLB entries. Accessing distant array elements in sequence (that is, elements that are not in the same page frame as previously accessed elements) requires frequent soft faults as new TLB entries are loaded and old TLB entries are discarded.

    Write a program that creates a two-dimensional array of numbers with a large size such as 10,000 rows by 10,000 columns. Iterate through the array in column-major order, assigning each element the sum of the row and column indices. Column-major means the column index (the second index) increments fastest. In other words, the column index increments in the inner loop.

Measure precisely how long this procedure takes. Note, you may need to take steps to ensure your programming language does not optimize away the entire calculation if the results from the array are not used later. It may suffice to print one of the array values after the timing is complete, or you may need to do something like sum all the array elements and print that result.

Repeat the process, including the timing, exactly as explained before, except change the inner loop to iterate over the row index (the first index) and the outer loop to iterate over the column index, making the access sequence row-major.

Since general-purpose computers perform many other tasks while running your code, you may need to perform both procedures multiple times to get a statistically valid result. You might start by running the experiment 10 times and averaging the times for column-major and row-major array access.

Are you able to determine a consistently superior array access method? Which order is fastest on your system using the language you selected? Note that the difference between the column-major and row-major access order may not be dramatic – it might be just a few percent.

Join our community Discord space

Join the book's Discord workspace for a monthly *Ask me Anything* session with the author:
https://discord.gg/7h8aNRhRuY

8
Performance-Enhancing Techniques

The fundamental aspects of processor and memory architectures discussed in previous chapters support the design of a complete and functional computer system. However, the performance of such a system would be poor compared to most modern processors without the addition of features to increase the speed of instruction execution.

Several performance-enhancing techniques are employed routinely in processor and system designs to achieve peak execution speed in real-world computer systems. These techniques do not alter what the processor does in terms of program execution and data processing; they just get it done faster.

After completing this chapter, you will understand the benefits of multilevel cache memory in computer architectures and the advantages and challenges associated with instruction pipelining. You'll also understand the performance improvement resulting from simultaneous multithreading and the purpose and applications of single-instruction, multiple-data processing.

The following topics will be covered in this chapter:

- Cache memory
- Instruction pipelining
- Simultaneous multithreading
- SIMD processing

Technical requirements

Files for this chapter, including answers to the exercises, are available at https://github.com/PacktPublishing/Modern-Computer-Architecture-and-Organization-Second-Edition.

Cache memory

Cache memory is a high-speed memory region (compared to the speed of main memory) that temporarily stores program instructions or data for future use. Usually, these instructions or data items have been retrieved from main memory recently and are likely to be needed again shortly.

The primary purpose of cache memory is to increase the speed of repeatedly accessing the same memory location and nearby memory locations. To be effective, accessing the cached items must be significantly faster than accessing the original source of the instructions or data, referred to as the **backing store**.

When caching is in use, each attempt to access a memory location begins with a search of the cache. If the requested item is present, the processor retrieves and uses it immediately. This is called a **cache hit**. If the cache search is unsuccessful (a **cache miss**), the instruction or data item must be retrieved from the backing store. In the process of retrieving the requested item, a copy is added to the cache for anticipated future use.

Cache memory is used for a variety of purposes in computer systems. Some examples of cache memory applications are:

- **Translation lookaside buffer (TLB)**: The TLB, as we saw in *Chapter 7, Processor and Memory Architectures*, is a form of cache memory used in processors supporting paged virtual memory. The TLB contains a collection of virtual-to-physical address translations that speed up access to page frames in physical memory. As instructions execute, each main memory access requires a virtual-to-physical translation. Successful searches of the TLB result in much faster instruction execution compared to the page table lookup process following a TLB miss. The TLB is part of the MMU and is not directly related to the varieties of processor cache memory discussed later in this section.

- **Disk drive caches**: Reading and writing the magnetized platters of rotating disk drives is orders of magnitude slower than accessing **dynamic RAM (DRAM)** devices. Disk drives generally implement cache memory to store the output of read operations and to temporarily hold data in preparation for writing. Drive controllers often store more data than the quantity originally requested in internal cache memory with the expectation that future reads will request data adjacent to the initial request.

If this turns out to be a correct assumption, which it often is, the drive can satisfy the second request immediately from cache without the delay associated with accessing the disk platters.

- **Web browser caches**: Web browsers generally store copies of recently accessed web pages in memory in anticipation of the user clicking the Back button to return to a previously viewed page. When this happens, the browser can retrieve some or all of the page content from its local cache and immediately redisplay the page without the need to access the remote web server and retrieve the same information again.
- **Processor instruction and data caches**: The following sections examine processor cache structures. The purpose of these caches is to improve the speed of access to instructions and data in comparison to the latency incurred when accessing DRAM modules.

Cache memory improves computer performance because many algorithms executed by operating systems and applications exhibit locality of reference. **Locality of reference** refers to the reuse of data that has been accessed recently (this is referred to as **temporal locality**) and to accessing data in physical proximity to data that has been accessed previously (this is called **spatial locality**).

Using the structure of the TLB as an example, temporal locality is exploited by storing a virtual-to-physical translation in the TLB for some time following initial access to a particular page frame. Any additional references to the same page frame in subsequent instructions will enjoy speedy access to the translation until it is eventually replaced in the cache by a different translation.

The TLB exploits spatial locality by referencing an entire page frame with a single TLB entry. Any subsequent accesses to different addresses on the same page will benefit from the presence of the TLB entry resulting from the first reference to the page.

As a rule, cache memory regions are small in comparison to the backing store. Cache memory devices are designed for maximum speed, which generally means they are more complex and costly per bit than the data storage technology used in the backing store. Due to their limited size, cache memory devices tend to fill quickly. When a cache does not have an available location to store a new entry, an older entry must be discarded. The cache controller uses a **cache replacement policy** to select which cache entry will be overwritten by the new entry.

The goal of processor cache memory is to maximize the percentage of cache hits over time, thus providing the highest sustained rate of instruction execution. To achieve this objective, the caching logic must determine which instructions and data will be placed into the cache and retained for future use.

A processor's caching logic does not have assurance that a cached data item will ever be used again once it has been inserted into the cache.

The logic of caching relies on the likelihood that, due to temporal and spatial locality, there is a very good chance the cached data will be accessed in the near future. In practical implementations on modern processors, cache hits typically occur on 95 to 97 percent of memory accesses. Because the latency of cache memory is a small fraction of the latency of DRAM, a high cache hit rate leads to a substantial performance improvement in comparison to a cache-free design.

The following sections discuss the multilevel caching technologies of modern processors and some of the cache replacement policies used in their implementations.

Multilevel processor caches

In the years since the introduction of personal computers, processors have undergone dramatic increases in the speed of instruction processing. The internal clocks of modern Intel and AMD processors are close to 1,000 times faster than the 8088 processor used in the first IBM PC. The speed of DRAM technology, in comparison, has increased at a much slower rate over time. Given these trends, if a modern processor were to access DRAM directly for all of its instructions and data, it would spend the vast majority of its time simply waiting for the DRAM to respond to each request.

To attach some approximate numbers to this topic, consider a modern processor capable of accessing a 32-bit data value from a processor register in 1 ns. Accessing the same value from DRAM might take 100 ns. Oversimplifying things somewhat, if each instruction requires a single access to a memory location, and the execution time for each instruction is dominated by the memory access time, we can expect a processing loop that accesses the data it requires from processor registers to run 100 times faster than the same algorithm accessing main memory on each instruction.

Now, assume cache memory is added to the system with an access time of 4 ns. By taking advantage of cache memory, the algorithm that accesses DRAM on each instruction will suffer the 100 ns performance penalty the first time a particular address is referenced, but subsequent accesses to the same and nearby addresses will occur at the cache speed of 4 ns. Although accessing the cache is four times slower than accessing registers, it is 25 times faster than accessing DRAM. This example represents the degree of execution speedup achievable through the effective use of cache memory in modern processors.

High-performance processors generally employ multiple cache levels with the goal of achieving peak instruction execution rate. Processor cache hardware is constrained in terms of size and performance by the economics of semiconductor technology. Selecting an optimally performing mix of processor cache types and sizes while achieving a price point acceptable to end users is a key goal of processor designers.

The two types of RAM circuits in common use as main memory and for processor internal storage are DRAM and static RAM. DRAM is inexpensive but has a comparatively slow access time, due largely to the time required to charge and discharge the bit cell capacitors during read and write operations.

Static RAM is much faster than DRAM but is much more costly, resulting in its use in smaller quantities in applications where performance is critical. DRAM designs are optimized for density, resulting in the largest possible number of bits stored on a single DRAM integrated circuit. Static RAM designs are optimized for speed, minimizing the time to read or write a location. Processor cache memory is generally implemented using static RAM.

Static RAM

Static RAM (SRAM) boasts a substantially faster access time than DRAM, albeit at the expense of significantly more complex circuitry. SRAM bit cells take up much more space on the integrated circuit die than the cells of a DRAM device capable of storing an equivalent quantity of data. As you will recall from *Chapter 4, Computer System Components*, a single DRAM bit cell consists of just one MOSFET and one capacitor.

The standard circuit for a single bit of SRAM contains six MOSFET transistors. Four of these transistors are used to form two NOT gates. These gates are based on the CMOS circuit shown in *Figure 4.3* in *Chapter 4, Computer System Components*, in the *Introducing the MOSFET* section. These gates are labeled G_1 and G_2 in the following diagram:

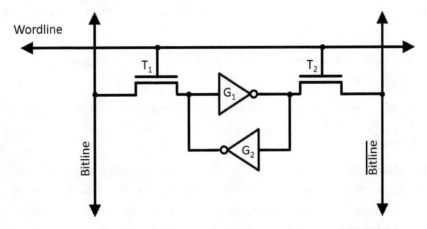

Figure 8.1: SRAM circuit diagram

The output of each of the NOT gates is connected to the input of the other, forming a flip-flop. Most of the time, the wordline is low, which turns off transistor switches T_1 and T_2, isolating the pair of gates. While the wordline is low (and power continues to be applied), the gates will persist on one of two states:

- **The stored bit is 0**: The input of G_1 is low and its output is high
- **The stored bit is 1**: The input of G_1 is high and its output is low

The access transistors (T_1 and T_2) function as switches that connect the bitlines to the cell for reading and writing. As with DRAM, driving the wordline high enables access to the bit cell by reducing the resistance across each access transistor to a very small value. To read the cell contents, the readout circuitry measures the voltage between the bitline pair labeled Bitline and $\overline{\text{Bitline}}$. (where the overbar represents the NOT operation). The two bitline signals always have opposing senses, forming a differential pair. Measuring the sign of the voltage difference between the two signals determines whether the cell contains a 0 or a 1.

When writing to the bit cell, the wordline is driven high and the bitline signals are driven to opposite voltage levels representing the desired value (0 or 1) to be written. The transistors writing the data to the bitlines must have substantially greater drive capability than the bit cell NOT gate transistors. This allows the desired value to be written to the cell, even if the flip-flop state must be overpowered to switch it to the state being written.

An SRAM bank is arranged in a rectangular grid of rows and columns in the same manner as DRAM. The wordline enables access to all SRAM bit cells along a single row. The bitlines connect to all columns of the grid of bit cells.

Unlike DRAM, SRAM does not require periodic refreshes to retain its data content. This is the reason it is referred to as *static* RAM.

In the next section, we will see how the first level of processor cache is constructed using SRAM.

Level 1 cache

In a multilevel cache architecture, the cache levels are numbered, beginning with 1. A level 1 cache (also referred to as an **L1 cache**) is the first cache the processor searches when requesting an instruction or data item from memory. Because it is the first stop in the cache search, an L1 cache is generally constructed using the fastest SRAM technology available and is physically located as close to the processor's logic circuitry as possible.

The emphasis on speed makes the L1 cache costly and power-hungry, which means it must be quite small in comparison to main memory, particularly in cost-sensitive applications. Even when it is modestly sized, a fast level 1 cache can provide a substantial performance boost in comparison to an otherwise equivalent processor that does not employ caching.

The processor (or the MMU, if present) transfers data between DRAM and cache in fixed-size data blocks called **cache lines**. Computers using DDR DRAM modules generally use a cache line size of 64 bytes. The same cache line size is commonly used at all cache levels.

Modern processors often divide the L1 cache into two sections of equal size, one for instructions and one for data. This configuration is referred to as a **split cache**. In a split cache, the level 1 instruction cache is referred to as the **L1 I-cache** and the level 1 data cache is the **L1 D-cache**. The processor uses separate buses to access each of the caches, thereby implementing a significant aspect of the Harvard architecture. This arrangement speeds instruction execution by enabling access to instructions and data in parallel, assuming L1 cache hits for both.

Modern processors employ a variety of strategies to organize cache memory and control its operation. The simplest cache configuration is direct mapping, introduced in the next section.

Direct-mapped cache

A **direct-mapped cache** is a block of memory organized as a one-dimensional array of cache sets, where each address in the main memory maps to a single set in the cache. Each **cache set** consists of the following items:

- A cache line, containing a block of data read from main memory
- A tag value, indicating the location in main memory corresponding to the cached data
- A valid bit, indicating whether the cache set contains data

This example represents an instruction cache, which is read-only. We'll examine read-write data caches in a later section.

There are times when the cache contains no data, such as immediately following processor power-on. The valid bit for each cache set is initially clear to indicate the absence of data in the set. When the valid bit is clear, use of the set for lookups is inhibited. Once data has been loaded into a cache set, the hardware sets the valid bit.

We will use a small L1 I-cache size of 512 bytes as an example. Because this is a read-only instruction cache, it need not support the ability to write to memory. The cache line size is 64 bytes. Dividing 512 bytes by 64 bytes per set results in 8 cache sets.

The 64 bytes in each set equals 2^6 bytes, which means the least significant 6 bits in the address select a location within the cache line. Three additional bits of an address are required to select one of the eight sets in the cache.

From this information, the following diagram shows the division of a 32-bit physical memory address into tag, set number, and cache line byte offset components:

Figure 8.2: Components of a 32-bit physical memory address

Each time the processor reads an instruction from DRAM (which is necessary any time the instruction is not already present in the cache), the MMU reads the 64-byte block containing the addressed location and stores it in the L1 I-cache set selected by the three **Set** bits of *Figure 8.2*. The upper 23 bits of the address are stored in the **Tag** field of the cache set, and the **Valid** bit is set.

As the processor fetches each subsequent instruction, the control unit uses the three Set bits in the instruction's address to select a cache set for comparison. The hardware compares the upper 23 bits of the executing instruction's address with the Tag value stored in the selected cache set. If the values match, a cache hit has occurred, and the processor reads the instruction from the cache line. If a cache miss occurs, the MMU reads the line from DRAM into the appropriate cache set (overwriting any existing data for that set) and provides the instruction to the control unit for execution.

The following diagram represents the organization of the entire 512-byte cache and its relation to the three fields in a 32-bit instruction address:

Chapter 8

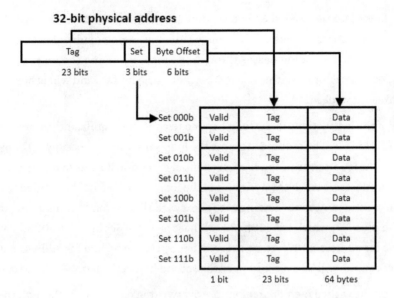

Figure 8.3: Relation of a 32-bit physical address to cache memory

To demonstrate why a direct-mapped cache can produce a high hit rate, assume we're running a program containing a loop starting at physical memory address 8000h (we're ignoring the upper 16 bits of the 32-bit address here for simplicity) and containing 256 bytes of code. The loop executes instructions sequentially from the beginning to the end of the 256 bytes and then branches back to the top of the loop for the next iteration.

Address 8000h contains 000b in its Set field, so this address maps to the first cache set, as shown in *Figure 8.3*. On the first pass through the loop, the MMU retrieves the 64-byte Set 000b cache line from DRAM and stores it in the first set of the cache. As the remaining instructions stored in the same 64-byte block execute, each will be retrieved directly from the cache.

As execution flows into the second 64 bytes, another read from DRAM is required. By the time the end of the loop is reached, Sets 000b through 011b have been populated with the loop's 256 bytes of code. For the remaining passes through the loop, assuming the thread runs without interruption, the processor will achieve a 100 percent cache hit rate and will achieve maximum instruction execution speed.

Alternatively, if the instructions in the loop happen to consume significantly more memory, the advantage of caching will be reduced. Assume the loop's instructions occupy 1,024 bytes, twice the cache size. The loop performs the same sequential execution flow from top to bottom. When the instruction addresses reach the midpoint of the loop, the cache has been filled with the first 512 bytes of instructions. At the beginning of the next cache line beyond the midpoint, the address will be 8000h plus 512, which is 8200h. 8200h has the same Set bits as 8000h, which causes the cache line for address 8000h to be overwritten by the cache line for address 8200h. Each subsequent cache line will be overwritten as the second half of the loop's code executes.

Even though all the cached memory regions are overwritten on each pass through the loop, the caching structure continues to provide a significant benefit because, once read from DRAM, each 64-byte line remains in the cache and is available for use as its instructions are executed. The downside in this example is the increased frequency of cache misses. This represents a substantial penalty because, as we've seen, accessing an instruction from DRAM may be 25 (or more) times slower than accessing the same instruction from the L1 I-cache.

> **VIRTUAL AND PHYSICAL ADDRESS TAGS IN CACHING**
>
> The example in this section assumes that cache memory uses physical memory addresses to tag cache entries. This implies the addresses used in cache searches are the output of the virtual-to-physical address translation process in systems using paged virtual memory. It is up to the processor designer to select whether to use virtual or physical addresses for caching purposes.
>
>
>
> In modern processor caches, it is not uncommon to use virtual address tags in one or more cache levels and physical address tags in the remaining levels. One advantage of using virtual address tagging is speed, since no virtual-to-physical translation is required during cache accesses. As we've seen, virtual-to-physical translation requires a TLB lookup and potentially a page table search in the event of a TLB miss. However, virtual address tagging introduces other issues such as **aliasing**, which occurs when the same virtual address refers to different physical addresses.
>
> As with many other aspects of processor performance optimization, this is a trade-off to be considered during cache system design.

This example was simplified by assuming instruction execution flows linearly from the top to the bottom of the loop without any detours. In real-world code, there are frequent calls to functions in different areas of application memory space and to system-provided libraries.

In addition to those factors, other system activities such as interrupt processing and thread context switching frequently overwrite information in the cache, leading to a higher cache miss frequency. This affects application performance because the main-line code must perform additional DRAM accesses to reload the cache following each diversion that caused a cache entry to be overwritten.

One way to reduce the effects of deviations from straight-line code execution is to set up multiple caches operating in parallel. This configuration is called a **set associative cache**, discussed next.

Set associative cache

In a two-way set associative cache, memory is divided into two equal-sized caches. Each of these has half the number of entries of a direct-mapped cache of the same total size. The hardware consults both caches in parallel on each memory access and a hit may occur in either one. The following diagram illustrates the simultaneous comparison of a 32-bit address tag against the tags contained in two L1 I-caches:

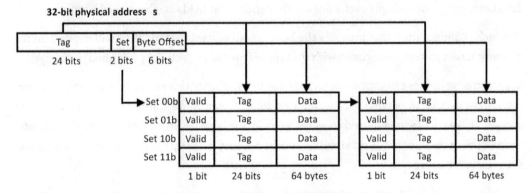

Figure 8.4: Set associative cache operation

The cache configuration shown here contains the same cache line size (64 bytes) as the cache in *Figure 8.3*, but only half as many sets per cache. The overall size of cache memory is the same as the previous example: 64 bytes per line times 4 rows times 2 caches equals 512 bytes. Because there are now four sets, the **Set** field in the physical address reduces to 2 bits, and the **Tag** field increases to 24 bits. Each set consists of two cache lines, one in each of the two caches.

When a cache miss occurs, the memory management logic must select which of the two cache tables to use as the destination for the data being loaded from DRAM. A common method is to track which of the relevant sets in the two tables has gone the longest without being accessed and overwrite that entry. This replacement policy, called **least-recently used** (LRU), requires hardware support to keep track of which cache line has been idle the longest. The LRU policy relies on the temporal locality heuristic that says data that has not been accessed for an extended time is less likely to be accessed again soon.

Another method for choosing between the two tables is to simply alternate between them on successive cache insertions.

The hardware to implement this replacement policy is simpler than the LRU policy, but there may be a performance impact due to the arbitrary selection of the line being invalidated. Cache replacement policy selection represents another area of trade-off between increased hardware complexity and incremental performance improvements.

In a two-way set associative cache, two cache lines from different physical locations with matching Set fields are present in the cache simultaneously, assuming both cache lines are valid. At the same time, in comparison to a direct-mapped cache of the same total size, each of the two-way set associative caches is half the size. This represents yet another design trade-off: the number of unique Set values that can be stored in a two-way set associative cache is reduced compared to direct mapping, but multiple cache lines with identical Set fields can be cached simultaneously.

The cache configuration that provides the best overall system performance will depend on the memory access patterns associated with the types of processing activity performed on the system.

Set associative caches can contain more than the two caches of this example. Modern processors often have 4, 8, or 16 parallel caches, referred to as 4-way, 8-way, and 16-way set associative caches. These caches are referred to as **set associative** because an address Tag field associates with all the cache lines in the set simultaneously. A direct-mapped cache implements a one-way set associative cache.

The advantage of multi-way set associative caches over direct-mapped caches is that they tend to have higher hit rates, resulting in better system performance than direct-mapped caches in most practical applications. If multi-way set associative caches provide better performance than one-way direct mapping, why not increase the level of associativity even further? Taken to the limit, this progression ends with fully associative caching.

Fully associative cache

Assuming the number of lines in the cache is a power of two, repetitively dividing the overall cache memory into a larger number of smaller parallel caches, until each cache contains only one line, results in a **fully associative cache**.

In our example of a 512-byte cache with 64 bytes per cache line, this process would result in eight parallel caches, each with only one set.

In this architecture, every memory access leads to a parallel comparison with the Tag values stored in all the cache lines. A fully associative cache using an effective replacement policy such as LRU can provide a very high hit rate, though at a substantial cost in circuit complexity and a corresponding consumption of integrated circuit die area.

In power-sensitive applications such as battery-powered mobile devices, the additional circuit complexity of a fully associative cache leads to increased battery drain. In desktop computers and cloud servers, excessive processor power consumption must be minimized to avoid the need for extraordinary cooling measures and to minimize the electric utility bill for cloud providers operating thousands of servers. Because of these costs, fully associative caches are rarely used as instruction and data caches in modern processors.

The concept behind the fully associative cache may seem familiar because it is the same concept employed in the TLB presented in *Chapter 7, Processor and Memory Architectures*. The TLB is typically a fully associative cache containing virtual-to-physical address translations. Although the use of fully associative caching in the TLB results in the drawbacks of circuit complexity, die area consumption, and power consumption described in the preceding paragraph, the performance benefit provided by the TLB is so substantial that full associativity is used in virtually all high-performance processors that implement paged virtual memory.

Our discussion up to this point has focused on instruction caching, which is normally a read-only process. The functional requirements for the data cache are much like those of the instruction cache, with one significant extension: in addition to reading from data memory, the processor must be free to write to data memory. This is the subject of the next section.

Processor cache write policies

In processors with split caches, the L1 D-cache is similar in structure to the L1 I-cache, except the circuitry must permit the processor to write to memory as well as read from it. Each time the processor writes a data value to memory, it must update the L1 D-cache line containing the data item, and, at some point, it must also update the DRAM location in physical memory containing the cache line. As with reading DRAM, writing to DRAM is a slow process compared to the speed of writing to the L1 D-cache.

The most common cache write policies in modern processors are:

- **Write-through:** This policy updates DRAM immediately every time the processor writes data to memory.
- **Write-back:** This policy holds the modified data in the cache until the line is about to be evicted from the cache. A write-back cache must provide an additional status bit associated with each cache line indicating whether the data has been modified since it was read from DRAM. This is called the **dirty bit**. If set, the dirty bit indicates the data in that line is modified and the system must write the data to DRAM before its cache line can be freed for reuse.

The write-back policy generally results in better system performance because it allows the processor to perform multiple writes to the same cache line without the need to access main memory on each write. In systems with multiple cores or multiple processors, the use of write-back caching introduces complexity, because the L1 D-cache belonging to a core that writes to a memory location will contain data that differs from the corresponding DRAM location, and from any copies in the caches of the other processors or cores. This memory content disagreement will persist until the cache line has been written to DRAM and refreshed in any processor caches it occupies.

It is generally unacceptable to have different processor cores retrieve different values when accessing the same memory location, so a solution must be provided to ensure that all processors always see the same data at each memory location. The challenge of maintaining identical views of memory contents across multiple processors is called the **cache coherence** problem.

A multi-core processor could resolve this issue by sharing the same L1 D-cache among all cores, but modern multi-core processors usually implement a separate L1 D-cache for each core to maximize access speed. In multiprocessor systems with shared main memory, where the processors are on separate integrated circuits, and in multi-core processors that do not share L1 cache, the problem is more complex.

Some multiprocessor designs perform **snooping** to maintain cache coherence. A processor snoops by monitoring memory write operations performed by the other processors. When a write occurs to a memory location present in the snooping processor's cache, the snooper takes one of two actions: it can invalidate its cache line by setting the **Valid** bit to `false`, or it can update its cache line with the data written by the other processor. If the cache line is invalidated, the next access to the line's address range will result in DRAM access, which picks up the data modified by the other processor.

Snooping can be effective in systems with a limited number of processors, but it does not scale well to systems containing dozens or hundreds of processors. This is because each processor must continuously monitor the write behavior of all the other processors. Other, more sophisticated, cache coherence protocols must be implemented in systems with large numbers of processors.

Level 2 and level 3 processor caches

The discussion to this point has focused on L1 instruction and data caches. These caches are designed to be as fast as possible, but the focus on speed restricts their size due to cache circuit complexity and power requirements.

Because of the great disparity in the latency performance of L1 cache and DRAM, it is reasonable to wonder whether providing an intermediate level of cache between L1 and DRAM could improve performance beyond that of a processor containing just an L1 cache. The answer is, yes: adding an L2 cache provides a substantial performance enhancement.

Modern high-performance processors generally contain a substantial bank of L2 cache on-chip. Unlike an L1 cache, an L2 cache typically combines instructions and data in a single memory region, representing the von Neumann architecture rather than the Harvard architectural characteristic of the split L1 cache structure.

An L2 cache is generally slower than an L1 cache but still much faster than direct accesses to DRAM. Although an L2 cache uses SRAM bit cells with the same basic circuit configuration shown in *Figure 8.1*, the L2 circuit design emphasizes a reduction in the per-bit die area and power consumption relative to the L1 design. These modifications permit an L2 cache to be much larger than an L1 cache, but they also make it significantly slower.

A typical L2 cache might be four or eight times as large as the combined L1 I- and D-caches, with an access time 2-3 times as long as L1. The L2 cache may or may not be inclusive of the L1 cache, depending on the processor design. An **inclusive L2 cache** always contains the cache lines contained in the L1 cache, plus others.

As we have seen, each time the processor accesses memory, it first consults the L1 cache. If an L1 cache miss occurs, the next place to check is the L2 cache. Because L2 is larger than L1, there is a significant chance the data will be found there. If the cache line is not in L2, DRAM access is required. Processors generally perform some of these steps in parallel to ensure a cache miss does not result in a lengthy sequential search for the requested data.

Each miss of both an L1 and L2 cache results in a DRAM read that populates the corresponding cache lines in L1 and L2, assuming L2 is inclusive. Each time the processor writes to data memory, both L1 and L2 must be updated eventually. An L1 D-cache implements a cache write policy (typically write-through or write-back) to determine when it must update the L2 cache. L2, similarly, implements its own write policy to determine when to write dirty cache lines to DRAM.

If using two cache levels helps with performance, why stop there? In fact, most modern high-performance processors implement three (or more!) levels of cache on-chip. As with the transition from L1 to L2, the transition from L2 to L3 involves a larger block of memory with slower access speed. Like L2, an L3 cache usually combines instructions and data in a single memory region. On a consumer-grade PC processor, an L3 cache typically consists of a few megabytes of SRAM with an access time 3-4 times slower than the L2 cache.

THE PROGRAMMER'S VIEW OF CACHE MEMORY

Although software developers do not need to take any steps to take advantage of cache memory, an understanding of the execution environment in which their software operates can help improve the performance of code running on processors containing multiple levels of cache. Where flexibility permits, sizing data structures and structuring code inner loops to operate within the confines of anticipated L1, L2, and L3 cache sizes can result in a significant increase in execution speed.

Since the performance of any piece of code is affected by many factors related to the processor architecture and system software behavior, the best way to determine the optimal algorithm among multiple viable alternatives is to carefully benchmark the performance of each one.

The multilevel cache hierarchy in modern processors results in a dramatic performance improvement in comparison to an otherwise equivalent system that lacks cache memory. Caching allows the speediest processors to run with minimal DRAM access delay a large percentage of the time. Although cache memory adds substantial complexity to processor designs and consumes a great deal of valuable die area and power, processor vendors have determined that the use of a multi-level cache memory architecture is well worth these costs.

Cache memory speeds up the execution of instructions by reducing memory access latency when retrieving instructions and the data referenced by those instructions.

The next area of performance enhancement that we'll look at is optimization opportunities within the processor to increase the rate of instruction execution. The primary method modern processors use to achieve this performance boost is pipelining.

Instruction pipelining

Before we introduce pipelining, we will first break down the execution of a single processor instruction into a sequence of discrete steps:

- **Fetch**: The processor control unit accesses the memory address of the next instruction to execute, as determined by the previous instruction, or from the predefined reset value of the program counter immediately after power-on, or in response to an interrupt. Reading from this address, the control unit loads the instruction opcode into the processor's internal instruction register.
- **Decode**: From the opcode, the control unit determines the actions to be taken during instruction execution. This may involve the ALU and may require read or write access to registers or memory locations.
- **Execute**: The control unit executes the requested operation, invoking an ALU operation if required.
- **Write-back**: The control unit writes the results of instruction execution to register or memory locations. The program counter is then updated to the address of the next instruction to be executed.

The processor performs the fetch, decode, execute, write-back cycle repetitively from the time of power application until the computer is turned off. In a relatively simple processor such as the 6502, the processor executes each of these steps as essentially isolated, sequential operations.

From the programmer's viewpoint, each instruction completes all these steps before the next instruction begins. After write-back has completed, the results and side effects of each instruction are available in registers, memory, and processor flags for immediate use by the next instruction. This is a straightforward execution model, and executing programs can safely assume that all the effects of previous instructions are completed when an instruction begins execution.

The following diagram is an example of instruction execution in a processor that requires one clock cycle for each of the fetch-decode-execute-writeback steps. Note that each step in this diagram is indicated by its initial: F, D, E, or W:

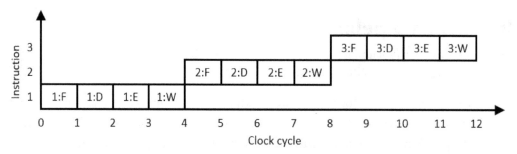

Figure 8.5: Sequential instruction execution

Each instruction requires four clock cycles to complete. At the hardware level, the processor represented in *Figure 8.5* consists of four execution subsystems, each of which becomes active during one of the four instruction clock cycles. The processing logic associated with each of these steps reads input information from memory and from processor registers and stores intermediate results in latches for use in later execution stages. After each instruction finishes, the next instruction begins execution on the next clock cycle.

The number of completed **instructions per clock (IPC)** provides a performance metric indicating how quickly a processor executes instructions relative to the processor clock speed. The processor in the example of *Figure 8.5* requires four clock cycles per instruction, resulting in an IPC of 0.25.

The potential for improved performance in this example arises from the fact that the circuitry involved in each of the four steps sits idle while the three remaining steps execute. Suppose that instead, as soon as the fetch hardware finishes fetching one instruction, it immediately begins fetching the next instruction. The following diagram shows the execution process when the hardware involved in each of the four instruction execution steps shifts from processing one instruction to the next instruction on each clock cycle:

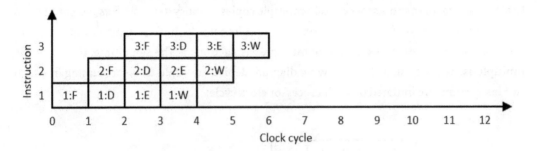

Figure 8.6: Pipelined instruction execution

This execution procedure is referred to as a **pipeline** because instructions enter and move through the execution stages from beginning to completion, like fluid moving through a physical pipeline. The processor pipeline contains multiple instructions at various stages of execution simultaneously. The reason for going to this trouble is evident in the preceding example: the processor is now completing one instruction per clock cycle, an IPC of 1.0, a four-fold speedup from the non-pipelined execution model of *Figure 8.5*. Similar levels of performance improvement are achieved in real-world processing using pipelining techniques.

In addition to overlapping the execution of sequential instructions via pipelining, there may be other opportunities to make productive use of processor subsystems that may otherwise be idle. Processor instructions fall into a few different categories that require different portions of the processor circuitry for their execution. Some examples include the following:

- **Branching instructions**: Conditional and unconditional branch instructions manipulate the program counter to set the address of the next instruction to be executed
- **Integer instructions**: Instructions involving integer arithmetic and bit manipulation access the integer portion of the ALU
- **Floating-point instructions**: Floating-point operations, on processors that provide hardware support for these operations, generally use circuitry that is separate from the integer ALU

By increasing the sophistication of the processor's instruction scheduling logic, it may be possible to initiate the execution of two instructions on the same clock cycle if they happen to use independent processing resources. For example, it has long been common for processors to dispatch instructions to the **floating-point unit (FPU)** for execution in parallel with non-floating-point instructions executing on the main processor core.

In fact, many modern processors contain multiple copies of subsystems such as integer ALUs to support the execution of multiple instructions simultaneously. In this architectural paradigm, the processor initiates the execution of multiple instructions at the same time, referred to as **multiple-issue** processing. The following diagram depicts multiple-issue processing in which two instructions are initiated on each processor clock cycle:

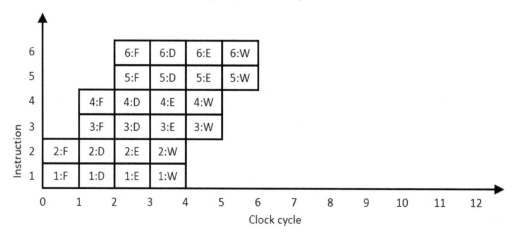

Figure 8.7: Multiple-issue pipelined instruction execution

This execution model doubles the number of instructions completed per clock cycle from the single-path pipeline of *Figure 8.6*, resulting in an IPC of 2.0. This represents a **superscalar processor**, which can issue (in other words, begin executing) more than one instruction per clock cycle. A **scalar processor**, in comparison, can issue only one instruction per clock cycle. To be clear, both *Figures 8.5* and *8.6* represent the behavior of a scalar processor, while *Figure 8.7* represents superscalar processing. A superscalar processor implements **instruction-level parallelism** (ILP) to increase execution speed. Virtually all modern high-performance, general-purpose processors are superscalar.

Superpipelining

Looking back at the scalar, non-pipelined processing model presented in *Figure 8.5*, we can consider how the processor clock speed is selected. In the absence of concerns about power consumption and heat dissipation, it is generally desirable to run the processor clock at the fastest possible speed. The upper limit on the clock speed for the processor of *Figure 8.5* is determined by the lengthiest timing path through each of the four subsystems involved in the execution stages. Different instructions may have drastically different execution time requirements. For example, an instruction to clear a processor flag requires very little execution time while a 32-bit division instruction may take much longer to produce its output.

It is inefficient to limit overall system execution speed based on the worst-case timing of a single instruction. Instead, processor designers look for opportunities to break the execution of complex instructions into a larger number of sequential steps. This approach is called **superpipelining**. Superpipelining consists of increasing the number of pipeline stages by breaking complex stages into multiple simpler stages. A superpipeline is, in essence, a processor pipeline with many stages, potentially numbering in the dozens. In addition to being superscalar, modern high-performance processors are generally superpipelined.

Breaking a pipeline into a larger number of superpipeline stages permits the simplification of each stage, reducing the time required to execute each stage. With faster-executing stages, it is possible to increase the processor clock speed. Superpipelining provides an instruction execution rate increase corresponding to the percentage increase in processor clock speed.

Reduced instruction set computer (RISC) processer instruction sets are designed to support effective pipelining. Most RISC instructions perform simple operations, such as moving data between registers and memory or adding two registers together. RISC processors usually have shorter pipelines compared to **complex instruction set computer (CISC)** processors. CISC processors, and their richer, more complex instruction sets, benefit from longer pipelines by breaking up long-running instructions into a series of sequential stages.

A big part of the challenge of efficiently pipelining processors based on legacy instruction sets such as x86 is that the original design of the instruction set did not fully consider the potential for later advances such as superscalar processing and superpipelining. As a result, modern x86-compatible processors devote a substantial proportion of their die area to complex logic implementing these performance-enhancing features for instructions that were not designed to operate in such an environment.

If breaking a pipeline into one or two dozen stages results in a substantial performance boost, why not continue by breaking the instruction pipeline into hundreds or even thousands of smaller stages to achieve even better performance? The answer: pipeline hazards.

Pipeline hazards

Implementing pipelining in a general-purpose processor is not as straightforward as the discussion to this point might imply. If the result of an instruction relies on the outcome of the previous instruction, the current instruction will have to wait until the result of the prior instruction becomes available. Consider this x86 code:

```
inc eax
add ebx, eax
```

Let's assume these two instructions are executed on a processor operating as in *Figure 8.6* (single-issue pipelining). The first instruction increments the eax register and the second adds this incremented value into the ebx register. The add instruction cannot execute the addition operation until the result of the increment operation in the previous instruction is available. If the second instruction's execute stage (labeled **2:E** in *Figure 8.6*) cannot execute until the first instruction's writeback stage has completed (**1:W** in the diagram), the 2:E stage has no choice but to wait for the 1:W stage to complete. Situations like this, where the pipeline cannot process the next stage of an instruction because of missing dependent information, are referred to as **pipeline hazards**.

One way in which processor pipelines deal with this issue is by implementing bypasses. After the first instruction's execute stage completes (labeled **1:E** in the diagram), the incremented value of the eax register has been computed, but it has not yet been written to the eax register. The second instruction's execute stage (**2:E** in the diagram) requires the incremented value of the eax register as an input. If the pipeline logic makes the result of the 1:E stage directly available to the 2:E stage without first writing to the eax register, the second instruction can complete execution without delay. The use of this type of shortcut to pass data between source and destination instructions without waiting for the completion of the source instruction is called a **bypass**. Bypasses are used extensively in modern processor designs to keep the pipeline working as efficiently as possible.

In some situations, a bypass is not possible because the necessary result simply cannot be computed before the destination instruction is ready to consume it. In this case, execution of the destination instruction must pause and await the delivery of the source instruction result. This causes the pipeline execution stage to become idle, which usually results in the propagation of the idle period through the remaining pipeline stages.

This propagating delay is called a **pipeline bubble**, analogous to an air bubble passing through a fluid pipeline. The presence of bubbles in the processing sequence reduces the effectiveness of pipelining.

Bubbles are bad for performance, so pipelined processor designers undertake substantial efforts to avoid them whenever possible. One way to do this is **out-of-order instruction execution**, referred to as **OoO**. Consider the two instructions listed in the earlier example, but now followed by a third instruction:

```
inc eax
add ebx, eax
mov ecx, edx
```

The third instruction does not depend on the results of the previous instructions. Instead of using a bypass to avoid a delay, the processor can employ OoO execution to avoid a pipeline bubble. The resulting instruction execution sequence would look like this:

```
inc eax
mov ecx, edx
add ebx, eax
```

The outcome of executing these three instructions is the same, but the separation in time between the first and third instructions has grown, reducing the likelihood of a pipeline bubble, or, if one still occurs, at least shortening its duration.

OoO execution involves detection of the presence of instruction dependencies and rearranges their execution order in a way that produces the same overall results, just not in the order originally coded. Some (typically CISC) processors perform this instruction reordering in real time during program execution. Other architectures (typically RISC) rely on intelligent programming language compilers to rearrange instructions during the software build process to minimize pipeline bubbles. The first approach requires a substantial investment in processing logic and the associated die real estate to perform reordering on the fly, while the second approach simplifies the processor logic but substantially complicates the job of assembly language programmers, who must now bear the burden of ensuring that pipeline bubbles do not excessively impair execution performance.

Some CISC processors use micro-operations and register renaming to achieve some of the performance benefits enjoyed by RISC architectures. These techniques are introduced in the next section.

Micro-operations and register renaming

The x86 instruction set architecture has presented a particular challenge for processor designers. Although several decades old, the x86 architecture remains a mainstay of personal and business computing. As a CISC configuration with only eight registers (in 32-bit x86), the techniques of highly efficient instruction pipelining and exploiting the large number of registers available in RISC architectures are much less helpful in the native x86 architecture.

To gain some of the benefits of the RISC methodology, x86 processor designers have taken the step of implementing the x86 instructions as sequences of micro-operations. A micro-operation, abbreviated **μop** (and pronounced *micro-op*), is a sub-step of a processor instruction. Simpler x86 instructions decompose to 1 to 3 μops, while more complex instructions require a larger number of μops. The decomposition of instructions into μops provides a finer level of granularity for evaluating dependencies upon the results of other μops and supports an increase in execution parallelism.

In tandem with the use of μops, modern processors generally provide additional internal registers, numbering in the dozens or hundreds, for storing intermediate μop results. These registers contain partially computed results destined for assignment to the processor's physical registers after an instruction's μops have all completed. The use of these internal registers is referred to as **register renaming**. Each renamed register has an associated tag value indicating the physical processor register it will eventually occupy. By increasing the number of renamed registers available for intermediate storage, the possibilities for instruction parallelism increase.

Several μops may be in various stages of execution at any given time. The dependencies of an instruction upon the result of a previous instruction will block the execution of a μop until its input is available, typically via the bypass mechanism described in the preceding section. Once all required inputs are available, the μop is scheduled for execution.

This mode of operation represents a **dataflow processing** model. Dataflow processing allows parallel operations to take place in a superscalar architecture by triggering μop execution when any data dependencies have been resolved.

High-performance processors perform the process of decoding instructions into μops after fetching each instruction. In some chip designs, the results of this decoding are stored in a Level 0 instruction cache, located between the processor and the L1 I-cache. The L0 I-cache provides the fastest possible access to pre-decoded μops for execution of code inner loops at the maximum possible speed. By caching the decoded μops, the processor avoids the need to repeat the instruction decoding pipeline stages for each subsequent access of the same instruction.

In addition to the hazards related to data dependencies between instructions, a second key source of pipeline hazards is the occurrence of conditional branching, discussed in the next section.

Conditional branches

Conditional branching introduces substantial difficulty into the pipelining process. The address of the next instruction following a conditional branch instruction cannot be confirmed until the branch condition has been evaluated.

There are two possibilities: if the branch condition is not satisfied, the processor will execute the instruction that follows the conditional branch instruction. If the branch condition is satisfied, the next instruction is at the address indicated in the instruction.

Several techniques are used to deal with the challenges presented by conditional branching in pipelined processors. Some of these are as follows:

- When possible, avoid branching altogether. Software developers can design algorithms with inner loops that minimize or eliminate conditional code. Optimizing compilers attempt to rearrange and simplify the sequence of operations to reduce the negative effects of conditional branching.
- The processor may delay fetching the next instruction until the branch condition has been evaluated. This typically introduces a bubble into the pipeline, degrading performance.
- The processor may perform the computation of the branch condition as early in the pipeline as possible. This identifies the correct branch more quickly, allowing execution to proceed with less delay.
- Some processors attempt to guess the result of the branch condition and begin tentatively executing instructions along that path. This is called **branch prediction**. If the guess turns out to be incorrect, the processor must clear the results of the incorrectly executed instructions from the pipeline (this is called **flushing the pipeline**) and start over with the first instruction on the correct path. Although an incorrect guess leads to a significant performance impact, correctly guessing the branch direction allows execution to proceed without delay. Some processors that perform branch prediction track the results of branch condition evaluations and use that information to improve the accuracy of future guesses when re-executing the same instruction.
- Upon encountering a conditional branch instruction, the processor can begin executing instructions along both branch paths simultaneously using its superscalar capabilities. This is referred to as **eager execution**.

    Once the branch condition has been determined, the results of execution along the incorrect path are discarded. Eager execution can only proceed if instructions avoid making changes that cannot be discarded. Writing data to main memory or to an output device is an example of a change that can't easily be undone, so eager execution would pause if such an action occurred while executing along a speculative path.

Modern high-performance processors devote a substantial portion of their logic resources to supporting effective pipelining under a wide variety of processing conditions. The combined benefits of multicore, superscalar, and superpipelined processing provide impressive performance enhancements in recent generations of sophisticated processor architectures. By performing instruction pipelining and executing instructions in a superscalar context, these features introduce parallelism into code sequences not originally intended to exploit parallel execution.

It is possible to further increase processing performance by introducing the parallel execution of multiple threads on a single processor core using simultaneous multithreading.

Simultaneous multithreading

As we learned in previous chapters, each executing process contains one or more threads of execution. When performing multithreading with time-slicing on a single-core processor, only one thread is in the running state at any moment in time. By rapidly switching between multiple ready-to-run threads, the processor creates the illusion (from the user's viewpoint) that multiple programs are running simultaneously.

This chapter introduced the concept of superscalar processing, which provides a single processing core with the ability to issue more than one instruction per clock cycle. The performance enhancement resulting from superscalar processing may be limited when the active sequence of instructions does not require a mixture of processor resources that aligns well with the capabilities of its superscalar functional units. For example, in a particular instruction sequence, integer processing units may be heavily used (resulting in pipeline bubbles), while address computation units remain mostly idle.

One way to increase the utilization of processor superscalar capabilities is to issue instructions from more than one thread on each clock cycle. This is called **simultaneous multithreading**. By simultaneously executing instructions from different threads, there is a greater likelihood that the instruction sequences will depend on disparate functional capabilities within the processor, thereby enabling increased execution parallelism and a higher rate of instruction execution.

Processors that support simultaneous multithreading must provide a separate set of registers for each simultaneously executing thread, as well as a complete set of renamed internal registers for each thread. The intent here is to provide more opportunities for utilization of the processor's superscalar capabilities.

Many modern, high-performance processors support the execution of two simultaneous threads, though some support as many as eight. As with most of the other performance enhancement techniques discussed in this chapter, increasing the number of simultaneous threads supported within a processor core eventually reaches a point of diminishing returns as the simultaneous threads compete for access to shared resources.

>
> **SIMULTANEOUS MULTITHREADING VERSUS MULTICORE PROCESSORS VERSUS MULTIPROCESSING**
>
> Simultaneous multithreading refers to a processor core with the ability to support the execution of instructions from different threads in the same pipeline stage on the same clock cycle. This differs from a multicore processor, in which each of the multiple processor cores on a single silicon die executes instructions independently of the others and only connects to the other cores through some level of cache memory.
>
> A multiprocessing computer contains multiple processor integrated circuits, each of which is usually contained in a separate integrated circuit package. Alternatively, a multiprocessor may be implemented as multiple processor chips assembled within a single package.

The performance optimization techniques discussed so far attempt to enhance the performance of scalar data processing, meaning that each instruction sequence operates on a small number of data values. Even superscalar processing, implemented with or without simultaneous multithreading, attempts to accelerate the execution of instructions that typically operate on one or two register-size data items at a time.

Vector data processing performs the same mathematical operation on each element in an array of data values simultaneously. Processor architectural features that improve execution parallelism for vector processing operations are the subject of the next section.

SIMD processing

Processors that issue a single instruction involving zero, one, or two data items per clock cycle are referred to as scalar processors. Processors capable of issuing multiple instructions per clock cycle, though not explicitly executing vector processing instructions, are called superscalar pro**cessors**. Some algorithms benefit from explicitly vectorized execution, which means performing the same operation on many data items simultaneously. Processor instructions tailored to such tasks are called **single-instruction, multiple-data (SIMD)** instructions.

The simultaneously issued instructions in superscalar processors generally perform different tasks on different data, representing a **multiple-instruction, multiple-data (MIMD)** parallel processing system. Some processing operations, such as the dot product operation used in digital signal processing described in *Chapter 6, Specialized Computing Domains*, perform the same mathematical operation on an array of values.

While the **multiply-accumulate (MAC)** operation described in *Chapter 6, Specialized Computing Domains*, performs scalar mathematical operations on each pair of vector elements in sequence, it is also possible to implement processor hardware and instructions capable of performing similar operations on more than a single pair of numbers at one time.

In modern processors, SIMD instructions are provided to perform tasks such as mathematics on numeric arrays, manipulation of graphics data, and substring searches in character strings.

The Intel implementation of **Streaming SIMD Extensions (SSE)** instructions, introduced in the Pentium III processors of 1999, provides a set of processor instructions and execution facilities for simultaneously operating on 128-bit data arrays. The data contained in the array can consist of integers or floating-point values. In the second generation of SSE (**SSE2**) instructions, the following data types can be processed in parallel:

- Two 64-bit floating-point values
- Four 32-bit floating-point values
- Two 64-bit integer values
- Four 32-bit integer values
- Eight 16-bit integer values
- Sixteen 8-bit integer values

SSE2 provides floating-point instructions for the familiar mathematical operations of addition, subtraction, multiplication, and division.

Instructions are also available for computing the square root, reciprocal, reciprocal of the square root, and returning the maximum value of the array elements. SSE2 integer instructions include comparison operations, bit manipulation, data shuffling, and data type conversion.

Later generations of the SSE instruction set have increased the data width and variety of supported operations. The latest iteration of SSE-type capabilities (as of 2021) is found in the AVX-512 instructions. **AVX** stands for **Advanced Vector Extensions** and provides register widths of 512 bits. AVX-512 includes, among other features, instructions optimized to support neural network algorithms.

One impediment to the widespread adoption of the different generations of SSE and AVX instructions is that, for end users to be able to take advantage of them, the instructions must be implemented in the processor, the operating system must support the instructions, and the compilers and other analytical tools used by the end users must take advantage of the SSE features. Historically, it has taken years following the introduction of new processor instructions before end users could easily take full advantage of their benefits.

The SIMD instructions available in modern processors have perhaps seen their most substantial application in the area of scientific computing. For researchers running complex simulations, machine learning algorithms, or sophisticated mathematical analyses, the availability of SSE and AVX instructions provides a way to achieve a substantial performance boost in code that performs extensive mathematical operations, character manipulations, and other vector-oriented tasks.

Summary

The majority of modern 32-bit and 64-bit processors combine most, if not all, of the performance-enhancing techniques presented in this chapter. A typical consumer-grade personal computer or smartphone contains a single main processor integrated circuit with four cores, each of which supports the simultaneous multithreading of two threads. This processor is superscalar, superpipelined, and contains three levels of cache memory. The processor decodes instructions into μops and performs sophisticated branch prediction.

Although the techniques presented in this chapter might seem overly complicated and arcane, in fact, each of us uses them routinely and enjoys the performance benefits that they yield each time we interact with any kind of computing device. The processing logic required to implement pipelining and superscalar operation is undeniably complex, but semiconductor manufacturers go to the effort of implementing these features for one simple reason: it pays off in the performance of their products and in the resulting value of those products as perceived by their customers.

This chapter introduced the primary performance-enhancing techniques employed in processor and computer architectures to achieve peak execution speed in real-world computing systems. These techniques do not change in any way what the processor produces as output; they just help get the work done faster. We examined the most important techniques for improving system performance including the use of cache memory, instruction pipelining, simultaneous multithreading, and SIMD processing.

The next chapter focuses on extensions commonly implemented at the processor instruction set level to provide additional system capabilities beyond generic computing requirements.

The extensions discussed in *Chapter 9*, *Specialized Processor Extensions*, include privileged processor modes, floating-point mathematics, power management, and system security management.

Exercises

1. Consider a direct-mapped L1 I-cache of 32 KB. Each cache line consists of 64 bytes and the system address space is 4 GB. How many bits are in the cache tag? Which bit numbers (bit 0 is the least significant bit) are they within the address word?

2. Consider an 8-way set associative L2 instruction and data cache of 256 KB, with 64 bytes in each cache line. How many sets are in this cache?

3. A processor has a 4-stage pipeline with maximum delays of 0.8, 0.4, 0.6, and 0.3 nanoseconds in stages 1-4, respectively. If the first stage is replaced with two stages that have maximum delays of 0.5 and 0.3 nanoseconds respectively, how much will the processor clock speed increase in percentage terms?

Join our community Discord space

Join the book's Discord workspace for a monthly *Ask me Anything* session with the author: `https://discord.gg/7h8aNRhRuY`

9
Specialized Processor Extensions

In the preceding chapters, we discussed the features of general-purpose computer architectures and some architectural specializations intended to address domain-specific requirements. This chapter focuses on extensions commonly implemented at the processor instruction set level and in other computer system hardware to provide additional system capabilities beyond generic computing needs.

After reading this chapter, you will understand the purpose of privileged processor modes and how they operate in multiprocessing and multiuser contexts. You will be familiar with the concepts of floating-point processing units and instruction sets, techniques for power management in battery-powered devices, and processor and system features that enhance system security.

We will discuss the following processor extensions and system features in this chapter:

- Privileged processor modes
- Floating-point mathematics
- Power management
- System security management

Technical requirements

The files for this chapter, including solutions to the exercises, are available at https://github.com/PacktPublishing/Modern-Computer-Architecture-and-Organization-Second-Edition.

Privileged processor modes

Most operating systems running on 32-bit and 64-bit processors control access to system resources using the concept of privilege levels. The primary reasons for managing access in this manner are to enhance system stability, prevent unauthorized interactions with system hardware, and prevent unauthorized access to data.

Privileged execution improves system stability by ensuring only trusted code is allowed to execute instructions that provide unrestricted access to resources such as processor configuration registers and I/O devices. The operating system kernel and related modules, including device drivers, require privileged access to perform their functions. Because any crash of a kernel process or a device driver is likely to halt the entire system immediately, these software components generally undergo a careful design process and rigorous testing before being released for general use.

Running the operating system in a privileged context prevents unauthorized applications from accessing system-managed data structures such as page tables and interrupt vector tables. Whether by accident or as a result of malicious intent, a user application may attempt to access data residing in system-owned memory or in memory belonging to a different user. The system access control mechanisms stop these attempts from succeeding and inform the misbehaving application of its transgression by initiating an exception, usually resulting in a program crash with an error message.

Chapter 3, *Processor Elements*, introduced the concepts associated with interrupts and how they are handled by the processor. Before getting into the details of privileged processor modes, we will first discuss interrupt and exception handling in more detail.

Handling interrupts and exceptions

Hardware interrupts, as we have seen, allow processors to respond promptly to requests for service from peripheral devices. A hardware interrupt notifies the processor of a need to take some action, usually involving data transfer to or from an external device.

Exceptions are similar to interrupts, with the key distinction that exceptions generally involve a response to some condition arising internally in the processor. One example of an internally generated exception is division by zero.

It is possible for user-level code to intentionally generate exceptions. In fact, this is a standard method used by unprivileged code to request system services provided by privileged code in the operating system kernel and device drivers.

The distinction between the terms **interrupt** and **exception** is not precisely defined. Interrupt processing and exception processing typically rely on the same or similar processor hardware resources operating in much the same manner. In fact, in the x86 architecture, the mnemonic for the instruction that initiates a software-generated interrupt (or exception) is int, short for interrupt.

Exceptions can result from error conditions occurring during software execution such as division by zero, as well as non-error situations such as page faults. Unless the interrupt or exception leads to the drastic response of terminating the application, an interrupt service routine or exception handler processes the event and returns control to the system scheduler, which eventually resumes execution of the interrupted thread.

When an interrupt or exception occurs, the processor transfers control to the corresponding handler in the following manner.

When an interrupt occurs, the processor permits the currently executing instruction to complete, then transfers control to an **interrupt service routine** (ISR). The code in the ISR performs the processing needed to complete the request initiated by the interrupt. After the ISR completes, the originally executing code resumes. This activity takes place without the awareness or approval of any threads that may have been executing at the time of the interrupt.

When responding to an exception, the processor transfers control to an exception handling routine, which is much like an ISR. An exception may arise during the execution of an instruction, preventing completion of the instruction's operations. If execution of an instruction is disrupted by an exception, the same instruction will be restarted after the exception handler completes and thread execution resumes. This mechanism allows page faults to occur at any point during program execution without affecting the results produced by the program (other than introducing a delay due to page fault processing).

Each type of interrupt or exception has a vector number referencing a row in the system interrupt vector table. The processor hardware consults the **interrupt vector table** (IVT) when an interrupt or exception occurs to select the appropriate ISR or exception handler. The IVT row indexed by the vector number contains the address of the ISR or handler for that vector.

When handling an interrupt or exception, the processor pushes any required context information onto the stack and transfers control to the handler. In the case of an interrupt, the handler services the device and clears its interrupt request. After processing the interrupt or exception, the handler returns control to the operating system. In the case of an exception, this may involve restarting the instruction that was executing when the exception occurred.

The following table summarizes some of the interrupt and exception types present in x86 processors running in protected mode:

Vector	Description	Cause
0	Divide error	A `div` or `idiv` instruction attempted an integer division by zero.
2	Non-maskable interrupt	The NMI hardware signal was asserted.
3	Breakpoint	The `int 3` instruction executed.
6	Invalid opcode	An attempt was made to execute a reserved opcode.
13	General protection	A prohibited access to memory or to a system resource was attempted.
14	Page fault	The MMU issued a request to resolve a virtual address.
32-255	User defined	These vectors are available for use by hardware interrupts or with the `int` instruction.

Table 9.1: Example x86 interrupt and exception types

Some interesting details related to the interrupts and exceptions in *Table 9.1* are as follows:

- The operation of the NMI input signal on vector 2 is similar to the NMI input of the 6502 processor, discussed in *Chapter 3, Processor Elements*.
- Although modern processors provide sophisticated, non-intrusive breakpoint capabilities, the x86 architecture retains the breakpoint functionality provided by the `int 3` instruction from the early days of the 8086. As we saw in *Chapter 3, Processor Elements*, the mechanism used by 6502 software debuggers to break program execution at a specified address is to temporarily replace the opcode at the break location with the 6502 BRK opcode. When execution reaches that location, the BRK handler takes control to allow user interaction with the system. The `int 3` instruction in the x86 architecture functions in the same manner.

In fact, unlike the x86 `int` instruction used with any of the other vector numbers, the `int 3` instruction is implemented as a single-byte opcode with the value CCh. A software interrupt with a different vector, such as `int 32`, is a 2-byte instruction. The `int 3` instruction enables breakpoint insertion by replacing a single byte of code with the value CCh.

- Vector 13, the general protection exception handler, activates on any attempt by an application to access memory or a system resource not allocated for its use. In modern operating systems, the system-provided handler for this vector terminates the running application and displays an error message.

- Vector 14, the page fault handler, triggers when an application attempts to access a valid page that is not present in physical memory. The handler attempts to locate the referenced page, which may be in a disk file or in the system swap file, loads the page into memory, updates the virtual-to-physical translation tables, and restarts the instruction that triggered the exception.

To summarize, hardware interrupts are initiated by I/O devices in need of data transfers or other types of servicing. Exceptions occur when the execution of an instruction sequence must be paused to handle a condition such as a page fault or attempted unauthorized memory access.

Hardware interrupts tend to occur at random times relative to the ongoing execution of program code, while behavior related to software-generated exceptions is often repeatable if the same code executes again while operating on the same data.

While some exceptions, such as a general protection fault, result in the termination of the process causing the exception, most interrupts and exceptions end with the interrupted thread resuming execution after the interrupt or exception handler has completed its processing.

EXCEPTIONS IN PROGRAMMING LANGUAGES

Many programming languages provide a facility for performing exception handling within an application. Programming language exceptions are significantly different from exceptions handled at the processor level. Exceptions in programming languages generally relate to error conditions at a much higher conceptual level than exceptions handled by processor hardware.

For example, C++ will generate (or "throw," in C++ terminology) an exception when a memory allocation request fails. This is not the same type of exception as system exceptions handled at the processor level. Be careful not to confuse exceptions in high-level programming languages with exceptions handled directly by the processor.

Protection rings

We can think of the protection strategy employed by modern processors and operating systems as similar in some ways to the defenses implemented in the design of medieval castles. A castle usually has a high wall surrounding the castle grounds, sometimes enhanced by the presence of a moat. This outer wall has a small number of well-defended access points, each of which is fortified against intruders by mechanisms such as drawbridges and around-the-clock guard details. Within the castle compound, the castle itself has sturdy walls and a small number of well-defended access points, further limiting access to the most sensitive areas.

The most privileged members of the castle hierarchy enjoy unrestricted access in and out of the castle and the outer wall. Less privileged members of society may have authorization to pass inward through the outer wall but are prohibited from accessing the castle directly. The least privileged members of the local population are prohibited from entering the castle under most conditions and may have to accept limitations on what they can do on occasions when they are granted access, such as being permitted to access only specified public areas.

The protections provided by this strategy can be represented as concentric rings, with the highest privilege required to access the innermost rings, while the outer ring represents the area requiring the least privilege, as in *Figure 9.1*:

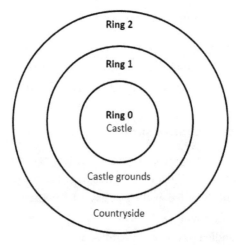

Figure 9.1: Protection ring example

This protection strategy implements three privilege levels that determine the types of access available to everyone in the system. Ring 0 requires the highest privilege level, while ring 2 requires no special privilege.

Modern processors, and the operating systems that run on them, use a very similar approach to prevent unauthorized access to critical resources while granting unprivileged users access to system capabilities approved for their use.

Although, in principle, it is possible to provide multiple intermediate levels between the highest and lowest privileged rings, most modern computer systems implement just two rings: a privileged ring, called the kernel or supervisor, and an unprivileged user ring. Some operating systems implement an intermediate ring containing device drivers, which grants access to the resources required to interact with I/O devices but does not provide the unfettered system-wide access of the kernel ring.

One reason operating systems such as Windows and Linux support only two privilege rings is because these systems have a design goal of portability across different processor architectures. Some processors support only two privilege rings, while others support a greater number. A portable operating system cannot implement more than two privilege rings if any of the underlying processor architectures do not support the desired number of rings. The x86 architecture, for example, supports up to four rings, but only two of these (ring 0, the most privileged, and ring 3, the least privileged) are used by Windows and Linux.

The following figure represents the ring organization of most operating systems that run on the x86 architecture:

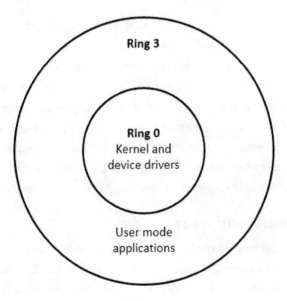

Figure 9.2: Protection rings in x86 operating systems

This ring-based privilege management system is the primary reason the infamous Windows "Blue Screen of Death," which was prevalent in the 1990s, appears so rarely for users of recent versions of Windows. User applications such as web browsers, email clients, and word processors will, on occasion, experience problems that cause the programs to crash. By virtue of the two-ring privilege enforcement mechanisms provided by modern operating systems such as Windows, Linux, macOS, and Android, the damage of an individual application crash is contained by the operating system, preventing the crash from affecting the operating system itself or any other program running on the system (at least, those programs that aren't dependent for correct operation upon the program that crashed).

Following an application crash, the operating system cleans up any resources in use by the crashed program, including allocated memory pages and open files. This allows computer systems such as web server hosts to remain operating continuously for hundreds of days despite the occasional crashes and restarts of software applications running on them.

In addition to protecting against application crashes, ring-based privilege control provides substantial security benefits against malicious actors. One type of attack a hacker might attempt is to insert some code into a system module that runs in ring 0. This insertion could occur in an executable file on disk, or the attacker may try to modify kernel code while it is running in memory. If successful, the attacker could then use this code to access data anywhere in the system because it is running in ring 0.

While achieving success with this type of attack is by no means impossible, modern processors and operating systems have implemented an extensive array of security measures and repaired many vulnerabilities present in earlier versions of operating systems. When system administrators and users take full advantage of the ring-based security features available in modern computer systems, there are very few feasible paths an attacker can take to access protected data. In fact, in most cases, a key element of successful hacks that make it into the news these days is a human-related security breakdown exploited by the attacker. We will discuss cybersecurity threats and methods for ensuring security in system and application software in the *System security management* section later in this chapter and in *Chapter 14, Cybersecurity and Confidential Computing Architectures*.

Supervisor mode and user mode

In a two-level protection ring hierarchy, the protection level of the currently executing thread is represented by a bit in a register. When operating in ring 0, the **supervisor mode** bit is 1, and when operating in **user mode** (ring 3 on x86) the supervisor mode bit is 0. The supervisor mode bit can only be modified by code running in supervisor mode.

The state of the supervisor mode bit determines which instructions are available for execution by the thread. Instructions that could interfere with system operation, such as the x86 `hlt` instruction, which halts processor instruction execution, are unavailable in user mode.

Any attempt to execute a prohibited instruction results in a general protection fault. In user mode, access by instructions to system memory regions and the memory allocated to other users is prohibited. In supervisor mode, all instructions are available for execution and all memory regions are accessible.

In the castle analogy, the supervisor mode bit represents the authorization presented to the castle guards that enables access to the castle grounds and to the castle itself. When set, the supervisor mode bit provides the keys to the kingdom.

System calls

All code belonging to the kernel and device drivers runs in ring 0, always. All user code runs in ring 3, always, even for users with enhanced operating system privileges such as system administrators. Code running in ring 3 is strictly controlled by the system and cannot directly do anything that involves allocating memory, opening a file, displaying information to the screen, or interacting with an I/O device. To access any of those system features, ring 3 user code must make a service request to the kernel.

The kernel service request must first pass through a gate (just like visitors entering our castle!) where the type of operation being requested, as well as any associated parameters, is scrutinized and validated before execution of the operation is allowed to proceed. The code performing the requested operation runs in supervisor mode at ring 0 and, when complete, returns control to user mode in ring 3.

In early versions of Windows (prior to Windows XP), an application used the software interrupt mechanism with vector 2eh to request system services. The system service calling protocol involved placing the parameters required by the requested service into processor registers and executing the `int 2eh` instruction, triggering a software interrupt. The handler ran in supervisor mode, resulting in a transition from ring 3 to ring 0. Upon completion of the handler, the system returned to ring 3 at the instruction following `int 2eh`.

One problem with the use of the `int 2eh` mechanism for requesting kernel services is that it is not very efficient. In fact, it takes over 1,000 processor clock cycles to get from the point at which the `int 2eh` instruction executes to the kernel code that begins to handle the exception. A busy system may request kernel services thousands of times per second.

To address this inefficiency, Intel implemented the sysenter and sysexit instructions in the x86 architecture beginning with the Pentium II processor in 1997. The purpose of these instructions is to accelerate the process of calling from ring 3 to ring 0, and later returning to ring 3. By using these instructions instead of int 2eh, entry into and exit from kernel mode speeds up by about a factor of three.

At around the time Intel began producing processors with the sysenter and sysexit instructions, AMD released the syscall and sysret instructions in their processor architectures, with the same performance objective. Unfortunately, the instructions in the Intel and AMD processor architectures are not compatible, which leads to a requirement for operating systems to differentiate between architectures when using accelerated kernel service calling instructions.

In the next section, we will look at the data formats and operations associated with floating-point arithmetic.

Floating-point arithmetic

Modern processors usually support integer data types in widths of 8, 16, 32, and 64 bits. Some smaller embedded processors may not directly support 64-bit or even 32-bit integers, while more sophisticated devices may support 128-bit integers. Integer data types are appropriate for use in a wide range of applications, but many areas of computing, particularly in the fields of science, engineering, and navigation, require the ability to represent fractional numbers with a high degree of accuracy.

As a simple example of the limitations of integer mathematics, suppose you need to divide 5 by 3. On a computer restricted to using integers, you can perform an integer calculation of this expression as follows, in C++:

```cpp
#include <iostream>
int main(void)
{
    int a = 5;
    int b = 3;
    int c = a / b;
    std::cout << "c = " << c << std::endl;
    return 0;
}
```

This program produces the following output:

```
c = 1
```

If you punch this expression into your pocket calculator, you'll find the result from the program is not very close to the actual result, which is approximately 1.6667. In computing applications where accurate calculations involving real numbers are required, the use of floating-point mathematics provides a practical solution.

Mathematically, the set of real numbers consists of all numbers, including all integers and fractions, along the number line from negative infinity to positive infinity. There is no limit to the number of digits in the integer and fractional parts of a real number.

Given the finite storage available in even the largest computer, it is clearly not possible to represent the entire range of real numbers in computer programs. A compromise is required if we are to represent real numbers in a useful manner. The standard solution is to represent real numbers with a mantissa and an exponent.

In areas of study involving the mathematics of very large or very small numbers, it is common to represent such numbers in terms of a mantissa and an exponent. For example, in physics, the gravitational constant is represented in the following format:

$$G = 6.674 \times 10^{-11} \frac{m^3}{kg\ s^2}$$

The **mantissa** represents the nonzero digits of the number after multiplication by a scaling factor that places those digits within a convenient range. The **exponent** represents the scaling factor by which the mantissa must be multiplied to produce the actual value.

In this example, the mantissa is *6.674* and the exponent is *-11*. This example uses a base-10 mantissa, which is convenient for manual calculation because multiplication by the scale factor *10^{-11}* can be performed by simply moving the position of the mantissa decimal point. In this example, an exponent of -11 requires moving the decimal point 11 places to the left, resulting in the equivalent value *G* = 0.00000000006674.

The use of floating-point notation avoids the error-prone use of lengthy sequences of zeros and allows both extremely large and small numbers to be represented in a compact format.

Any number along the real number line can be represented in floating-point notation. However, there can be no limit on the number of digits in the mantissa or exponent of the notation if the goal is to truly represent all real numbers.

In computer representations of floating-point numbers, both the mantissa and exponent are limited to predefined bit widths. These ranges have been chosen to fit floating-point numbers within standard data widths, typically 32 or 64 bits, while providing an adequate number of mantissa and exponent digits for a wide variety of applications.

Increasing the number of mantissa digits increases the precision of the numerical values representable in floating-point format. Increasing the number of exponent digits increases the range of numbers that can be represented. Because of the finite lengths of the mantissa and exponent, the result of a floating-point calculation often does not precisely equal the actual real-valued result. Instead, the best we can expect is that the floating-point result will be the nearest representable value to the correct result.

Modern processors commonly support 32-bit and 64-bit representations of floating-point numbers. Rather than the base-10 exponents described in the previous example, computers work with base-2 exponents. The general format of a computer floating-point number is:

$$\text{sign} \times \text{mantissa} \times 2^{exponent}$$

The sign is simply +1 or -1. In the binary representation, a positive number has a sign bit of 0 and a negative number has a sign bit of 1. Having separated the sign from the rest of the number, the remaining value is nonnegative. For any value other than zero (which is handled as a special case), this number can be scaled into the range of greater than or equal to 1 and less than 2 by multiplying by some power of 2.

Continuing the example of the gravitational constant, because the sign is +1, the value after removing the sign is unchanged: 6.674×10^{-11}. The mantissa for this number can be scaled into the range ($1 \leq m < 2$), where m represents the mantissa, by multiplying by 2^{34}. The result of this scaling is:

$$G = +1 \times 1.1465845 \times 2^{-34}$$

We multiplied the original number by 2^{34} to get the mantissa into the desired range, so the floating-point representation of the number must be multiplied by 2^{-34} to undo the scaling operation.

Since our computers operate on binary numbers, we must next convert the mantissa to a binary representation. The format used in floating-point processing is to represent the range between 1 and 2 with an unsigned integer.

For example, if we assume the binary mantissa is 16 bits wide, the mantissa of 1.0 is represented by 0000h and the value just below 2.0 (actually 1.99998474) is represented by FFFFh. A decimal mantissa m is converted to a 16-bit binary mantissa using the expression $(m - 1) \times 2^{16}$ and rounding the result to the nearest integer. A 16-bit binary mantissa m is converted to decimal with the expression $1 + m \times 2^{-16}$.

In our example, the decimal mantissa of 1.1465845 converts to a 16-bit binary mantissa of 0010010110000111b, or 2587h.

Using the scaling procedure described in this section, floating-point numbers can be represented in binary form with any desired bit widths for the mantissa and exponent.

For the purposes of cross-processor compatibility, it is helpful to define a limited number of binary floating-point formats capable of supporting a wide variety of use cases and adopt those formats industry-wide.

The IEEE 754 standard for the computer representation of floating-point numbers was adopted in 1985 for this reason. Before looking at key features of the standard, we'll explore the design of the Intel 8087 floating-point coprocessor, which served as the source of many of the concepts later enshrined in the IEEE 754 standard.

The 8087 floating-point coprocessor

Modern processors with floating-point hardware generally implement a set of instructions for floating-point computations and perform these operations in a dedicated functional unit. In a superscalar processor, the main processor core continues to execute other categories of instructions while the **floating-point unit (FPU)** executes floating-point calculations in parallel.

Recall from *Chapter 1, Introducing Computer Architecture*, that the original IBM PC of 1981 contained a socket for an 8087 floating-point coprocessor. The 8087 performs floating-point computations in hardware at speeds roughly 100 times faster than a functionally equivalent software implementation running on the host processor.

Because installation of the 8087 was optional, most PC software applications that wished to take advantage of its capabilities first tested for the presence of the 8087 and, if it was not found, reverted to a library of much slower floating-point code.

The 8087 supports the following data types for numeric processing:

- 16-bit two's complement integer
- 32-bit two's complement integer
- 64-bit two's complement integer
- 18-digit signed packed **binary coded decimal (BCD)**
- 32-bit signed short real with a 24-bit mantissa and an 8-bit exponent
- 64-bit signed long real with a 53-bit mantissa and an 11-bit exponent
- 80-bit signed temporary real with a 64-bit mantissa and a 15-bit exponent

Each data type is stored as a series of bytes in memory. The formats used for the real data types are shown in the following figure:

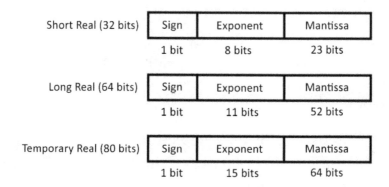

Figure 9.3: 8087 coprocessor floating-point data formats

The short and long real formats use an implied 1 bit as the most significant bit of the mantissa, and do not include this bit in their binary representations. As a special case, the value zero is represented in these formats by setting both the mantissa and exponent to zero.

The temporary real format is used internally in the 8087 to store intermediate results. This format has extended precision in comparison to the long real format to minimize the propagation of rounding errors over a series of calculations.

Each of the real number formats can be represented as $(-1)^s(2^{E-bias})(m)$, where S is the sign bit, E is the exponent, and m is the mantissa. This *bias* is 127 in the short real format, 1,023 in the long real format, and 16,383 in the temporary real format.

Subtraction of the *bias* term converts the unsigned integer stored in the exponent field to a signed value.

Our example real number, G, with a decimal mantissa of 1.1465845 and a binary scale of 2^{-34}, is represented in the short real format with a sign bit of 0, an exponent of (-34 + 127) = 5Dh, and a mantissa of $(1.1465845 - 1) \times 2^{23} = 12C348h$. Combining all three components, the 32-bit representation of 6.674×10^{-11} is $(\text{sign bit} = 0) \times 2^{31} + biased\ exponent = 5Dh \times 2^{23} + (mantissa = 12C348h)$. The resulting 32-bit single-precision floating-point value is 2E92C348h.

The 8087 adds 68 opcode mnemonics to the 8086/8088 instruction set for performing arithmetic, trigonometric, exponential, and logarithmic functions. In a PC program using the 8087, the code consists of a single stream of 8088 and 8087 instructions retrieved in the usual sequential manner by the 8088 processor. The 8087 passively monitors the address and data buses as the processor executes instructions and only becomes active when an 8087 opcode appears. The 8088 treats 8087 instructions as no-operation (or *nop*) instructions and ignores them.

When the 8087 begins executing an instruction, it can take control of the host bus to transfer data between its internal registers and system memory using DMA cycles. The 8087 and 8088 do not directly transfer data between themselves and can only share data by storing it in memory for use by the other processor.

8087 instruction execution proceeds independently of the 8088, making an 8087-equipped PC a truly parallel processing system. The 8087 has a BUSY output signal for use by the 8088 to determine whether the 8087 is currently processing an instruction.

When the 8088 requires the results of an 8087 operation, the 8088 must wait for the 8087 BUSY signal to de-assert, at which point the 8088 is free to access the memory locations containing the output of the 8087 instruction.

The IEEE 754 floating-point standard

The most widely used formats for representing floating-point numbers in modern computer systems are those defined in the IEEE 754 standard. **IEEE**, the **Institute of Electrical and Electronic Engineers**, publishes a wide variety of standards related to electricity, electronics, and computing. The original version of the IEEE 754 standard was adopted in 1985, based largely on the data types and mathematical operations of the Intel 8087.

The 8087 was not entirely compliant with the initial IEEE 754 standard, which was published several years after the introduction of the 8087. Later Intel floating-point coprocessors, beginning with the 80387 in 1987, were fully standard-compliant. Today's 32-bit and 64-bit processors generally implement an IEEE 754-compliant floating-point unit on the same integrated circuit die as the main processor.

The IEEE 754 standard was updated in 2008 and again in 2019. The current version is IEEE 754-2019 and contains definitions for base-2 floating-point number formats with widths of 16, 32, 64, 128, and 256 bits. It also contains base-10 floating-point number formats with widths of 32, 64, and 128 bits. The 32-bit and 64-bit base-2 floating-point formats are generally supported in programming languages that include floating-point operations. Support for the remaining IEEE 754 data types tends to be more limited and nonstandard across processors, programming languages, and operating systems.

The next section presents features many modern processor architectures implement to manage system power consumption.

Power management

For users of portable battery-powered devices such as smartphones, tablets, and laptop computers, the ability to operate for long time periods without recharging is an important feature. Designers of portable systems place great emphasis on ensuring battery power consumption is minimized under all operating conditions.

Some techniques designers use to reduce power consumption are:

- Placing idle subsystems in a low-power state or turning them off completely when they are not needed. This technique may not be possible for peripherals that must be available to respond to incoming requests, such as network interfaces.
- Reducing integrated circuit supply voltages and clock speeds during periods when execution speed is not critical.
- When possible, saving system state information and turning the processor power off. Users of laptop computers are familiar with the two options for reducing power consumption when the system is not in use: standby and hibernate. In standby mode, system RAM continues to be powered while the rest of the system is turned off. Standby mode enables very fast startup when the user resumes using the system. This responsiveness comes at a cost: keeping RAM powered consumes a significant amount of energy.

In hibernate mode, the entire system state is written to disk storage and the system powers down completely. Hibernate mode requires essentially zero power, though it normally takes quite a bit longer than standby mode to resume operation.

- When periodic processing is required, such as in real-time applications, placing the processor in a low-power state each time processing completes. The processor remains in that state until a timer interrupt (or some other type of interrupt) wakes the processor for the next iteration. Many embedded processors provide a low-power idle mode in which the instruction pipeline is halted but remains ready to instantly resume execution in response to an interrupt. Some RTOS implementations support this concept by idling the processor when all tasks are blocked waiting for an interrupt or other event.

Smartphones and other battery-powered devices make extensive use of these methods to drive battery usage to a minimum while remaining instantly responsive to user actions and to external inputs such as incoming calls and social media notifications.

Modern processors in high-performance computers and in embedded devices generally support the ability to adjust processor clock speed during code execution, and in some cases provide the ability to alter the processor power supply voltage as well. This combination of power management techniques is addressed in the next section.

Dynamic voltage frequency scaling

Optimization of processor clock frequency and supply voltage to minimize power consumption is called **dynamic voltage frequency scaling (DVFS)**. When performing DVFS, a processor regularly evaluates the performance requirements of its current operating state and adjusts the processor clock frequency and power supply voltage to minimize battery usage while continuing to perform at an acceptable level.

In a CMOS chip containing many transistors, a reasonable estimate of power consumption is,

$$P = \sum_{i=1}^{N} C_i v_i^2 f_i$$

In this expression, P represents the total power consumption of the chip, which contains N MOS transistors. C_i is the capacitance driven by the ith transistor, v_i is the transistor's supply voltage, and f_i is its operating frequency.

Examining this formula, we see that circuit power consumption is proportional to the square of the supply voltage and is directly proportional to capacitance and operating frequency. Because capacitance in the circuit plays a significant role in power consumption, it is beneficial to implement the device using a fabrication process that reduces the size of circuit gates, thus keeping the capacitance to a minimum. This is one reason a transition to an integrated circuit fabrication process supporting smaller feature sizes results in devices with reduced power consumption.

DVFS attempts to minimize overall power consumption by reducing both the supply voltage and processor clock frequency as much as possible, all the time.

Reducing the circuit supply voltage can provide a dramatic reduction in power consumption. However, any reduction of the processor supply voltage must be carefully coordinated with adjustments to the clock frequency. When the supply voltage is reduced, CMOS transistors switch more slowly. This is because the capacitance driven by each transistor remains unchanged, but it charges and discharges at a slower rate because there is less voltage driving the charge/discharge process. As the system voltage is reduced, the clock frequency must be reduced in tandem for the circuit to function properly.

Recalling the hydraulic system analogy from *Chapter 2, Digital Logic*, the effect of reducing the CMOS supply voltage is equivalent to reducing the water pressure that fills the balloon attached to the side of the pipe: with reduced system pressure, the balloon fills more slowly.

Reducing the processor supply voltage also reduces the circuit's noise immunity. When this happens, it becomes easier for external interference, such as the electrical field emanating from a motor starting up in a household appliance, to disrupt the internal operation of the processor. Any processor voltage reduction must be carefully managed to ensure continued reliable operation.

Reducing the processor clock frequency (in addition to any clock slowing required by supply voltage reduction) reduces power consumption by an approximately proportional amount. Reducing the operating frequency may also increase system reliability because each gate transition has more time to propagate and reach its final state before being latched by the gates it is driving.

A complex embedded device such as a smartphone often needs to transition rapidly between higher-and lower-power states as its various inputs stimulate it with user actions, timer-triggered events, and information arriving over wireless connections. While it is beneficial for the system to nimbly switch between lower-and higher-power modes as circumstances change, it is also important to limit the rate at which such transitions occur because the mere act of switching between power modes consumes some power itself.

Users expect their computers and other sophisticated digital devices to consume no more power than necessary, and they also expect those systems to be secure. The next section introduces the key aspects of system security management.

System security management

We have seen how the separation of privilege levels between kernel and user modes supports the effective separation of applications started by one user from those of other users and from system processes. This represents security at the level of executing software.

This is fine as far as it goes, but what about systems that must remain secure even when untrusted users have unrestricted physical access to them? Additional measures must be implemented at the hardware level to prevent curious or malicious users from accessing protected code, data, and hardware resources.

Before getting into the details of hardware-level security features, it is helpful to list some of the categories of information and other resources that must be protected in digital systems:

- **Personal information**: Information such as government identification numbers, passwords for accessing bank accounts, contact lists, emails, and text messages must be protected even if a portable device containing that information is lost or stolen.
- **Business information**: Trade secrets, customer lists, research products, and strategic plans are some categories of confidential business data that may have great value in the hands of competitors or criminals. Businesses also collect a great deal of personal information about their customers and are required to undertake substantial efforts to keep this information secure.
- **Government information**: Government organizations maintain a tremendous quantity of personal information about their citizens and must ensure it can only be used for authorized purposes. Governments also develop vast amounts of information related to national security that requires extensive security protocols.

Beyond the obvious physical security measures of storing sensitive information in a sturdy, access-controlled facility with effective alarm systems, a few measures can be taken to ensure a system is secure against a wide variety of attacks.

Consider the smartphone. A technically capable individual may be able to disassemble the phone's case and gain access to its circuit-level hardware.

If this person can monitor the external electrical signals of the processor and its communication paths to other system components, what kinds of information might be gathered? The answer depends on the types and quantity of hardware security implemented in the system design.

A first step in secure system design is to avoid inserting security vulnerabilities during development. During the development of a system containing an embedded processor, it is quite useful to provide a hardware debugger interface. A **hardware debugger** enables the connection of a PC to the device using a special interface cable. Using this connection, developers can reprogram flash memory, set breakpoints in the code, display the values of registers and variables, and single-step through code. If the debugger connection remains in the circuit board in the released version of the design, it may be possible for users to connect their own debugger to the system and work with it in the same manner as the developers.

This is clearly undesirable for any system intended to operate securely. Because the ability to connect a debugger continues to be quite useful even after a product is released, developers sometimes attempt to leave the debugger signals present in the circuit but camouflage them in some manner to make their functions less obvious. While this approach may be effective to some degree, dedicated hackers have demonstrated the ability to ferret out the presence of even the most cleverly disguised debugging interfaces and leverage them to access the processor's internals. Leaving an accessible hardware debugging interface present in a released product is a serious security vulnerability.

Many processor vendors have begun implementing security features to prevent even dedicated, technically capable attackers from breaching system security protections. To be effective, system developers must enable and fully leverage these capabilities in their system designs. Some examples of security technologies include the following:

- **Password-protected hardware debugger interface**: Some processor families support the addition of a password to the standard hardware debugging interface. In these systems, an initial authentication process must take place in which the connected system provides a strong password (such as a 256-bit number) before the processor enables debug functionality. This is an effective approach that retains the ability to securely troubleshoot issues that arise after the product is released.

- **Internal encryption engine with key storage**: Some processors provide encryption and decryption capabilities and store secret keys for use during operation. The keys must be stored in the processor during system manufacture and are not externally accessible after they have been stored.

This combination of encryption engine and stored keys allows secure communication with authorized external devices. The use of high-speed, hardware-based encryption and decryption capabilities allows secure full-speed communication between physically separated subsystems such as those within an automobile.

- **Device memory protection:** Many processor families provide several options for the protection of memory regions. For example, a ROM bank containing code can be locked after programming to ensure it cannot be reprogrammed later. Code regions can also be blocked from being read as data while still allowing access for execution. Processors that lack a full memory management unit often have a subsystem called the **memory protection unit (MPU)** to manage the security requirements of the various processor memory regions.

Several capabilities necessary for ensuring secure system operation have been combined in a device called the Trusted Platform Module, discussed in the next section.

Trusted Platform Module

Modern PCs contain a subsystem called the **Trusted Platform Module (TPM)** that provides a variety of security features. A TPM is a tamper-resistant processor specifically designed to support cryptographic operations. The TPM in a modern PC may be a small module that plugs into the motherboard, or it is sometimes implemented as a separate functional unit within the main processor integrated circuit.

The general categories of functionality provided by the TPM are:

- **Random number generation:** When it is necessary to pick a large number that will be very hard for someone else to guess, the best approach is to use a truly random method to pick the number. While most programming languages provide a mechanism for generating random-looking number sequences, in most cases these sequences are often at least somewhat predictable once a series of samples is available. The TPM provides a *true* random number generation capability suitable for generating cryptographic keys capable of providing strong protection for secure data. A true random number generator uses specialized hardware to generate an unpredictable sequence of 0 and 1 bits.
- **Cryptographic key generation:** The TPM can create cryptographic keys in standard formats for use internal to the system and upon request by user applications. These keys are commonly employed for purposes such as email digital signatures and for securing access to web applications.

- **Cryptographic key storage**: In addition to generating cryptographic keys, the TPM can securely store multiple keys for various purposes. **Public-key cryptography** uses keys generated in pairs. One key of the pair is the **public key**, which can be freely shared with anyone. The other is the **private key**, which must be protected for use only by the key owner. Data encrypted with the public key can only be decrypted using the private key. Similarly, data encrypted with the private key can only be decrypted by the corresponding public key. Public-key cryptography enables many security capabilities such as email encryption and document digital signatures. The TPM can be instructed to lock a private key permanently within it, ensuring that it cannot be discovered by malicious software or even by a sophisticated attacker with full physical access to the system.
- **System integrity verification**: The TPM is intimately involved in the secure boot process. This process ensures that only trusted firmware and software components are allowed to execute during the boot process. The secure boot process involves computing and verifying the digital signature of each firmware and software component used during boot prior to its execution.
- **System health monitoring**: The TPM can ensure system security features such as disk encryption and secure boot requirements continue to remain active continuously during system operation.
- **Authentication services**: The TPM provides a programming interface offering cryptographic functions on an as-needed basis for use by applications at the unprivileged level. This represents a secure mechanism for binding a user's identity to a digital device. The private key contained in the TPM provides strong, unforgeable identification of the user.

The TPM enables modern PCs and other digital devices to securely engage in financial and privacy-sensitive interactions over insecure communication channels including the internet.

The overarching purpose of security mechanisms such as the TPM is to disrupt the efforts of cyberattackers who are attempting to steal information or financial assets belonging to computer system users. The range of attacks performed and some of the defenses against them are discussed in the next section.

Thwarting cyberattackers

It is not sufficient to design a computer system with a standard set of security features and assume hackers will be unable to penetrate those defenses. Attackers have demonstrated extraordinary cleverness and will use any available avenue to gain insight into the inner workings of a system they find interesting, whether for reasons of intellectual curiosity or for criminal motives.

Some areas to consider beyond standard security techniques are:

- **Power, timing, and emission analysis**: By using fine-grained monitoring of system power consumption, the timing of external signal activity, or even the radio frequency emissions generated during algorithm execution, it may be possible to reverse engineer the activities going on inside the processor. Attacks based on these methods have been successful at extracting secret encryption keys during cryptographic algorithm execution, for example.
- **Power disruption**: Attacks that induce power supply voltage fluctuations or dropouts during system power-up or operation have been shown to leave many systems in a vulnerable state in which some security features are ineffective. A robust system design must anticipate such behavior, whether natural or intentional, and revert to a safe and secure state whenever the power supply is not performing within established parameters.
- **Physical alteration**: An attacker may replace some components in the system in an effort to gain an enhanced level of control or access to sensitive functions or data. For example, replacing the boot ROM in an embedded device may allow the system to operate normally while also enabling the attacker to gain unrestricted system access using code in the replacement ROM. A growing number of processor families support the use of digital signatures to verify the authenticity of ROM code. To verify the contents of the ROM device during startup, the processor runs a cryptographic **hash algorithm** over the ROM contents. The results of the hash (the signature) must match the signature preloaded in internal processor storage before the ROM code is allowed to execute. The hash algorithm is designed to make it essentially impossible to alter the ROM data content while still producing the expected signature. For additional security, the ROM data content can also be encrypted, which prevents the attacker from analyzing the code it contains.

This section has described modern approaches that are widely used to implement security in digital devices and examined some of the more esoteric security vulnerabilities that have been exploited by dedicated attackers. The design of a secure system must begin with a firm grounding in the basic elements of security, but also requires a great deal of ingenuity in considering the many ways a determined attacker may attempt to breach its defenses.

Summary

Building upon the preceding chapters, this chapter introduced computer architecture features addressing domain-specific functional requirements. We focused on extensions commonly implemented at the processor instruction set level and in components external to the processor that provide additional system capabilities beyond generic computing requirements.

You should now have a good understanding of privileged processor modes and how they are used in multiprocessing and multiuser contexts, the concepts of floating-point processors and instruction sets, techniques for power management in battery-powered devices, and the processor and system features intended to enhance computer system security.

This background prepares us for the next chapter, where we will examine the most popular processor architectures and instruction sets currently used in personal computing, business computing, and smart portable devices. These architectures are the x86, the x64, and the 32-bit and 64-bit variants of ARM.

Exercises

Using a programming language that allows access to the byte representation of floating-point data types (such as C or C++), write a function that accepts a 32-bit single-precision value as input. Extract the sign, exponent, and mantissa from the bytes of the floating-point value and display them. Remove the bias term from the exponent before displaying its value and display the mantissa as a decimal number. Test the program with the values 0, -0, 1, -1, 6.674e-11, 1.0e38, 1.0e39, 1.0e-38, and 1.0e-39. The numeric values listed here containing *e* are using the C/C++ text representation of floating-point numbers. For example, 6.674e-11 means 6.674×10^{-11}.

1. Modify the program from *Exercise 1* to also accept a double-precision floating-point value, and print the sign, exponent (with the bias removed), and mantissa from the value. Test it with the same input values as in *Exercise 1*, and with the values 1.0e308, 1.0e309, 1.0e-308, and 1.0e-309.
2. Search the internet for information about the NXP Semiconductors i.MX RT1060 processor family. Download the product family datasheet and answer the following questions about these processors.
3. Do the i.MX RT1060 processors support the concept of supervisor mode instruction execution? Explain your answer.
4. Do the i.MX RT1060 processors support the concept of paged virtual memory? Explain your answer.
5. Do the i.MX RT1060 processors support floating-point operations in hardware? Explain your answer.
6. What power management features do the i.MX RT1060 processors support?
7. What security features do the i.MX RT1060 processors support?

Join our community Discord space

Join the book's Discord workspace for a monthly *Ask me Anything* session with the author: https://discord.gg/7h8aNRhRuY

10

Modern Processor Architectures and Instruction Sets

Most modern personal computers contain a processor supporting either the Intel or AMD version of the x86 32-bit and x64 64-bit architectures. In contrast, almost all smartphones, smartwatches, tablets, and many embedded systems contain ARM 32-bit or 64-bit processors. This chapter takes a detailed look at the registers and instruction sets of these processor families.

After completing this chapter, you will understand the high-level architectures and unique attributes of the x86, x64, 32-bit ARM, and 64-bit ARM registers, instruction sets, assembly languages, and key aspects of legacy features supported in these architectures.

This chapter covers the following topics:

- x86 architecture and instruction set
- x64 architecture and instruction set
- 32-bit ARM architecture and instruction set
- 64-bit ARM architecture and instruction set

Technical requirements

The files for this chapter, including the answers to the exercises, are available at https://github.com/PacktPublishing/Modern-Computer-Architecture-and-Organization-Second-Edition.

x86 architecture and instruction set

For the purposes of this discussion, the term **x86** refers to the 16-bit and 32-bit instruction set architecture of the series of processors that began with the Intel 8086, introduced in 1978. The 8088, released in 1979, is functionally very similar to the 8086, except it has an 8-bit data bus instead of the 16-bit bus of the 8086. The 8088 was the central processor in the original IBM PC.

Subsequent generations of this processor series were named 80186, 80286, 80386, and 80486, leading to the term "x86" as shorthand for members of the family. Subsequent generations dropped the numeric naming convention and received the names Pentium, Core, i Series, Celeron, and Xeon.

Advanced Micro Devices (AMD), a semiconductor manufacturing company that competes with Intel, has been producing x86-compatible processors since 1982. Some recent AMD x86 processor generations have been named Ryzen, Opteron, Athlon, Turion, Phenom, and Sempron.

Code execution compatibility between Intel and AMD processors is good in many aspects. There are some key differences between processors from the two vendors, including the chip pin configuration and chipset compatibility.

In general, Intel processors only work in motherboards and with chipsets designed for Intel chips, and AMD processors only work in motherboards and with chipsets designed for AMD chips. We will highlight some other differences between Intel and AMD processors later in this section.

The 8086 and 8088 are 16-bit processors, despite the 8-bit data bus of the 8088. Internal registers in these processors are 16 bits wide and the instruction set operates on 16-bit data values. The 8088 transparently executes two bus cycles to transfer each 16-bit value between the processor and memory.

The 8086 and 8088 do not support the more sophisticated features of modern processors such as paged virtual memory and protection rings. These early processors also have only 20 address lines, limiting the addressable memory to 1 MB. A 20-bit address cannot fit in a 16-bit register, so it is necessary to use a somewhat complicated system of segment registers and offsets to access the full 1 MB address space.

In 1985, Intel released the 80386 with enhancements that mitigate many of these limitations. The 80386 introduced these features:

- **32-bit architecture**: Addresses, registers, and the ALU are 32 bits wide and instructions operate natively on operands up to 32 bits wide.

- **Protected mode**: This mode implements the multilevel privilege mechanism consisting of ring numbers 0 to 3 that we examined in *Chapter 9, Specialized Processor Extensions*. In Windows and Linux, ring 0 is kernel mode, and ring 3 is user mode. Rings 1 and 2 are not used in these operating systems.
- **On-chip MMU**: The 80386 MMU supports a flat memory model enabling any location in the 4 GB space to be accessed with a 32-bit address. Manipulation of segment registers and offsets is no longer required. The MMU supports paged virtual memory.
- **3-stage instruction pipeline**: The pipeline accelerates instruction execution, as discussed in *Chapter 8, Performance-Enhancing Techniques*.
- **Hardware debug registers:** The debug registers support setting up to four breakpoints that stop code execution at a specified virtual address when the address is accessed and a selected condition is satisfied. The available break conditions are code execution, data write, and data read or write. These registers are only available for use by code running in ring 0.

Modern x86 processors boot into the 16-bit operating mode of the original 8086, which is now called **real mode**. This mode retains compatibility with software written for the 8086/8088 environment, such as the MS-DOS operating system.

In most modern systems running on x86 processors, a transition to protected mode occurs during system startup. Once in protected mode, the operating system remains in protected mode until the computer shuts down.

MS-DOS ON A MODERN PC

Although the x86 processor in a modern PC is compatible at the instruction level with the original 8088, running an old copy of MS-DOS on a modern computer system is unlikely to be a straightforward process. The peripheral devices and their interfaces in modern PCs are not compatible with the corresponding interfaces in PCs from the 1980s. MS-DOS would need a driver that understands how to interact with the USB-connected keyboard of a modern motherboard, for example.

These days, the primary use for 16-bit mode in x86 processors is to serve as a bootloader for a protected mode operating system. Because most developers of computerized devices and the software that runs on them are unlikely to be involved in implementing such a capability, the remainder of our x86 discussion in this chapter will address protected mode and the associated 32-bit flat memory model.

The x86 architecture supports unsigned and signed two's complement integer data types with widths of 8, 16, 32, 64, and 128 bits. The names assigned to these data types are as follows:

- **Byte:** 8 bits
- **Word:** 16 bits
- **Doubleword:** 32 bits
- **Quadword:** 64 bits
- **Double quadword:** 128 bits

In most cases, the x86 architecture does not mandate the storage of these data types on natural boundaries. The **natural boundary** of a data type is any address evenly divisible by the size of the data type in bytes.

Storing any of the multi-byte types at unaligned boundaries is allowed but is discouraged because it causes a negative performance impact: instructions operating on unaligned data consume additional clock cycles. A few instructions that operate on double quadwords require naturally aligned storage and will generate a general protection fault if unaligned access is attempted.

x86 natively supports floating-point data types in widths of 16, 32, 64, and 80 bits. The 32-bit, 64-bit, and 80-bit formats are those presented in *Chapter 9, Specialized Processor Extensions*. The 16-bit format is called **half-precision** floating-point and has an 11-bit mantissa, an implied leading 1 bit, and a 5-bit exponent. The half-precision floating-point format is used extensively in GPU processing.

In the next section, we will look at the x86 register set in detail.

The x86 register set

The x86 architecture protected mode has eight 32-bit wide general-purpose registers, a flags register, and an instruction pointer. There are also six segment registers and additional processor model-specific configuration registers. The segment registers and model-specific registers are configured by system software during startup and are, in general, not relevant to the developers of applications and device drivers. For these reasons, we will not discuss the segment registers and model-specific registers further.

The 16-bit general-purpose registers in the original 8086 architecture are named AX, CX, DX, BX, SP, BP, SI, and DI. The reason for listing the first four registers in this non-alphabetic order is because this is the sequence in which these eight registers are pushed onto the stack by a pushad (push all registers) instruction.

Chapter 10

In the transition to the 32-bit architecture of the 80386, each register grew to 32 bits. The 32-bit version of a register's name is prefixed with the letter "E" to indicate this extension.

It is possible to access portions of 32-bit registers in smaller bit widths. For example, the lower 16 bits of the 32-bit EAX register are referenced as AX. The AX register can be further accessed as individual bytes using the names AH (high-order byte) and AL (low-order byte). The following diagram shows the register names and the subsets of each:

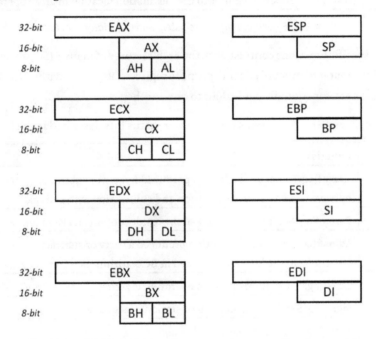

Figure 10.1: Register names and subsets

Writing to a portion of a 32-bit register, for example, the AL register, affects only the bits in that portion. In the case of AL, loading an 8-bit value modifies the lowest 8 bits of EAX, leaving the other 24 bits unaffected.

In keeping with the x86's CISC architecture, several functions associated with various instructions are tied to specific registers. *Table 10.1* provides a description of the functions associated with each of the x86 general-purpose registers:

Register	Name	Function
EAX	Accumulator	Arithmetic operations
ECX	Counter	Loop counter and shift/rotate counter

EDX	Data	Arithmetic and I/O operations
EBX	Base	Pointer to data
ESP	Stack pointer	Pointer to the top of the stack
EBP	Base pointer	Pointer to the stack base within a function
ESI	Source index	Pointer to the source location in array operations
EDI	Destination index	Pointer to the destination location in array operations

Table 10.1: x86 general-purpose registers and associated functions

These register-specific functions contrast with the architectures of many RISC processors, which tend to provide a greater number of general-purpose registers. Registers within a RISC processor are, for the most part, functionally equivalent to one another.

The x86 flags register, `EFLAGS`, contains the processor status bits described in *Table 10.2*:

Bit	Name	Function
0	CF	Carry flag: Indicates if addition produced a carry or subtraction produced a borrow. Used as input by addition and subtraction instructions.
2	PF	Parity flag: Set if the low 8 bits of the result contain an even number of 1 bits.
4	AF	Adjust flag: Indicates if addition produced a carry or subtraction produced a borrow from the lower 4 bits. Used in BCD arithmetic.
6	ZF	Zero flag: Set if the result of an operation is zero.
7	SF	Sign flag: Set if the result of an operation is negative.
8	TF	Trap flag: Used in single-step debugging.
9	IF	Interrupt enable flag: Setting this bit enables hardware interrupts.
10	DF	Direction flag: Controls the direction of string processing. When clear, the order is lowest to highest addresses. When set, the order is highest to lowest addresses.
11	OF	Overflow flag: Set if an operation resulted in a signed overflow.
12-13	IOPL	I/O privilege level: The privilege level of the currently executing thread. IOPL 0 is kernel mode, and 3 is user mode.
14	NT	Nested task flag: Controls the chaining of interrupts.
16	RF	Resume flag: Used for processing exceptions during debugging.

17	VM	Virtual 8086 mode flag: If set, 8086 compatibility mode is active. This mode allows some MS-DOS applications to be run in the context of a protected mode operating system.
18	AC	Alignment check flag: If set, memory alignment checking is active. For example, if the AC flag is set, storing a 16-bit value to an odd address triggers an Alignment Check exception. x86 processors can perform unaligned memory accesses when this flag is not set, but the number of instruction cycles required may increase.
19	VIF	Virtual interrupt flag: Virtual version of the IF flag in virtual 8086 mode.
20	VIP	Virtual interrupt pending flag: Set when an interrupt is pending in virtual 8086 mode.
21	ID	ID flag: If this bit can be set, the cpuid instruction is supported. cpuid returns processor identification and feature information.

Table 10.2: x86 flags' register bits

All bits in the EFLAGS register that are not listed in *Table 10.2* are reserved and are unused.

The 32-bit instruction pointer, EIP, contains the address of the next instruction to execute, unless a branch is taken. When a branch is taken, the address of the branch destination is loaded into EIP and execution continues from there.

The x86 architecture is little-endian, meaning multi-byte values are stored in memory with the least significant byte at the lowest address and the most significant byte at the highest address.

x86 addressing modes

As one would expect for a CISC architecture, x86 supports a variety of addressing modes. There are several rules associated with addressing source and destination operands that must be followed to create valid instructions. For instance, the sizes of the source and destination operands of a mov instruction must be equal. The assembler will attempt to select a suitable size for an operand that has an ambiguous size (for example, an immediate value of 7) to match the width of a destination location (such as the 32-bit register EAX). In cases where the size of an operand cannot be inferred, size keywords such as byte ptr must be provided.

The assembly language in these examples uses *Intel syntax*, which places the operands in destination-source order. Intel syntax is used primarily in the Windows and MS-DOS contexts. An alternative notation, known as *AT&T syntax*, places operands in source-destination order. AT&T syntax is used in Unix-based operating systems. All examples in this book will use the Intel syntax.

The x86 architecture supports a variety of addressing modes, which we will look at next. Comments in assembly code begin with a semicolon and continue to the end of the line.

Implied addressing

In this addressing mode, the register is implied by the instruction opcode. For example:

```
clc ; Clear the carry flag (CF in the EFLAGS register)
```

Register addressing

One or both source and destination registers are encoded in the instruction:

```
mov eax, ecx ; Copy the contents of register ECX to EAX
```

Registers may be used as the first operand, the second operand, or both operands.

Immediate addressing

An immediate value is provided as an instruction operand:

```
mov eax, 7 ; Move the 32-bit value 7 into EAX
mov ax, 7 ; Move the 16-bit value 7 into AX (the lower 16 bits of EAX)
```

When using Intel syntax, it is not necessary to prefix immediate values with the # character.

Direct memory addressing

The address of the value is provided as an instruction operand:

```
mov eax, [078bch] ; Copy the 32-bit value at hex address 78BC to EAX
```

In x86 assembly code, square brackets surrounding an expression indicate the expression is an address. When performing moves or other operations are performed on square-bracketed operands, the value being operated upon is the data at the specified address. The exception to this rule is the LEA (load effective address) instruction, which we'll examine later.

Register indirect addressing

The operand is a register containing the address of the data value:

```
mov eax, [esi] ; Copy the 32-bit value at the address contained in ESI to
             ; EAX
```

This mode is equivalent to using a pointer to reference a variable in C or C++.

Indexed addressing

The operand indicates a register plus offset that combine to provide the address of the data value:

```
mov eax, [esi + 0bh] ; Copy the 32-bit value at the address (ESI + 0bh) to
                     ; EAX
```

This mode is useful for accessing the elements of a data structure. In this scenario, the ESI register contains the address of the structure, and the added constant is the byte offset of the element from the beginning of the structure.

Based indexed addressing

The operand indicates a base register, an index register, and an offset that sum together to calculate the address of the data value:

```
mov eax, [ebx + esi + 10] ; Copy the 32-bit value starting at the address
                          ; (EBX + ESI + 10) to EAX
```

This mode is useful for accessing individual data elements within an array of data structures. In this example, the EBX register contains the address of the beginning of the structure array, ESI contains the offset of the referenced structure within the array, and the constant value (10) is the offset of the desired element from the beginning of the selected structure.

Based indexed addressing with scaling

The operand is composed of a base register, an index register multiplied by a scale factor, and an offset that sum together to calculate the address of the data value:

```
mov eax, [ebx + esi*4 + 10] ; Copy the 32-bit value starting at the
                            ; address (EBX + ESI*4 + 10) to EAX
```

In this addressing mode, the value in the index register can be multiplied by 1 (the default), 2, 4, or 8 before being summed with the other components of the operand address. There is no performance penalty associated with using the scaling multiplier. This feature is helpful when iterating over arrays containing elements with sizes of 2, 4, or 8 bytes.

Most of the general-purpose registers can be used as the base or index register in the based addressing modes.

The following diagram shows the possible combinations of register usage and scaling in the based addressing modes:

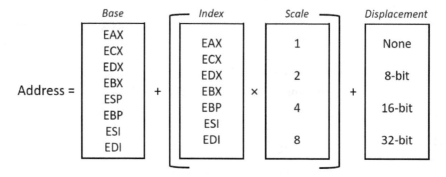

Figure 10.2: Based addressing mode

All eight general-purpose registers are available for use as the base register. Of those eight, only ESP is unavailable for use as the index register.

x86 instruction categories

The x86 instruction set was introduced with the Intel 8086 and has been extended several times over the years. Some of the most significant changes relate to the extension of the architecture from 16 to 32 bits, which added protected mode and paged virtual memory. In almost all cases, the new capabilities have been added while retaining full backward compatibility.

The full x86 instruction set contains several hundred instructions. We will not discuss all of them in this chapter. This section will provide brief summaries of the more important and commonly encountered instructions applicable to user-mode applications and device drivers.

This subset of x86 instructions can be divided into a few general categories: data movement; stack manipulation; arithmetic and logic; conversions; control flow; string and flag manipulation; input/output; and protected mode. We will also cover some miscellaneous instructions that do not fall into any specific category.

Data movement

Data movement instructions do not affect the processor flags. The following instructions perform data movement:

- mov: Copies the data value referenced by the second operand to the location provided as the first operand.

- cmovcc: Conditionally moves the second operand's data to the register provided as the first operand if the *cc* condition is true. The condition is determined from one or more of the following processor flags: CF, ZF, SF, OF, and PF. The condition codes are e (equal), ne (not equal), g (greater), ge (greater or equal), a (above), ae (above or equal), l (less), le (less or equal), b (below), be (below or equal), o (overflow), no (no overflow), z (zero), nz (not zero), s (SF=1), ns (SF=0), cxz (register CX is zero), and ecxz (the ECX register is zero).
- movsx, movzx: These are variants of the mov instruction performing sign extension and zero extension, respectively. The source operand must be a smaller size than the destination.
- lea: Computes the address provided by the second operand and stores it at the location given in the first operand. The second operand is surrounded by square brackets. Unlike the other data movement instructions, the computed address is stored in the destination rather than the data value located at that address.

Stack manipulation

Stack manipulation instructions do not affect the processor flags. These instructions are:

- push: Decrements ESP by 4, and then places the 32-bit operand into the stack location pointed to by ESP.
- pop: Copies the 32-bit data value pointed to by ESP to the operand location (a register or memory address), and then increments ESP by 4.
- pushfd, popfd: Pushes or pops the EFLAGS register.
- pushad, popad: Pushes or pops the EAX, ECX, EDX, EBX, ESP, EBP, ESI, and EDI registers, in that order.

Arithmetic and logic

The arithmetic and logic instructions modify the processor flags. The following instructions perform arithmetic and logic operations:

- add, sub: Perform integer addition or subtraction. When subtracting, the second operand is subtracted from the first. Both operands can be registers, or one operand can be a memory location and the other a register. One operand can be a constant.
- adc, sbb: Performs integer addition or subtraction using the CF flag as a carry input (for addition) or as a borrow input (for subtraction).
- cmp: Subtracts the two operands and discards the result while updating the OF, SF, ZF, AF, PF, and CF flags based on the result.
- neg: Negates the operand.

- `inc`, `dec`: Increments or decrements the operand by one.
- `mul`: Performs unsigned integer multiplication. The size of the product depends on the size of the operand. A byte operand is multiplied by `AL` and the result is placed in `AX`. A word operand is multiplied by `AX` and the result is placed in `DX:AX`, with the upper 16 bits in `DX`. A doubleword is multiplied by `EAX` and the result is placed in `EDX:EAX`.
- `imul`: Performs signed integer multiplication. The first operand must be a register and receives the result of the operation. There may be a total of two or three operands. In the two-operand form, the first operand multiplies the second operand, and the result is stored in the first operand (a register). In the three-operand form, the second operand multiplies the third operand, and the result is stored in the first operand register. In the three-operand form, the third operand must be an immediate value.
- `div`, `idiv`: Performs unsigned (`div`) or signed (`idiv`) division. The size of the result depends on the size of the operand. A byte operand is divided into `AX`, the quotient is placed in `AL`, and the remainder is placed in `AH`. A word operand is divided into `DX:AX`, the quotient is placed in `AX`, and the remainder is placed in `DX`. A doubleword is divided into `EDX:EAX`, the quotient is placed in `EAX`, and the remainder is placed in `EDX`.
- `and`, `or`, `xor`: Performs the corresponding logical operation on the two operands and stores the result in the destination operand location.
- `not`: Performs a logical `NOT` (bit inversion) operation on a single operand.
- `sal`, `shl`, `sar`, `shr`: Performs a logical (`shl` and `shr`) or arithmetic (`sal` and `sar`) shift of the byte, word, or doubleword argument left or right by 1 to 31 bit positions. `sal` and `shl` place the last bit shifted out into the carry flag and insert zeros into the vacated least significant bits. `shr` places the last bit shifted out into the carry flag and inserts zeros into the vacated most significant bits. `sar` differs from `shr` by propagating the sign bit into the vacated most significant bits.
- `rol`, `rcl`, `ror`, `rcr`: Performs a left or right rotation by 0 to 31 bits, optionally through the carry flag. `rcl` and `rcr` rotate through the carry flag, while `rol` and `ror` do not.
- `bts`, `btr`, `btc`: Reads a specified bit number (provided as the second operand) within the bits of the first operand into the carry flag, then either sets (`bts`), resets (`btr`), or complements (`btc`) that bit. These instructions may be preceded by the `lock` keyword to make the operation atomic.
- `test`: Performs a logical `AND` operation of two operands and updates the `SF`, `ZF`, and `PF` flags based on the result.

Conversions

Conversion instructions extend a smaller data size to a larger size. These instructions are:

- `cbw`: Converts a byte (`AL` register) into a word (`AX`).
- `cwd`: Converts a word (`AX` register) into a doubleword (`DX:AX`).
- `cwde`: Converts a word (`AX` register) into a doubleword (`EAX`).
- `cdq`: Converts a doubleword (`AX` register) into a quadword (`EDX:EAX`).

Control flow

Control flow instructions conditionally or unconditionally transfer execution to an address. The control flow instructions are:

- `jmp`: Transfers control to the instruction at the address provided as the operand.
- `jcc`: Transfers control to the instruction at the address provided as the operand if the condition *cc* is true. The condition codes were described previously in the `cmovcc` instruction description. The condition is determined from one or more of the following processor flags: `CF`, `ZF`, `SF`, `OF`, and `PF`.
- `call`: Pushes the current value of `EIP` onto the stack and transfers control to the instruction at the address provided as the operand.
- `ret`: Pops the top-of-stack value and stores it in `EIP`. If an operand is provided, it pops the given number of bytes from the stack to clear parameters.
- `loop`: Decrements the loop counter in `ECX` and, if not zero, transfers control to the instruction at the address provided as the operand.

String manipulation

String manipulation instructions may be prefixed by the `rep` keyword to repeat the operation the number of times given by the `ECX` register, incrementing or decrementing the source and destination location on each iteration, depending on the state of the `DF` flag. The operand size processed on each iteration can be a byte, word, or doubleword. The source address of each string element is given by the `ESI` register and the destination by the `EDI` register. These instructions are:

- `mov`: Moves a string element
- `cmps`: Compares elements at corresponding locations in two strings
- `scas`: Compares a string element to the value in `EAX`, `AX`, or `AL`, depending on the operand size

- `lods`: Loads the string into `EAX`, `AX`, or `AL`, depending on the operand size
- `stos`: Stores `EAX`, `AX`, or `AL`, depending on the operand size, to the address in `EDI`

Flag manipulation

Flag manipulation instructions modify bits in the `EFLAGS` register. The flag manipulation instructions are:

- `stc, clc, cmc`: Sets, clears, or complements the carry flag, `CF`
- `std, cld`: Sets or clears the direction flag, `DF`
- `sti, cli`: Sets or clears the interrupt flag, `IF`

Input/output

Input/output instructions read data from or write data to peripheral devices. The input/output instructions are:

- `in, out`: Moves 1, 2, or 4 bytes between `EAX`, `AX`, or `AL` and an I/O port, depending on the operand size
- `ins, outs`: Moves a data element between memory and an I/O port in the same manner as the string instructions
- `rep ins, rep outs`: Moves blocks of data between memory and an I/O port in the same manner as the string instructions

Protected mode

The following instructions access the features of protected mode:

- `sysenter, sysexit`: Transfers control from ring 3 to ring 0 (sysenter) or from ring 0 to ring 3 (sysexit) in Intel processors.
- `syscall, sysret`: Transfers control from ring 3 to ring 0 (syscall) or from ring 0 to ring 3 (sysret) in AMD processors. In x86 (32-bit) mode, AMD processors also support sysenter and sysexit.

Miscellaneous instructions

These instructions do not fit into the categories previously listed:

- `int`: Initiates a software interrupt. The operand is the interrupt vector number.
- `nop`: No operation.
- `cpuid`: Provides information about the processor model and its capabilities.

Other instruction categories

The instructions listed in this section are some of the more common instructions you will come across in x86 applications and device drivers beyond those listed in the preceding sections. The x86 architecture contains a wide variety of instruction categories, including the following:

- **Floating-point instructions**: These instructions are executed by the x87 floating-point unit.
- **SIMD instructions**: This category includes the MMX, SSE, SSE2, SSE3, SSE4, AVX, AVX2, and AVX-512 instructions. Some of the instruction sets in this category were introduced in the *SIMD processing* section of *Chapter 8, Performance-Enhancing Techniques*.
- **AES instructions**: These instructions support encryption and decryption using the **Advanced Encryption Standard (AES)**.
- **MPX instructions**: The **memory protection extensions (MPX)** enhance memory integrity by preventing errors such as buffer overruns.
- **SMX instructions**: The **safer mode extensions (SMX)** improve system security in the presence of user trust decisions.
- **TSX instructions**: The **transactional synchronization extensions (TSX)** enhance the performance of multithreaded execution using shared resources.
- **VMX instructions**: The **virtual machine extensions (VMX)** support the secure and efficient execution of virtualized operating systems.

Additional processor registers are provided for use by the floating-point and SIMD instructions.

There are even more categories of x86 instructions beyond those listed here, a few of which have been retired in later generations of the architecture.

Common instruction patterns

Listed below are some examples of instruction usage patterns you will come across frequently in compiled code. The techniques used in these examples produce the desired result while minimizing code size and the number of clock cycles required:

```
xor  reg, reg ; Set reg to zero
test reg, reg ; Test if reg contains zero
add  reg, reg ; Shift reg left by one bit
```

x86 instruction formats

Individual x86 instructions are of variable length and can range in size from 1 to 15 bytes. The components of a single instruction, including any optional bytes, are laid out in memory in the following sequence:

- **Prefix bytes**: One or more optional prefix bytes provide auxiliary opcode execution information. For example, the lock prefix performs bus locking in a multiprocessor system to enable atomic test-and-set type operations. rep and related prefixes enable string instructions to perform repeated operations on string elements in a single instruction. Other prefixes are available to provide hints for conditional branch instructions or to override the default size of an address or operand.
- **Opcode bytes**: An x86 opcode, consisting of 1 to 3 bytes, follows any prefix bytes. For some opcodes, an additional 3 opcode bits are encoded in a *ModR/M* byte following the opcode.
- **ModR/M byte**: Not all instructions require this byte. The *ModR/M* byte contains three information fields providing an address mode and operand register information. The upper two bits of this byte (the *Mod* field) and the lower three bits (the *R/M* field) combine to form a 5-bit field with 32 possible values. Of these, 8 values identify register operands, and the other 24 values specify addressing modes. The remaining 3 bits (the *reg/opcode* field) either indicate a register or provide three additional opcode bits, depending on the instruction.
- **Address displacement bytes**: 0, 1, 2, or 4 bytes provide an address displacement used in computing the operand address.
- **Immediate value bytes**: If the instruction includes an immediate value, it is in the last 1, 2, or 4 bytes of the instruction.

The variable-length nature of x86 instructions makes the process of instruction decoding quite complex. It is also challenging for debugging tools to disassemble a sequence of instructions in reverse order, perhaps to display the code leading up to a breakpoint.

This difficulty arises because it is possible for a trailing subset of bytes within a lengthy instruction to form a complete, valid instruction. This complexity is a notable difference from the more regular instruction formats used in RISC architectures.

x86 assembly language

It is possible to develop programs of any level of complexity in assembly language.

Most modern applications, however, are largely or entirely developed in high-level languages. Assembly language tends to be used in cases where the use of specialized instructions is required, or a level of extreme optimization is necessary that is unachievable with an optimizing compiler.

Regardless of the language used in application development, all code must ultimately execute as processor instructions. To fully understand how code executes on a computer system, there is no substitute for examining the state of the system following the execution of each individual instruction. A good way to learn to understand and operate in this environment is to write some assembly code.

The x86 assembly language example in the following listing is a complete x86 application that runs in a Windows command console, printing a text string and then exiting:

```
.386
.model FLAT,C
.stack 400h

.code
includelib libcmt.lib
includelib legacy_stdio_definitions.lib

extern printf:near
extern exit:near

public main
main proc
    ; Print the message
    push    offset message
    call    printf

    ; Exit the program with status 0
    push    0
    call    exit
main endp

.data
message db "Hello, Computer Architect!",0

end
```

A description of the contents of this assembly language file follows:

- The `.386` directive indicates the instructions in this file should be interpreted as applying to 80386 and later-generation processors.
- The `.model FLAT,C` directive specifies a 32-bit flat memory model and the use of C language function calling conventions.
- The `.stack 400h` directive specifies a stack size of 400 h (1,024) bytes.
- The `.code` directive indicates the start of executable code.
- The `includelib` and `extern` directives reference system-provided libraries and the functions within them to be used by the program.
- The `public` directive indicates that the function name, `main`, is an externally visible symbol.
- The lines between `main proc` and `main endp` are the assembly language instructions making up the `main` function.
- The `.data` directive indicates the start of data memory. The `message db` statement defines the message string as a sequence of bytes, followed by a zero byte.
- The end directive marks the end of the program.

This file, named `hello_x86.asm`, can be assembled and linked to form the executable `hello_x86.exe` program with the following command, which runs the Microsoft Macro Assembler:

```
ml /Fl /Zi /Zd hello_x86.asm
```

The components of this command are:

- `ml` runs the assembler (`ml.exe`)
- `/Fl` creates a listing file
- `/Zi` includes symbolic debugging information in the executable file
- `/Zd` includes line number debugging information in the executable file
- `hello_x86.asm` is the name of the assembly language source file

This is a portion of the `hello_x86.1st` listing file generated by the assembler:

```
.386
.model FLAT,C
.stack 400h
```

```
00000000                        .code
                                includelib libcmt.lib
                                includelib legacy_stdio_definitions.lib

                                extern printf:near
                                extern exit:near

                                public main
00000000                        main proc
                                    ; Print the message
00000000  68 00000000 R         push    offset message
00000005  E8 00000000 E         call    printf

                                    ; Exit the program with status 0
0000000A  6A 00                 push    0
0000000C  E8 00000000 E         call    exit
00000011                        main endp


00000000                        .data
00000000  48 65 6C 6C 6F        message db "Hello, Computer Architect!",0
          2C 20 43 6F 6D
          70 75 74 65 72
          20 41 72 63 68
          69 74 65 63 74
          21 00
```

This listing displays the address offsets from the beginning of the main function in the left column. On lines containing instructions, the opcode follows the address offset. Address references in the code (for example, offset message) are displayed as 00000000 in the listing because these values are determined during linking, and not during assembly, which is when this listing is generated.

This is the output displayed when running this program:

```
C:\>hello_x86.exe
Hello, Computer Architect!
```

Next, we will look at the extension of the 32-bit x86 architecture to the 64-bit x64 architecture.

x64 architecture and instruction set

The original specification for a processor architecture extending the x86 processor and instruction set to 64 bits, named **AMD64**, was introduced by AMD in 2000. The first AMD64 processor, the Opteron, was released in 2003. Intel found itself following AMD's lead and developed an AMD64-compatible architecture, eventually given the name **Intel 64**. The first Intel processor that implemented the 64-bit architecture was the Xeon, introduced in 2004. The name of the architecture shared by AMD and Intel came to be called **x86-64**, reflecting the evolution of x86 to 64 bits, and, in popular usage, this term has been shortened to **x64**.

The first Linux version supporting the x64 architecture was released in 2001, well before the first x64 processors were even available. Windows began supporting the x64 architecture in 2005.

Processors implementing the AMD64 and Intel 64 architectures are largely compatible at the instruction set level of user-mode programs. There are a few differences between the architectures, the most significant of which is the difference in support of the sysenter/sysexit Intel instructions and the syscall/sysret AMD instructions we saw earlier.

In general, operating systems and programming language compilers manage these differences, making them rarely an issue of concern to software and system developers. Developers of kernel software, drivers, and assembly code must take these differences into account.

The principal features of the x64 architecture are:

- x64 is a mostly compatible 64-bit extension of the 32-bit x86 architecture. Most software, particularly user-mode applications, written for the 32-bit environment, should execute without modification in a processor running in 64-bit mode. 64-bit mode is also referred to as **long mode**.
- The eight 32-bit general-purpose registers of x86 are extended to 64 bits in x64. The register name prefix R indicates 64-bit registers. For example, in x64, the extended x86 EAX register is called RAX. The x86 register subcomponents EAX, AX, AH, and AL continue to be available in x64.
- The instruction pointer, RIP, is now 64 bits. The flags register, RFLAGS, also extends to 64 bits, though the upper 32 bits are reserved. The lower 32 bits of RFLAGS are the same as EFLAGS in the x86 architecture.
- Eight 64-bit general-purpose registers have been added, named R8 through R15.
- 64-bit integers are supported as a native data type.

- x64 processors retain the option of running in x86 compatibility mode. This mode enables the use of 32-bit operating systems and allows any application built for x86 to run on x64 processors. In 32-bit compatibility mode, the 64-bit extensions are unavailable.

Virtual addresses in the x64 architecture are 64 bits wide, supporting an address space of 16 **exabytes (EB)**, equivalent to 2^{64} bytes. Current processors from AMD and Intel, however, support only 48 bits of virtual address space. This restriction reduces processor hardware complexity while still supporting up to 256 **terabytes (TB)** of virtual address space. Current-generation processors also support a maximum of 48 bits of physical address space. This permits a processor to address 256 TB of physical RAM, though modern motherboards do not support the number of DRAM devices such a system would require.

The x64 register set

In the x64 architecture, the extension of x86 register lengths to 64 bits and the addition of registers R8 through R15 results in the register map shown in *Figure 10.3*:

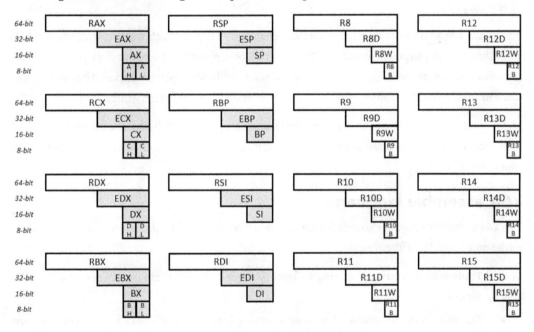

Figure 10.3: x64 registers

In *Figure 10.3*, the x86 registers described in the preceding section (and present in x64) are shaded. The x86 registers have the same names and are the same sizes when operating in 64-bit mode.

The 64-bit extended versions of the x86 registers have names starting with the letter **R**. The new 64-bit registers (R8 through R15) can be accessed in smaller widths using the appropriate suffix letter:

- Suffix **D** accesses the lower 32 bits of the register: R11D
- Suffix **W** accesses the lower 16 bits of the register: R11W
- Suffix **B** accesses the lower 8 bits of the register: R11B

Unlike the x86 registers, the new registers in the x64 architecture are truly general purpose and do not perform any special functions at the processor instruction level.

x64 instruction categories and formats

The x64 architecture implements essentially the same instruction set as x86, with 64-bit extensions. When operating in 64-bit mode, the x64 architecture uses a default address size of 64 bits and a default operand size of 32 bits. A new opcode prefix byte, rex, specifies the use of 64-bit operands.

The format of x64 instructions in memory matches that of the x86 architecture, with some exceptions that, for our purposes, are minor. The addition of support for the rex prefix byte is the most significant variation from the x86 instruction format. Address displacements and immediate values within some instructions can be 64 bits wide, in addition to all the bit widths supported in x86.

Although it is possible to define instructions that are longer than 15 bytes, the processor instruction decoder will raise a general protection fault if an attempt is made to decode an instruction longer than 15 bytes.

x64 assembly language

The x64 assembly language source file for the hello program is like the x86 version of this code, with some notable differences:

- There is no directive specifying a memory model because there is only one x64 memory model.
- The Windows x64 **application programming interface** (**API**) uses a calling convention that stores the first four arguments to a called function in the RCX, RDX, R8, and R9 registers, in that order. This differs from the default x86 calling convention, which pushes parameters onto the stack. Both library functions called by this program (printf and exit) take a single argument, passed in RCX.

- The calling convention requires the caller of a function to allocate stack space to hold at least the number of arguments passed to the called functions, with a minimum reservation space for four arguments, even if fewer are being passed. Because the stack grows downward in memory, this requires a subtraction from the stack pointer. The `sub rsp, 40` instruction performs this stack allocation. Normally, after the called function returns, it would be necessary to adjust the stack pointer to remove this allocation. Our program calls the `exit` function, terminating program execution, which makes this step unnecessary.

The code for the 64-bit version of the `hello` program is as follows:

```
.code
includelib libcmt.lib
includelib legacy_stdio_definitions.lib

extern printf:near
extern exit:near

public main
main proc
    ; Reserve stack space
    sub     rsp, 40

    ; Print the message
    lea     rcx, message
    call    printf

    ; Exit the program with status 0
    xor     rcx, rcx
    call    exit
main endp

.data
message db "Hello, Computer Architect!",0

end
```

This file, named hello_x64.asm, is assembled and linked to form the executable hello_x64.exe program with the following call to the Microsoft Macro Assembler (x64 version):

```
ml64 /Fl /Zi /Zd hello_x64.asm
```

The components of this command are:

- ml64 runs the 64-bit assembler
- /Fl creates a listing file
- /Zi includes symbolic debugging information in the executable file
- /Zd includes line number debugging information in the executable file
- hello_x64.asm is the name of the assembly language source file

This is a portion of the hello_x64.lst listing file generated by the assembler command:

```
00000000                            .code
                                    includelib libcmt.lib
                                    includelib legacy_stdio_definitions.lib

                                    extern printf:near
                                    extern exit:near

                                    public main
00000000                            main proc
                                    ; Reserve stack space
00000000  48/ 83 EC 28                   sub     rsp, 40

                                    ; Print the message
00000004  48/ 8D 0D                      lea     rcx, message
          00000000 R
0000000B  E8 00000000 E                  call    printf

                                    ; Exit the program with status 0
00000010  48/ 33 C9                      xor     rcx, rcx
00000013  E8 00000000 E                  call    exit
00000018                            main endp

00000000                            .data
00000000 48 65 6C 6C 6F             message db "Hello, Computer Architect!",0
```

```
        2C 20 43 6F 6D
        70 75 74 65 72
        20 41 72 63 68
        69 74 65 63 74
        21 00
```

The output of running this program is as follows:

```
C:\>hello_x64.exe
Hello, Computer Architect!
```

This completes our brief introduction to the x86 and x64 architectures. There is a great deal more to be learned, and indeed the *Intel 64 and IA-32 Architectures Software Developer's Manual, Volumes 1 through 4*, contains nearly 5,000 pages of detailed documentation on these architectures. We have just scratched the surface in this chapter.

Next, we will take a similar top-level tour of the ARM 32-bit and 64-bit architectures.

32-bit ARM architecture and instruction set

The ARM architectures define a family of RISC processors suitable for use in a wide variety of applications. Processors based on ARM architectures are preferred in designs where a combination of high performance, low power consumption, and small physical size is needed.

ARM Holdings, a British semiconductor and software company, developed the ARM architectures and licenses them to other companies who implement processors in silicon. Many applications of the ARM architectures are **system-on-chip** (**SoC**) designs combining a processor with specialized hardware to support functions such as cellular radio communications in smartphones.

ARM processors are employed in a broad spectrum of applications, from tiny battery-powered devices to supercomputers. ARM processors serve as embedded processors in safety-critical systems such as automotive anti-lock brakes and as general-purpose processors in smartwatches, portable phones, tablets, laptop computers, desktop computers, and servers. As of 2021, over 180 billion ARM processors have been manufactured.

ARM processors are true RISC systems with a large set of general-purpose registers and single-cycle execution of most instructions. Standard ARM instructions have a fixed width of 32 bits, though a separate variable-length instruction set named **T32** (formerly called **Thumb**) is available for applications where memory is at a premium. The T32 instruction set uses a mixture of 16- and 32-bit instructions.

Current-generation ARM processors support both the ARM and T32 instruction sets and can switch between the two sets on the fly. Most operating systems and applications prefer to use the T32 instruction set over the ARM set because code density is improved.

ARM is a **load/store architecture**, requiring data to be loaded from memory to a register before any processing such as an ALU operation can take place with it. A subsequent instruction stores the result back to memory. While this might seem like a step back from the x86 and x64 architectures, which operate directly on operands in memory in a single instruction, in practice, the load/store approach permits several sequential operations to be performed at high speed on an operand once it has been loaded into one of the many processor registers.

ARM processors are bi-endian. A configuration setting is available to select the little-endian or big-endian byte order for multi-byte values. The default setting is little-endian, which is the configuration commonly used by operating systems.

The ARM architecture natively supports these data types:

- **Byte**: 8 bits
- **Halfword**: 16 bits
- **Word**: 32 bits
- **Doubleword**: 64 bits

WHAT'S IN A WORD?

There is a potentially confusing difference between the data type names of the ARM architecture and those of the x86 and x64 architectures: in x86 and x64, a word is 16 bits and a doubleword is 32 bits. In ARM, a word is 32 bits and a doubleword is 64 bits.

ARM processors support eight distinct execution privilege levels. These levels, and their abbreviations, are as follows:

- User (USR)
- Supervisor (SVC)
- Fast interrupt request (FIQ)
- Interrupt request (IRQ)
- Monitor (MON)
- Abort (ABT)

- **Undefined (UND)**
- **System (SYS)**

For the purposes of operating systems and user applications, the most important privilege levels are USR and SVC. The two interrupt request modes, FIQ and IRQ, are used by device drivers for processing interrupts.

In most operating systems running on ARM, including Windows and Linux, the kernel mode runs in ARM SVC mode, equivalent to ring 0 on x86/64. ARM USR mode is equivalent to ring 3 on x86/x64. Applications running under Linux on ARM processors use software interrupts to request kernel services, which involves a transition from USR mode to SVC mode.

The ARM architecture provides system capabilities beyond those of the main processor via the concept of coprocessors. Each coprocessor implements a specialized category of functionality in support of the main processor. Up to 16 coprocessors can be implemented in a system, with predefined functions assigned to four of them.

Coprocessor 15 implements the MMU and other system functions. If present, coprocessor 15 must support the instruction opcodes, register set, and behaviors specified for the MMU. Coprocessors 10 and 11 combine to provide floating-point functionality in processors equipped with that feature. Coprocessor 14 provides debugging functions.

The ARM architectures have evolved through several versions over the years. The architectural variant currently in wide use is **ARMv8-A**. ARMv8-A supports 32-bit and 64-bit operating systems and applications. 32-bit applications can run under a 64-bit ARMv8-A operating system.

Virtually all high-end smartphones and portable electronic devices produced since 2016 are designed around processors or SoCs based on the ARMv8-A architecture. The description that follows will focus on ARMv8-A 32-bit mode. We will look at the differences in ARMv8-A 64-bit mode in a later section in this chapter.

The ARM register set

In USR mode, the ARM architecture has 16 general-purpose 32-bit registers named R0 through R15. The first 13 registers are truly general-purpose, while the last three have the following defined functions:

- R13 is the stack pointer, also named SP in assembly code. This register points to the top of the stack.

- R14 is the link register, also named LR. This register holds the return address while in a called function. The use of a link register differs from x86/x64, which pushes the return address onto the stack.

  The reason for using a register to hold the return address is because it is significantly faster to resume execution at the address in LR at the end of a function than it is to pop the return address from the stack and resume execution at that address.

- R15 is the program counter, also named PC. Due to pipelining, the value contained in PC is usually two instructions ahead of the currently executing instruction. Unlike x86/x64, it is possible for user code to directly read and write the PC register. Writing an address to PC causes execution to immediately jump to the newly written address.

The **current program status register (CPSR)** contains status and mode control bits, similar to EFLAGS/RFLAGS in the x86/x64 architectures.

Bit	Name	Function
0-3	M	Mode: The current execution privilege level (USR, SVC, and so on).
4	T	Thumb: Set if the T32 (Thumb) instruction set is active. If clear, the ARM instruction set is active. User code can set and clear this bit.
9	E	Endianness: Setting this bit enables big-endian mode. If clear, little-endian mode is active. Most code uses little-endian mode.
27	Q	Cumulative saturation flag: Set if, at some point in a series of operations, an overflow or saturation occurred.
28	V	Overflow flag: Set if the operation resulted in a signed overflow.
29	C	Carry flag: Indicates whether addition produced a carry, or subtraction produced a borrow.
30	Z	Zero flag: Set if the result of an operation is zero.
31	N	Negative flag: Set if the result of an operation is negative.

Table 10.3: Selected CPSR bits

CPSR bits not listed in *Table 10.3* are either reserved or represent functions not discussed in this chapter.

By default, most instructions do not affect the flags. The S suffix must be used with, for example, an addition instruction (adds) to cause the result to affect the flags. Comparison instructions are the exception to this rule; they update the flags automatically.

ARM addressing modes

In true RISC fashion, the only ARM instructions that can access system memory are those that perform register loads and stores.

The `ldr` instruction loads a register from memory, while `str` stores a register to memory. A separate instruction, `mov`, transfers the contents of one register to another or moves an immediate value into a register.

When computing the target address for a load or store operation, ARM starts with a base address provided in a register and adds an increment to arrive at the target memory address. There are three methods for determining the increment that will be added to the base register in register load and store instructions:

- **Offset:** A signed constant is added to the base register. The offset is stored as part of the instruction. For example, `ldr r0, [r1, #10]` loads r0 with the word at the address r1+10. As shown in the following addressing mode examples, pre- or post-indexing can optionally update the base register to the target address before or after the memory location is accessed.
- **Register:** An unsigned increment stored in a register can be added to or subtracted from the value in a base register. For example, `ldr r0, [r1, r2]` loads r0 with the word at the address r1+r2. Either of the registers can be thought of as the base register.
- **Scaled register:** An increment in a register is shifted left or right by a specified number of bit positions before being added to or subtracted from the base register value. For example, `ldr r0, [r1, r2, lsl #3]` loads r0 with the word at the address r1+(r2×8). The shift can be a logical left or right shift, `lsl` or `lsr`, inserting zero bits in the vacated bit positions, or an arithmetic right shift, `asr`, that replicates the sign bit in the vacated positions.

The addressing modes available for specifying source and destination operands in ARM instructions are presented in the following sections.

Immediate

An immediate value is provided as part of the instruction. The possible immediate values consist of an 8-bit value, coded in the instruction, rotated through an even number of bit positions. A full 32-bit value cannot be specified because the instruction itself is, at most, 32 bits wide. To load an arbitrary 32-bit value into a register, the `ldr` instruction must be used instead to load the value from memory:

```
mov r0, #10 // Load the 32-bit value 10 decimal into r0
mov r0, #0xFF000000 // Load the 32-bit value FF000000h into r0
```

The second example contains the 8-bit value FFh in the instruction opcode. During execution, it is rotated left by 24-bit positions into the most significant 8 bits of the word.

Register direct

This mode copies one register to another:

```
mov r0, r1 // Copy r1 to r0
mvn r0, r1 // Copy NOT(r1) to r0
```

Register indirect

The address of the operand is provided in a register. The register containing the address is surrounded by square brackets:

```
ldr r0, [r1] // Load the 32-bit value at the address given in r1 to r0
str r0, [r3] // Store r0 to the address in r3
```

Unlike most instructions, str uses the first operand as the source and the second as the destination.

Register indirect with offset

The address of the operand is computed by adding an offset to the base register:

```
ldr r0, [r1, #32] // Load r0 with the value at the address [r1+32]
str r0, [r1, #4] // Store r0 to the address [r1+4]
```

Register indirect with offset, pre-incremented

The address of the value is determined by adding an offset to the base register. The base register is updated to the computed address and this address is used to load the destination register:

```
ldr r0, [r1, #32]! // Load r0 with [r1+32] and update r1 to (r1+32)
str r0, [r1, #4]!  // Store r0 to [r1+4] and update r1 to (r1+4)
```

Register indirect with offset, post-incremented

The base address is first used to access the memory location. The base register is then updated to the computed address:

```
ldr r0, [r1], #32 // Load [r1] to r0, then update r1 to (r1+32)
str r0, [r1], #4  // Store r0 to [r1], then update r1 to (r1+4)
```

Double register indirect

The address of the operand is the sum of a base register and an increment register. The register names are surrounded by square brackets:

```
ldr r0, [r1, r2] // Load r0 with [r1+r2]
str r0, [r1, r2] // Store r0 to [r1+r2]
```

Double register indirect with scaling

The address of the operand is the sum of a base register and an increment register shifted left or right by the given number of bits. The register names and the shift information are surrounded by square brackets:

```
ldr r0, [r1, r2, lsl #5] // Load r0 with [r1+(r2*32)]
str r0, [r1, r2, lsr #2] // Store r0 to [r1+(r2/4)]
```

The next section introduces the general categories of ARM instructions.

ARM instruction categories

The instructions described in this section are from the T32 instruction set.

Load/store

These instructions move data between registers and memory:

- ldr, str: Copies an 8-bit (suffix b for byte), 16-bit (suffix h for halfword), or 32-bit value between a register and a memory location. ldr copies the value from memory to a register, while str copies a register to memory. ldrb copies 1 byte into the lower 8 bits of a register.
- ldm, stm: Loads or stores multiple registers. Copies 1 to 16 registers to or from memory. For example, the instruction ldm r1, {r0, r2, r4-r11} loads registers r0, r2, and r4 through r11 from contiguous memory beginning at the address provided in r1. Any subset of registers can be loaded from, or stored to, memory using these instructions.

Stack manipulation

These instructions store data to, and retrieve data from, the stack:

- push, pop: Pushes or pops any subset of the registers to or from the stack, for example, push {r0, r2, r4-r11}. These instructions are variants of the ldm and stm instructions.

Register movement

These instructions transfer data between registers:

- mov, mvn: Moves a register (mov), or its bit-inversion (mvn), to the destination register.

Arithmetic and logic

These instructions mostly have one destination register and two source operands. The first source operand is a register, while the second can be a register, a shifted register, or an immediate value.

Including the s suffix causes these instructions to set the condition flags. For example, adds performs addition and sets the condition flags:

- add, sub: Adds or subtracts two numbers. For example, add r0, r1, r2, lsl #3 is equivalent to the expression $r0 = r1 + (r2 \times 2^3)$. The lsl operator performs a logical shift left of the second operand, r2.
- adc, sbc: Adds or subtracts two numbers with carry or borrow.
- neg: Negates a number.
- and, orr, eor: Performs logical AND, OR, or XOR operations.
- orn, eon: Performs logical OR or XOR operations between the first operand and the bitwise-inverted second operand.
- bic: Clears selected bits in a register.
- mul: Multiplies two numbers.
- mla: Multiplies two numbers and accumulates the result. This instruction has an additional operand to specify the accumulator register.
- sdiv, udiv: Signed and unsigned division, respectively.

Comparisons

These instructions compare two values and set the condition flags based on the result of the comparison. The s suffix is not needed with these instructions to set the condition codes:

- cmp: Subtracts two numbers, discards the result, and sets the condition flags. This is equivalent to a subs instruction, except the result is discarded.
- cmn: Adds two numbers, discards the result, and sets the condition flags. This is equivalent to an adds instruction, except the result is discarded.
- tst: Performs a bitwise AND, discards the result, and sets the condition flags. This is equivalent to an ands instruction, except the result is discarded.

Control flow

These instructions transfer control conditionally or unconditionally to a target address:

- b: Performs an unconditional branch to the target address.
- b*cc*: Branches based on one of these condition codes as *cc*: eq (equal), ne (not equal), gt (greater than), lt (less than), ge (greater or equal), le (less or equal), cs (carry set), cc (carry clear), mi (minus: N flag = 1), pl (plus: N flag = 0), vs (V flag set), vc (V flag clear), hi (higher: C flag set and Z flag clear), or ls (lower or same: C flag clear and Z flag clear).
- bl: Branches to the specified address and stores the address of the next instruction in the link register (r14, also called lr). The called function returns to the calling code with the mov pc, lr instruction.
- bx: Branches and selects the instruction set. If bit 0 of the target address is 1, T32 mode is entered. If bit 0 is clear, ARM mode is entered. Bit 0 of instruction addresses must always be zero due to ARM's address alignment requirements. This frees bit 0 to select the instruction set.
- blx: Branches with a link and selects the instruction set. This instruction combines the functions of the bl and bx instructions.

Supervisor mode

This instruction allows user-mode code to initiate a call to supervisor mode:

- svc (**supervisor call**): Initiates a software interrupt that causes the supervisor mode exception handler to process a system service request.

Breakpoint

This instruction is used by debuggers during software development:

- bkpt (**trigger a breakpoint**): This instruction takes a 16-bit operand for use by debugging software to identify the breakpoint.

Conditional execution

Many ARM instructions support conditional execution, which uses the same condition codes as the branch instructions to determine whether individual instructions are executed. If an instruction's condition evaluates false, the instruction is processed as a no-op. The condition code is appended to the instruction mnemonic. This conditional execution mechanism is formally known as **predication**.

For example, this function converts a **nibble** (the lower 4 bits of a byte) into an ASCII character version of the nibble:

```
// Convert the low 4 bits of r0 to an ascii character in r0
nibble2ascii:
and r0, #0xF
cmp r0, #10
addpl r0, r0, #('A' - 10)
addmi r0, r0, #'0'
mov pc, lr
```

The `cmp` instruction subtracts 10 from the nibble in `r0` and sets the N flag if `r0` is less than 10. If `r0` is greater than or equal to 10, the N flag is clear.

If N is clear, the `addpl` instruction executes (`pl` means "plus," as in "not negative"), and the `addmi` instruction does not execute. If N is set, the `addpl` instruction does not execute and the `addmi` instruction executes. After this sequence completes, `r0` contains a character in the range 0-9 or A-F.

The use of conditional instruction execution helps keep the instruction pipeline flowing efficiently by avoiding branches.

Other instruction categories

ARM processors optionally support a range of SIMD and floating-point instructions. Additional instructions are provided that are generally only used during system configuration.

32-bit ARM assembly language

The ARM assembly example in this section uses the syntax of the GNU Assembler, provided with the Android Studio **integrated development environment** (**IDE**). Other assemblers may use a different syntax. As with the Intel syntax for the x86 and x64 assembly languages, the operand order for most instructions is the destination followed by the source.

The ARM assembly language source file for the `hello` program is as follows:

```
.text
.global _start

_start:
    mov     r0, #1          // int fd 1 (stdout)
    ldr     r1, =message    // const void *buf
```

```
        mov     r2, #count      // size_t count
        mov     r7, #4          // syscall 4 (sys_write)
        svc     0
        mov     r0, #0          // int status (0=OK)
        mov     r7, #1          // syscall 1 (sys_exit)
        svc     0

.data
message:
    .ascii      "Hello, Computer Architect!"
count = . - message
```

This file, named `hello_arm.s`, is assembled and linked to form the executable program `hello_arm` with the following commands. These commands use the development tools provided with the **Android Studio Native Development Kit (NDK)**. The commands assume the Windows PATH environment variable has been set to include the NDK tools directory:

```
arm-linux-androideabi-as -al=hello_arm.lst -o hello_arm.o hello_arm.s
arm-linux-androideabi-ld -o hello_arm hello_arm.o
```

The components of these commands are:

- `arm-linux-androideabi-as` runs the assembler
- `-al=hello_arm.lst` creates a listing file named `hello_arm.lst`
- `-o hello_arm.o` creates an object file named `hello_arm.o`
- `hello_arm.s` is the name of the assembly language source file
- `arm-linux-androideabi-ld` runs the linker
- `-o hello_arm` creates an executable file named `hello_arm`
- `hello_arm.o` is the name of the object file provided as input to the linker

This is a portion of the `hello_arm.lst` listing file generated by the assembler command:

```
1                   .text
2                   .global _start
3
4                   _start:
5 0000 0100A0E3         mov     r0, #1          // int fd 1 (stdout)
6 0004 14109FE5         ldr     r1, =message    // const void *buf
7 0008 1A20A0E3         mov     r2, #count      // size_t count
```

```
 8 000c 0470A0E3           mov      r7, #4        // syscall 4 (sys_write)
 9 0010 000000EF           svc      0
10
11 0014 0000A0E3           mov      r0, #0        // int status (0=OK)
12 0018 0170A0E3           mov      r7, #1        // syscall 1 (sys_exit)
13 001c 000000EF           svc      0
14
15                .data
16                message:
17 0000 48656C6C           .ascii    "Hello, Computer Architect!"
17      6F2C2043
17      6F6D7075
17      74657220
17      41726368
18                count = . - message
```

You can run this program on an Android device with **Developer options** enabled. We won't go into the procedure for enabling those options here, but you can learn more about that topic with an internet search.

This is the output displayed when running this program on an Android ARM device connected to the host PC with a USB cable:

```
C:\>adb push hello_arm /data/local/tmp/hello_arm
C:\>adb shell chmod +x /data/local/tmp/hello_arm
C:\>adb shell /data/local/tmp/hello_arm
Hello, Computer Architect!
```

These commands use the **Android Debug Bridge (adb)** tool included with Android Studio. Although the hello_arm program runs on the Android device, output from the program is sent back to the PC and appears in the command window.

The next section introduces the 64-bit ARM architecture, an extension of the 32-bit ARM architecture.

64-bit ARM architecture and instruction set

The 64-bit version of the ARM architecture, named **AArch64**, was announced in 2011. This architecture has 31 general-purpose 64-bit registers, 64-bit addressing, a 48-bit virtual address space, and a new instruction set named **A64**.

The 64-bit instruction set is a superset of the 32-bit instruction set, allowing existing 32-bit code to run unmodified on 64-bit processors.

Instructions are 32 bits wide, and most operands are 32 or 64 bits. The A64 register functions differ in some respects from 32-bit mode: the program counter is no longer directly accessible as a register and an additional register is provided that always returns an operand value of zero.

At the user privilege level, most A64 instructions have the same mnemonics as the corresponding 32-bit instructions. The assembler determines whether an instruction operates on 64-bit or 32-bit data based on the operands provided. The following rules determine the operand length and register size used by an instruction:

- 64-bit register names begin with the letter X; for example, x0
- 32-bit register names begin with the letter W; for example, w1
- 32-bit registers occupy the lower 32 bits of the corresponding 64-bit register number

When working with 32-bit registers, the following rules apply:

- Register operations such as right shifts behave the same as in the 32-bit architecture. A 32-bit arithmetic right shift uses bit 31 as the sign bit, not bit 63.
- Condition codes for 32-bit operations are set based on the result in the lower 32 bits.
- Writes to a W register set the upper 32 bits of the corresponding X register to zero.

The A64 is a load/store architecture with the same instruction mnemonics for memory operations (ldr and str) as 32-bit mode. There are some differences and limitations in comparison to the 32-bit load and store instructions:

- The base register must be an X (64-bit) register.
- An address offset can be any of the same types as in 32-bit mode, as well as an X register. A 32-bit offset can be zero-extended or sign-extended to 64 bits.
- Indexed addressing modes can only use immediate values as an offset.
- A64 does not support the ldm or stm instructions for loading or storing multiple registers in a single instruction. Instead, A64 adds the ldp and stp instructions for loading or storing a pair of registers in a single instruction.
- A64 only supports conditional execution for a small subset of instructions.

Stack operations are significantly different in A64. Perhaps the biggest difference in this area is that the stack pointer must maintain 16-byte alignment when accessing data.

64-bit ARM assembly language

This is the 64-bit ARM assembly language source file for the `hello` program:

```
.text
.global _start

_start:
    // Print the message to file 1 (stdout) with syscall 64
    mov     x0, #1
    ldr     x1, =msg
    mov     x2, #msg_len
    mov     x8, #64
    svc     0

    // Exit the program with syscall 93, returning status 0
    mov     x0, #0
    mov     x8, #93
    svc     0

.data
msg:
    .ascii      "Hello, Computer Architect!"
msg_len = . - msg
```

This file, named `hello_arm64.s`, is assembled and linked to form the executable `hello_arm64` program with the following commands. These commands use the 64-bit development tools provided with the Android Studio NDK. The use of these commands assumes the Windows PATH environment variable has been set to include the tools directory:

```
aarch64-linux-android-as -al=hello_arm64.lst -o hello_arm64.o ^
hello_arm64.s
aarch64-linux-android-ld -o hello_arm64 hello_arm64.o
```

The components of these commands are:

- `aarch64-linux-android-as` runs the assembler
- `-al=hello_arm64.lst` creates a listing file named `hello_arm64.lst`
- `-o hello_arm64.o` creates an object file named `hello_arm64.o`

- `hello_arm64.s` is the name of the assembly language source file
- `aarch64-linux-android-ld` runs the linker
- `-o hello_arm64` creates an executable file named `hello_arm64`
- `hello_arm64.o` is the name of the object file provided as input to the linker

This is a portion of the `hello_arm64.lst` listing file generated by the assembler:

```
    1                        .text
    2                        .global _start
    3
    4                    _start:
    5                            // Print the message to file 1 (stdout) with syscall 64
    6 0000 200080D2         mov     x0, #1
    7 0004 E1000058         ldr     x1, =msg
    8 0008 420380D2         mov     x2, #msg_len
    9 000c 080880D2         mov     x8, #64
   10 0010 010000D4         svc     0
   11
   12                            // Exit the program with syscall 93, returning status 0
   13 0014 000080D2         mov     x0, #0
   14 0018 A80B80D2         mov     x8, #93
   15 001c 010000D4         svc     0
   16
   17                        .data
   18                    msg:
   19 0000 48656C6C         .ascii      "Hello, Computer Architect!"
   19      6F2C2043
   19      6F6D7075
   19      74657220
   19      41726368
   20                        msg_len = . - msg
```

You can run this program on an Android device with **Developer options** enabled, as described earlier. This is the output displayed when running this program on an Android ARM device connected to the host PC with a USB cable:

```
C:\>adb push hello_arm64 /data/local/tmp/hello_arm64
```

```
C:\>adb shell chmod +x /data/local/tmp/hello_arm64
C:\>adb shell /data/local/tmp/hello_arm64
Hello, Computer Architect!
```

This completes our introduction to the 32-bit and 64-bit ARM architectures.

Summary

Having completed this chapter, you should have a good understanding of the high-level architectures and features of the x86, x64, 32-bit ARM, and 64-bit ARM registers, instruction sets, and assembly languages.

The x86 and x64 architectures represent a mostly CISC approach to processor design, using variable-length instructions that can take many cycles to execute, a lengthy pipeline, and (in x86) a limited number of processor registers.

The ARM architectures, on the other hand, implement RISC processors with mostly single-cycle instruction execution, a large register set, and (somewhat) fixed-length instructions. Early versions of ARM had pipelines as short as three stages, though later generations have considerably more stages.

Is one of these architectures better than the other, in a general sense? It may be that each is better in some ways, and system designers must make their selection of processor architecture based on the specific needs of the system under development. Of course, there is a great deal of inertia behind the use of x86/x64 processors in personal computing, business computing, and server applications. Similarly, there is much history behind the dominance of ARM processors in smart personal devices and embedded systems. Many factors beyond raw performance must be considered in the processor selection process when designing a new computer or smart device.

In the next chapter, we'll look at the RISC-V architecture. RISC-V was developed from a clean sheet, incorporating lessons learned from the history of processor development and without any of the baggage required to maintain support for decades-old legacy designs.

Exercises

1. Install the free Visual Studio Community edition, available at https://visualstudio.microsoft.com/vs/community/, on a Windows PC. Once installation is complete, open the Visual Studio IDE and select **Get Tools and Features...** under the **Tools** menu. Install the **Desktop development with C++** workload.

In the Windows search box in the Task bar, begin typing `Developer Command Prompt for VS 2022`. When the app appears in the search menu, select it to open Command Prompt.

Create a file named `hello_x86.asm` with the content shown in the source listing in the *x86 assembly language* section of this chapter.

Build the program using the command shown in the *x86 assembly language* section of this chapter and run it. Verify that the output **Hello, Computer Architect!** appears on the screen.

2. Write an x86 assembly language program that computes the following expression and prints the result as a hexadecimal number: $[(129 - 66) \times (445 + 136)] \div 3$. As part of this program, create a callable function to print 1 byte as two hex digits.

3. In the Windows search box in the Task bar, begin typing `x64 Native Tools Command Prompt for VS 2022`. When the app appears in the search menu, select it to open Command Prompt.

   Create a file named `hello_x64.asm` with the content shown in the source listing in the *x64 assembly language* section of this chapter.

   Build the program using the command shown in the *x64 assembly language* section of this chapter and run it. Verify that the output **Hello, Computer Architect!** appears on the screen.

4. Write an x64 assembly language program that computes the following expression and prints the result as a hexadecimal number: $[(129 - 66) \times (445 + 136)] \div 3$. As part of this program, create a callable function to print 1 byte as two hex digits.

5. Install the free Android Studio IDE, available at https://developer.android.com/studio/. Once installation is complete, open the Android Studio IDE, create a new project, and select **SDK Manager** under the **Tools** menu. Select the **SDK Tools** tab and check the **NDK** option, which may say **NDK (Side by side)**. Complete the installation of the NDK.

   Locate the following files under the SDK installation directory (the default location is under `%LOCALAPPDATA%\Android`) and add their directories to your `PATH` environment variable: `arm-linux-androideabi-as.exe` and `adb.exe`. Hint: The following command works for one version of Android Studio (your path may vary):

   ```
   set PATH=%PATH%;%LOCALAPPDATA%\Android\Sdk\ndk\23.0.7599858\toolchains\llvm\prebuilt\windows-x86_64\bin
   ```

Create a file named hello_arm.s with the content shown in the source listing in the *32-bit ARM assembly language* section of this chapter.

Build the program using the commands shown in the *32-bit ARM assembly language* section of this chapter.

Enable **Developer Options** on an Android phone or tablet. Search the internet for instructions on how to do this.

Connect your Android device to the computer with a USB cable.

Copy the program executable image to the phone using the commands shown in the *32-bit ARM assembly language* section of this chapter and run the program. Verify that the output **Hello, Computer Architect!** appears on the host computer screen.

Disable **Developer Options** on your Android phone or tablet.

6. Write a 32-bit ARM assembly language program that computes the following expression and prints the result as a hexadecimal number: $[(129 - 66) \times (445 + 136)] \div 3$. As part of this program, create a callable function to print 1 byte as two hex digits.

7. Locate the following files under the Android SDK installation directory (the default location is under %LOCALAPPDATA%\Android) and add their directories to your PATH environment variable: aarch64-linux-android-as.exe and adb.exe. Hint: The following command works for one version of Android Studio (your path may vary):

```
set PATH=%PATH%;%LOCALAPPDATA%\Android\Sdk\ndk\23.0.7599858\
toolchains\llvm\prebuilt\windows-x86_64\bin;%LOCALAPPDATA%\Android\
Sdk\platform-tools
```

Create a file named hello_arm64.s with the content shown in the source listing in the *64-bit ARM assembly language* section of this chapter.

Build the program using the commands shown in the *64-bit ARM assembly language* section of this chapter.

Enable **Developer Options** on an Android phone or tablet.

Connect your Android device to the computer with a USB cable.

Copy the program executable image to the phone using the commands shown in the *64-bit ARM assembly language* section of this chapter and run the program. Verify that the output **Hello, Computer Architect!** appears on the host computer screen.

Disable **Developer Options** on your Android phone or tablet.

8. Write a 64-bit ARM assembly language program that computes the following expression and prints the result as a hexadecimal number: $[(129 - 66) \times (445 + 136)] \div 3$. As part of this program, create a callable function to print 1 byte as two hex digits.

Join our community Discord space

Join the book's Discord workspace for a monthly *Ask me Anything* session with the author: https://discord.gg/7h8aNRhRuY

11
The RISC-V Architecture and Instruction Set

This chapter introduces the exciting, relatively new RISC-V (pronounced *risk five*) processor architecture and instruction set. RISC-V is a completely open-source specification for a reduced instruction set processor. Complete user-mode (non-privileged) and privileged instruction set specifications have been released and a wide variety of hardware implementations of this architecture are currently available. There are specifications for a number of instruction set extensions to support general-purpose computing, high-performance computing, and embedded applications. Commercially available processors implement many of these extensions.

The following topics will be covered in this chapter:

- The RISC-V architecture and applications
- The RISC-V base instruction set
- RISC-V extensions
- RISC-V variants
- 64-bit RISC-V
- Standard RISC-V configurations
- RISC-V assembly language
- Implementing RISC-V in a **field-programmable gate array (FPGA)**

After completing this chapter, you will understand the architecture and capabilities of the RISC-V processor and its optional extensions.

You will have learned the basics of the RISC-V instruction set and will understand how RISC-V can be tailored to support a variety of computer architectures, from low-end embedded systems to warehouse-scale cloud server farms. You will also see how to implement a RISC-V processor in a low-cost FPGA board.

Technical requirements

Files for this chapter, including answers to the exercises, are available at `https://github.com/PacktPublishing/Modern-Computer-Architecture-and-Organization-Second-Edition`.

The RISC-V architecture and applications

The RISC-V architecture, publicly announced in 2014, was developed at the University of California, Berkeley, by Yunsup Lee, Krste Asanović, David A. Patterson, and Andrew Waterman. This effort followed four previous major RISC architectural design projects at UC Berkeley, leading to the name RISC-V, where *V* represents the Roman numeral five.

The RISC-V project began as a clean sheet with several major goals:

- Design a RISC **instruction set architecture (ISA)** suitable for use across a wide spectrum of applications, from micro-power embedded devices to high-performance cloud server multiprocessors.

- Provide an ISA that is free to use by anyone, for any application. This contrasts with the ISAs of almost all other commercially available processors, which are the carefully guarded intellectual property of the company that owns them.

- Incorporate lessons learned from previous decades of processor design, avoiding wrong turns and suboptimal features that other architectures must retain in newer generations to maintain compatibility with previous, sometimes ancient in technological terms, generations.

- Provide a small but complete base architecture suitable for use in embedded devices. The base ISA is the minimal set of capabilities any RISC-V processor must implement. The base RISC-V is a 32-bit processor with 31 general-purpose registers.

All instructions are 32 bits long. The base ISA supports integer addition and subtraction but does not include integer multiplication and division. This avoids forcing minimal processor implementations to include relatively expensive multiplication and division hardware for applications that do not require those operations.

- Provide optional ISA extensions to support floating-point mathematics, atomic memory operations, and multiplication and division.
- Provide additional ISA extensions to support privileged execution modes, similar to the x86, x64, and ARM privileged implementations discussed in *Chapter 10, Modern Processor Architectures and Instruction Sets*.
- Support a compressed instruction set, implementing 16-bit versions of many 32-bit instructions. In processors implementing this extension, 16-bit instructions may be freely interspersed with 32-bit instructions.
- Provide optional ISA extensions to support 64-bit, and even 128-bit, processor word sizes and paged virtual memory on single- and multi-core processors and in multiprocessing configurations.

RISC-V processors are available on the market today at competitive prices and, given the sophistication of the ISA design and the advantages of its free-to-use nature, we can expect the market share of RISC-V processors to grow rapidly in the coming years. RISC-V Linux distributions are available, which include all the software development tools needed to build and run applications on RISC-V-based computers and smart devices.

Some applications where RISC-V processors have begun to make significant inroads are:

- Artificial intelligence and machine learning
- Embedded systems
- Extreme-scale computing
- Ultra-low-power processing
- Processing at the IoT edge

We will look at some implementations of the RISC-V architecture for these application areas later in this chapter.

Figure 11.1 presents the RISC-V base ISA register set.

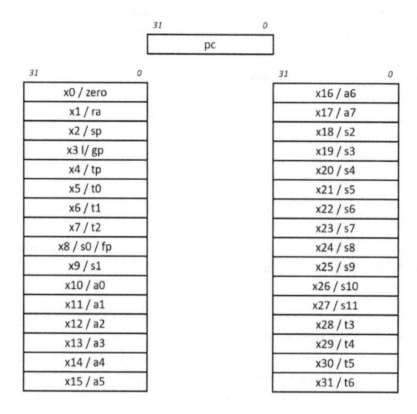

Figure 11.1: RISC-V base ISA register set

The registers are 32 bits wide. General-purpose registers x1 through x31 can be used without any restrictions and with no special functions assigned by the processor hardware. The x0 register is hardwired to return zero when it is read and will discard any value written to it. We will see some interesting uses of the x0 register shortly.

Each register has one or two alternate names, shown in *Figure 11.1*. These names correspond to the usage of registers in the standard RISC-V **application binary interface (ABI)**. The ABI is necessary because registers x1-x31 are functionally interchangeable and software compatibility requires that we specify which register should serve as the stack pointer, which registers should contain arguments to functions, which should contain return values, and so forth. The register designations are:

- **ra**: Function return address.
- **sp**: Stack pointer.

- **gp**: Global data pointer.
- **tp**: Thread-local data pointer.
- **t0-t6**: Temporary storage.
- **fp**: Frame pointer for function-local stack data (this usage is optional).
- **s0-s11**: Saved registers (if the frame pointer is not in use, x8 becomes s0).
- **a0-a7**: Arguments passed to functions. Any additional arguments are passed on the stack. Function return values are passed in a0 and a1.

The pc register contains the 32-bit program counter, holding the address of the current instruction.

You may be surprised to see that there are no processor flags registered in the RISC-V ISA. Some operations that modify flags in other processor architectures instead store their results in a RISC-V register. For example, the signed (slt) and unsigned (sltu) RISC-V comparison instructions subtract two operands and set a destination register to 0 or 1 depending on the sign of the result. A subsequent conditional branch instruction can use the value in that register to determine which code path to take.

Some of the flags found in other processors must be computed in RISC-V. For example, there is no carry flag. To determine whether an addition resulted in a carry, it is necessary to perform a comparison between the sum and one of the operands of the addition instruction. If the sum is greater than or equal to the addend (either addend can be used for the comparison), a carry did not occur; otherwise, the addition produced a carry.

Most of the base ISA computational instructions use a three-operand format in which the first operand is a destination register, the second operand is a source register, and the third operand is either a source register or an immediate value. This is an example of a three-operand instruction:

```
add x1, x2, x3
```

This instruction adds the x2 register to the x3 register and stores the result in the x1 register.

To avoid introducing instructions that are not strictly necessary, many instructions take on extra duties that are performed by dedicated instructions in other processor architectures. For example, RISC-V contains no instruction that simply moves one register to another. Instead, a RISC-V addition instruction adds a source register and an immediate value of zero and stores the result in a destination register, producing the same result. The instruction to transfer the x2 register to the x1 register is therefore add x1, x2, 0, assigning the value (x2 + 0) to x1.

The RISC-V assembly language provides several pseudo-instructions using familiar terminology to implement such functions. For example, the assembler translates the mv x1, x2 pseudo-instruction to a literal add x1, x2, 0 instruction.

The RISC-V base instruction set

The RISC-V base instruction set is composed of just 47 instructions. Eight are system instructions that perform system calls and access performance counters. The remaining 39 instructions fall into the categories of computational instructions, control flow instructions, and memory access instructions. We will examine each of these categories in turn.

Computational instructions

All the computational instructions except lui and auipc use the three-operand form. The first operand is the destination register, the second is a source register, and the third is either a second source register or an immediate value. Instruction mnemonics using an immediate value (except for auipc) end with the letter i. These are the instructions and their functions:

- add, addi, sub: Perform addition and subtraction. The immediate value in the addi instruction is a 12-bit signed value. The sub instruction subtracts the second source operand from the first. There is no subi instruction because addi can add a negative immediate value.
- sll, slli, srl, srli, sra, srai: Perform logical left and right shifts (sll and srl), and arithmetic right shifts (sra). Logical shifts insert zero bits into vacated locations. Arithmetic right shifts replicate the sign bit into vacated locations. The number of bit positions to shift is taken from the lowest 5 bits of the second source register or from the 5-bit immediate value.
- and, andi, or, ori, xor, xori: Perform the indicated bitwise operation on the two source operands. Immediate operands are 12 bits.
- slt, slti, sltu, sltui: The *set if less than* instructions set the destination register to 1 if the first source operand is less than the second source operand. This comparison is in terms of two's complement (slt) or unsigned (sltu) operands. Immediate operand values are 12 bits.
- lui: Load upper immediate. This instruction loads bits 12-31 of the destination register with a 20-bit immediate value. Setting a register to an arbitrary 32-bit immediate value requires two instructions: First, lui sets bits 12-31 to the upper 20 bits of the value. Then, addi adds in the lower 12 bits to form the complete 32-bit result. lui has two operands: the destination register and the immediate value.

- auipc: Add upper immediate to PC and store the result in the destination register. This instruction adds a 20-bit immediate value to the upper 20 bits of the program counter. This instruction enables PC-relative addressing in RISC-V.

    To form a complete 32-bit PC-relative address, auipc forms a partial result, then an addi instruction adds in the lower 12 bits.

Control flow instructions

The conditional branching instructions perform comparisons between two registers and, based on the result, may transfer control within the range of a signed 12-bit address offset from the current PC. Two unconditional jump instructions are available, one of which (jalr) provides access to the entire 32-bit address range:

- beq, bne, blt, bltu, bge, bgeu: Branch if equal (beq), not equal (bne), less than (blt), less than unsigned (bltu), greater or equal (bge), or greater or equal, unsigned (bgeu). These instructions perform the designated comparison between two registers and, if the condition is satisfied, transfer control to the address offset provided in the 12-bit signed immediate value.
- jal: Jump and link. Transfer control to the PC-relative address provided in the 20-bit signed immediate value and store the address of the next instruction (the return address) in the destination register.
- jalr: Jump and link, register. Compute the target address as the sum of the source register and a signed 12-bit immediate value, then jump to that address and store the address of the next instruction in the destination register. When preceded by the auipc instruction, the jalr instruction can perform a PC-relative jump anywhere in the 32-bit address space.

Memory access instructions

The memory access instructions transfer data between a register and a memory location. The first operand is the register to be loaded or stored. The second operand is a register containing a memory address. A signed 12-bit immediate value is added to the address in the register to produce the final address used for the load or store.

The load instructions perform sign extension for signed values or zero extension for unsigned values. The sign or zero extension operation fills in all 32 bits in the destination register when a smaller data value (a byte or halfword) is loaded.

Unsigned loads are specified by a trailing u in the mnemonic:

- lb, lbu, lh, lhu, lw: Load an 8-bit byte (lb), a 16-bit halfword (lh), or a 32-bit word (lw) into the destination register. For byte and halfword loads, the instruction will either sign-extend (lb and lh) or zero-extend (lbu and lhu) to fill the 32-bit destination register. For example, the lw x1, 16(x2) instruction loads the word at the address (x2 + 16) into register x1.
- sb, sh, sw: Store a byte (sb), halfword (sh), or word (sw) to a memory range matching the size of the data value.
- fence: Enforce memory access ordering in a multithreaded context. The purpose of this instruction is to ensure a coherent view of cached data across threads. This instruction takes two operands: The first specifies the types of memory accesses that must complete prior to the fence instruction. The second specifies the types of memory accesses controlled following the fence. The operation types ordered by this instruction are memory reads and writes (r and w) and I/O device inputs and outputs (i and o). For example, the fence rw, rw instruction will guarantee that all loads and stores involving memory addresses occurring before the fence instruction will complete before any subsequent memory loads or stores take place. This instruction ensures that any values present in processor caches are properly synchronized with memory or the I/O device.
- fence.i: This instruction ensures that any stores to instruction memory have completed before the fence.i instruction completes. This instruction is primarily useful in the context of self-modifying code.

System instructions

Of the eight system instructions, one invokes a system call, one initiates a debugger breakpoint, and the remaining six read and write system **control and status registers (CSRs)**. The CSR manipulation instructions read the current value of the selected CSR into a register, then update the CSR by either writing a new value, clearing selected bits, or setting selected bits. The source value for the CSR modification is provided in a register or as an immediate 5-bit value. CSRs are identified by a 12-bit address. Each CSR instruction performs the read and write of the CSR as an atomic operation:

- ecall: Invoke a system call. Registers used for passing parameters into and returning from the call are defined by the ABI, not by processor hardware.
- ebreak: Initiate a debugger breakpoint.

- `csrrw`, `csrrwi`, `csrrc`, `csrrci`, `csrrs`, `csrrsi`: Read the specified CSR into a destination register and either write a source operand value to the register (`csrrw`), clear any 1 bit in the source operand in the register (`csrrc`), or set any 1 bit in the source operand in the register (`csrrs`).

    These instructions take three operands: the first is the destination register receiving the value read from the CSR, the second is the CSR address, and the third is a source register or a 5-bit immediate value (i suffix).

Six CSRs are defined in the base RISC-V architecture, all read-only. To execute any of the CSR access instructions in read-only mode, the `x0` register must be provided as the third operand. These registers define three 64-bit performance counters:

- `cycle`, `cycleh`: The lower (`cycle`) and upper (`cycleh`) 32 bits of the 64-bit count of elapsed system clock cycles since a reference time—typically, system startup. The frequency of the system clock may vary if **dynamic voltage and frequency scaling (DVFS)** is active.
- `time`, `timeh`: These are the lower (`time`) and upper (`timeh`) 32 bits of the 64-bit count of elapsed fixed-frequency real-time clock cycles since a reference time—typically, system startup.
- `instret`, `instreth`: The lower (`instret`) and upper (`instreth`) 32 bits of the 64-bit count of processor instructions retired. Retired instructions are those that have completed execution.

The two 32-bit halves of each performance counter cannot be read in a single atomic operation. To prevent erroneous readings, the following procedure must be used to reliably read each of the 64-bit counters:

1. Read the upper 32 bits of the counter into a register.
2. Read the lower 32 bits of the counter into another register.
3. Read the upper 32 bits into yet another register.
4. Compare the first and second reads of the upper 32 counter bits. If they differ, jump back to *step 1*.

This procedure will read a valid count value, even though the counter continues to run between the reads. In general, execution of this sequence should require, at most, one backward jump in *step 4*.

Pseudo-instructions

The RISC-V architecture has a truly reduced instruction set, lacking several types of instructions present in the instruction sets we have investigated in earlier chapters. The functions of many of those more familiar instructions can be performed with RISC-V instructions, though perhaps not in an immediately intuitive manner.

The RISC-V assembler supports several pseudo-instructions, each of which translates to one or more RISC-V instructions providing a type of functionality one might expect in a general-purpose processor instruction set. *Table 11.1* presents a few of the most useful RISC-V pseudo-instructions.

Pseudo-instruction	RISC-V instruction(s)	Function
nop	addi x0, x0, 0	No operation
mv rd, rs	addi rd, rs, 0	Copy rs to rd
not rd, rs	xori rd, rs, -1	rd = NOT rs
neg rd, rs	sub rd, x0, rs	rd = -rs
j offset	jal x0, offset	Unconditional jump
jal offset	jal x1, offset	Near function call (20-bit offset)
call offset	auipc x1, offset [31:12] + offset [11] jalr x1, offset [11:0] (x1)	Far function call (32-bit offset)
ret	jalr x0, 0 (x1)	Return from function
beqz rs, offset	beq rs, x0, offset	Branch if equal to zero
bgez rs, offset	bge rx, x0, offset	Branch if greater than or equal to zero
bltz rs, rt, offset	blt rs, x0, offset	Branch if less than zero
bgt rs, rt, offset	blt rt, rs, offset	Branch if greater than
ble rs, rt, offset	bge rt, rs, offset	Branch if less than or equal
fence	fence iorw, iorw	Fence all memory and I/O accesses
csrr rd, csr	csrrw rd, csr, x0	Read a CSR
li rd, immed	addi rd, x0, immed	Load 12-bit immediate

li rd, immed	lui rd, immed [31:12] + immed [11] addi rd, x0, immed [11:0]	Load 32-bit immediate
la rd, symbol	auipc rd, delta [31:12] + delta [11] addi rd, rd, delta [11:0]	Load the address of symbol, where delta=(symbol-pc)
lw rd, symbol	auipc rd, delta [31:12] + delta [11] lw rd, rd, delta [11:0] (rd)	Load word at symbol, where delta=(symbol-pc)
sw rd, symbol, rt	auipc rt, delta [31:12] + delta [11] sw rd, rd, delta [11:0] (rt)	Store word at symbol, where delta=(symbol-pc)

Table 11.1: RISC-V pseudo-instructions

In *Table 11.1*, rd is the destination register, rs is the source register, csr is a control and status register, symbol is an absolute data address, and offset is a PC-relative instruction address.

Instructions combining the upper 20 bits of an address or immediate value with an immediate value containing the lower 12 bits must perform a step to reverse the effect of the sign extension of bit 11 of the lower 12-bit value in the second instruction of each sequence. This is necessary because the immediate value in the addi instruction is always treated as signed. The most significant bit of the 12-bit immediate value will be sign-extended through bit 31 before being added to the upper 20 bits.

The following example demonstrates the problem and the solution. Assume we want to load the value 0xFFFFFFFF into a register using lui and addi and naively add the upper and lower portions, as shown here:

```
lui x1, 0xFFFFF # x1 now equals 0xFFFFF000
addi x1, x1, 0xFFF
```

One point to note regarding assembly code in this chapter is that the RISC-V assembler uses the # character to begin a comment.

The `addi` instruction sign-extends 0xFFF to 0xFFFFFFFF before adding it to 0xFFFFF000. The result of the addition is then 0xFFFFEFFF, which is not what we want. Adding bit 11 of the lower 12 bits to the upper 20 bits will fix this, as shown in the following code block:

```
lui x1, 0xFFFFF+1 # Add bit 11; x1 now equals 0x00000000
addi x1, x1, 0xFFF
```

The result is now 0xFFFFFFFF, the correct value. This procedure will work for any other numeric value. If bit 11 happens to be 0, nothing will be added to the upper 20 bits.

Privilege levels

The RISC-V architecture defines three privilege levels at which a thread can run:

- User (U) privilege level
- Supervisor (S) privilege level
- Machine (M) privilege level

All RISC-V implementations must support M mode, the most privileged level, which can access all system features. M mode is active following system reset. The code in a simple embedded system can run entirely in M mode.

In a slightly more sophisticated use case, a secure boot process might run at the M privilege level, loading, verifying, and starting an application that runs in U mode. This approach is appropriate for a secure embedded solution.

In addition to the mandatory M level, a RISC-V processor may implement U or both S and U privilege levels. A system running a general-purpose operating system uses S mode and U mode in the same manner as the kernel and user modes of the processors and operating systems discussed in previous chapters.

RISC-V U-mode applications request system services with the `ecall` (environment call) instruction, generating an exception handled at the S level. The privilege architecture of RISC-V directly supports modern operating systems such as Linux.

Separate collections of CSRs are defined to enable configuration, control, and monitoring of the system at each of the three privilege levels. Depending on the privilege level of the running thread and the level of a CSR, the thread may have read-write, read-only, or no access to the CSR. Threads at higher privilege levels can access CSRs at lower privilege levels.

The RISC-V S privilege level supports paged virtual memory with a 32-bit address space divided into 4 KB pages. A 32-bit virtual address is separated into a 20-bit virtual page number and a 12-bit page offset.

Two additional virtual memory configurations are defined for the RISC-V 64-bit environment. The first is a 39-bit address space, supporting 512 GB of virtual memory.

For applications requiring even more virtual addresses, a 48-bit address space is available, supporting 256 TB of virtual memory. Although the 48-bit configuration offers far more memory than the 39-bit configuration, it also requires additional storage for page tables and consumes more processing time traversing those tables.

The following instructions support privileged execution levels:

- `mret, sret, uret`: These instructions return from the exception handler initiated by an `ecall` instruction. Each of these instructions can be executed at the privilege level indicated by the first letter of the instruction or higher. Executing one of these instructions referencing a privilege level lower than that of the current thread will return from the exception initiated at the lower level.
- `wfi`: Wait for interrupt. This instruction requests the current thread to stall until an interrupt becomes available for servicing. The RISC-V specification only requires that this instruction serves as a hint, so a particular implementation may process a `wfi` instruction as a no-op rather than stalling the thread. Because it is possible for the processor to handle `wfi` as a no-op, the code that follows a `wfi` instruction must explicitly check for the presence of pending interrupts in need of processing. This sequence typically occurs inside a loop.
- `sfence.vma`: Flush virtual memory page table data from cache to memory. The leading s in the instruction mnemonic indicates that this instruction is targeted for use at the supervisor privilege level.

RISC-V defines additional instructions and CSRs supporting virtualization and the hypervisor that manages the virtual environment. RISC-V virtualization will be covered in *Chapter 12, Processor Virtualization*.

RISC-V extensions

The instruction set described previously is named **RV32I**, which stands for the **RISC-V 32-bit integer instruction set**. Although the RV32I ISA provides a complete and useful instruction set for many purposes, it lacks several functions and features available in other popular processors such as x86 and ARM.

The RISC-V extensions provide a mechanism for adding capabilities to the base instruction set in an incremental and compatible manner. Implementors of RISC-V processors can selectively include extensions in their design to optimize trade-offs between chip size, system capability, and performance.

These flexible design options are also available to developers of low-cost FPGA-based systems. We'll see more about implementing a RISC-V processor in an FPGA later in this chapter. The major extensions we will cover now are named M, A, C, F, and D. We'll also mention some other available extensions.

The M extension

The RISC-V M extension adds integer multiplication and division functionality to the base RV32I instruction set. The following instructions are included in this extension:

- mul: Multiply two 32-bit registers and store the lower 32 bits of the result in the destination register.
- mulh, mulhu, mulhsu: Multiply two 32-bit registers and store the upper 32 bits of the result in the destination register. Treat the multiplicands as both signed (mulh), both unsigned (mulhu), or signed rs1 times unsigned rs2 (mulhsu). rs1 is the first source register in the instruction and rs2 is the second.
- div, divu: Perform division of two 32-bit registers, rounding the result toward zero, on signed (div) or unsigned (divu) operands.
- rem, remu: Return the remainder corresponding to the result of a div or divu instruction on the operands.

Division by zero does not raise an exception. To detect division by zero, code should test the divisor and branch to an appropriate handler if it is zero.

The A extension

The RISC-V A extension provides atomic read-modify-write operations to support multithreaded processing in shared memory.

The load-reserved (lr.w) and store-conditional (sc.w) instructions work together to perform a memory read followed by a write to the same location in an atomic sequence. The load-reserved instruction places a reservation on the memory address during the load. If another thread writes to the same location while the reservation is in effect, the reservation is canceled.

When the store-conditional instruction executes, it returns a value indicating if it successfully completed the atomic operation. If the reservation remains valid (in other words, no intervening write occurred to the target address), the store-conditional instruction writes the register to memory and returns zero, indicating success.

If the reservation was canceled, the store-conditional instruction does not alter the memory location and returns a nonzero value indicating that the store operation failed. The following instructions implement the load-reserved and store-conditional operations:

- `lr.w`: Load a register from a memory location and place a reservation on the address.
- `sc.w`: Store a register to a memory location conditionally. Set the destination register to zero if the operation succeeded and the memory location was written, or set the destination register to a nonzero value if the reservation was canceled. If the reservation was canceled, the memory location is not modified by this instruction.

The **atomic memory operation (AMO)** instructions atomically load a word from a memory location into the destination register, perform a binary operation between the value that was loaded and `rs2`, and store the result back to the memory address. The following instructions implement the AMO operations:

- `amoswap.w`: Atomically swap `rs2` into the `rs1` memory location.
- `amoadd.w`: Atomically add `rs2` into the `rs1` memory location.
- `amoand.w`, `amoor.w`, `amoxor.w`: Atomically perform AND, OR, or XOR operations with `rs2` into the `rs1` memory location.
- `amomin.w`, `amominu.w`, `amomax.w`, `amomaxu.w`: Atomically perform minimum or maximum selection of signed or unsigned (instructions with the u suffix) values with `rs2` into the `rs1` memory location.

The C extension

The RISC-V C extension implements compressed instructions with the goals of minimizing the amount of memory consumed by instruction storage and reducing the amount of bus traffic required to fetch instructions.

All RV32I instructions discussed previously are 32 bits in length. The C extension provides alternate 16-bit representations of many of the most frequently used RV32I instructions. Each compressed instruction is equivalent to one full-length instruction. No mode switching is necessary, meaning programs can freely intermix 32-bit RV32I instructions and compressed 16-bit instructions.

In fact, assembly language programmers do not even need to take steps to specify whether an instruction should be generated in compressed form. The assembler and linker are capable of transparently emitting compressed instructions where possible to minimize code size, in most cases with no execution performance penalty.

When working with processors and software development toolsets supporting the RISC-V C extension, the benefits of compressed instructions are immediately available to developers working in assembly language as well as to those working with higher-level languages.

The F and D extensions

The RISC-V F and D extensions provide hardware support for single-precision (F) and double-precision (D) floating-point arithmetic in accordance with the IEEE 754 standard. The F extension adds 32 floating-point registers named f0-f31 and a control and status register named fcsr to the architecture. These registers are all 32 bits. This extension includes a set of floating-point instructions that complies with the IEEE 754-2008 single-precision standard.

Most floating-point instructions operate on the floating-point registers. Data transfer instructions are provided to load floating-point registers from memory, store floating-point registers to memory, and move data between floating-point registers and integer registers.

The D extension widens f0-f31 to 64 bits. In this configuration, each f register can hold a 32-bit value or a 64-bit value. Double-precision floating-point instructions are added, in compliance with the IEEE 754-2008 double-precision standard. The D extension requires the F extension to be present.

Other extensions

Several additional extensions to the RISC-V architecture, detailed in the following list, have been defined, are in development, or are at least under consideration for future development:

- **RV32E architecture**: This is not actually an extension; rather, it is a modified architecture intended to reduce processor hardware requirements below those of the RV32I instruction set for the smallest embedded systems. The only difference between RV32I and RV32E is the reduction in the number of integer registers to 15. This change is expected to reduce processor die area and power consumption by about 25% compared to an otherwise equivalent RV23I processor. x0 remains a dedicated zero register. Halving the number of registers frees up 1 bit in each register specifier in an instruction. These bits are guaranteed to remain unused in future revisions and are thus available for use in custom instruction extensions.

- **Q extension:** The Q extension supports 128-bit quad-precision floating-point mathematics, as defined in the IEEE 754-2008 standard.
- **L extension:** The L extension supports decimal floating-point arithmetic, as defined in the IEEE 754-2008 standard.
- **B extension:** The B extension supports bit manipulations such as inserting, extracting, and testing individual bits.
- **J extension:** The J extension supports dynamically translated languages such as Java and JavaScript.
- **T extension:** The T extension supports memory transactions composed of atomic operations across multiple addresses.
- **P extension:** The P extension provides packed **single instruction, multiple data (SIMD)** instructions for floating-point operations in small RISC-V systems.
- **V extension:** The V extension supports data-parallel, or vector, operations. The V extension does not specify the lengths of data vectors; that decision is left to the implementers of a RISC-V processor design. A typical implementation of the V extension might support 512-bit data vectors, though implementations with up to 4,096-bit vector lengths are currently available.
- **N extension:** The N extension provides support for handling interrupts and exceptions at the U privilege level.
- **Zicsr extension:** The Zicsr extension performs atomic read-modify-write operations on the system CSRs. These instructions are described earlier in this chapter in the *System instructions* section.
- **Zifencei extension:** The Zifencei extension defines the fence.i instruction, described in the *Memory access instructions* section.

The next section covers some of the currently available RISC-V processor variants.

RISC-V variants

Some of the wide range of applications where variants of the RISC-V architecture are gaining significant usage are:

- **Artificial intelligence and machine learning:** Esperanto Technologies has developed a **system-on-chip (SoC)** containing over 1,000 RISC-V processors and placed 6 of these chips on a single PCIe card. This design is optimized for high-performance machine learning recommendation workloads in large data centers while consuming minimum power.

- **Embedded systems**: The Efinix VexRiscv core is a soft CPU intended for implementation in an FPGA. The VexRiscv uses the RV32I ISA with M and C extensions.

  A complete SoC implementation includes the CPU, memory, and a selectable set of I/O interfaces such as general-purpose I/O, timers, serial interfaces, and chip-to-chip communication interfaces such as **Serial Peripheral Interface (SPI)**.

  Extreme-scale computing: Several efforts are underway to develop **High-Performance Computing (HPC)** systems, commonly known as supercomputers, based on RISC-V processors. Specific goals of these projects include increasing the energy efficiency of the processor cores and enhancing the vector processing capabilities implemented in processor hardware.

- **Ultra-low power processing**: Micro Magic, Inc. offers a 64-bit RISC-V core with a claimed power consumption of only 0.01 Watt when running at a 1 GHz clock frequency. This rate of energy usage will allow battery-powered devices to operate for days, weeks, or months on a single charge.

  Processing at the IoT edge: Smart devices operating at the so-called **internet of things (IoT)** edge may have high processing loads for tasks such as artificial intelligence inference and data encryption while being tightly constrained in terms of power consumption by factors such as battery lifetime and limited cooling capabilities. Some examples of IoT devices are doorbell cameras, voice command processors such as Amazon Alexa, and Wi-Fi-enabled power outlets. As in the previous examples, the goals for these processor designs are focused on maximum processing performance while limiting power consumption to an acceptable level.

These are just a few of the application areas where RISC-V is making inroads. We can expect the growth of RISC-V usage in these and other areas to continue.

The next section covers the extension of the base RISC-V ISA to 64 bits.

64-bit RISC-V

This chapter has discussed the 32-bit RV32I architecture and instruction set and several important extensions. The RV64I instruction set expands RV32I to a 64-bit architecture. As in RV32I, instructions are 32 bits wide. In fact, the RV64I instruction set is almost entirely the same as RV32I, except for these significant differences:

- Integer registers are widened to 64 bits.
- Addresses are widened to 64 bits.
- Bit shift counts in instruction opcodes increase in size from 5 to 6 bits.

- Several new instructions are provided to operate on 32-bit values in a manner equivalent to RV32I. These instructions are necessary because most instructions in RV64I operate on 64-bit values and there are many situations in which it is necessary to operate efficiently on 32-bit values. These word-oriented instructions have an opcode mnemonic suffix of W. The W-suffix instructions produce signed 32-bit results. These 32-bit values are sign-extended (even if they are unsigned values) to fill the 64-bit destination register. In other words, bit 31 of each result is copied into bits 32-63.

The following new instructions are defined in RV64I:

- `addw, addiw, subw, sllw, slliw, srlw, srliw, sraw, sraiw`: These instructions perform almost identically to the RV32I instruction with the same mnemonic minus the W suffix. They work with 32-bit operands and produce 32-bit results. The result is sign-extended to 64 bits.
- `ld, sd`: Load and store a 64-bit **doubleword**. These are the 64-bit versions of the `lw` and `sw` instructions in the RV32I instruction set.

The remaining RV32I instructions perform the same functions in RV64I, except addresses and registers are 64 bits in length. The same opcodes, both in assembly source code and in binary machine code, are used in both instruction sets.

In the next section, we will examine some standard 32-bit and 64-bit RISC-V configurations that are commercially available. Each of these consists of a base ISA plus selected extensions.

Standard RISC-V configurations

The RV32I and RV64I instruction sets provide a base set of capabilities that is useful mainly in smaller embedded system designs. Systems intended to support multithreading, multiple privilege levels, and general-purpose operating systems require several of the RISC-V extensions to operate correctly and efficiently.

The minimum RISC-V configuration recommended for establishing an application development target consists of a base RV32I or RV64I instruction set architecture augmented with the I, M, A, F, D, Zicsr, and Zifencei extensions. The abbreviation for this combination of features is G, as in RV32G or RV64G. Many G configurations additionally support the compressed instruction extension, with the names RV32GC and RV64GC.

In embedded applications, a common configuration is RV32IMAC, providing the base instruction set plus multiply/divide functionality, atomic operations, and compressed instruction support.

Marketing materials for RISC-V processors frequently use these shorthand descriptions of processor capabilities.

The following section presents a complete program in RISC-V assembly language.

RISC-V assembly language

The following RISC-V assembly language example is a complete application that runs on a RISC-V processor:

```
.section .text
.global main

main:
    # Reserve stack space and save the return address
    addi    sp, sp, -16
    sd      ra, 0(sp)

    # Print the message using the C library puts function
1:  auipc   a0, %pcrel_hi(msg)
    addi    a0, a0, %pcrel_lo(1b)
    jal     ra, puts

    # Restore the return address and sp, and return to caller
    ld      ra, 0(sp)
    addi    sp, sp, 16
    jalr    zero, ra, 0

.section .rodata
msg:
    .asciz "Hello, Computer Architect!\n"
```

This program prints the following message in a console window and then exits:

```
Hello, Computer Architect!
```

The following are some points of interest within the assembly code:

- The %pcrel_hi and %pcrel_lo directives select the high 20 bits (%pcrel_hi) or low 12 bits (%pcrel_lo) of the PC-relative address of the label provided as an argument. The combination of the auipc and addi instructions places the address of the message string in a0.

- 1: is a local label. When referencing a local label, the letter b is appended to reference a label earlier in the code (backward), or f is appended to reference a label later in the code (forward). The %pcrel_hi and %pcrel_lo directives are paired: the 1: local label resolves the lower 12 bits of the offset to the msg address.

In the next section, we will run some code in a fully functional RISC-V processor implemented in an FPGA.

Implementing RISC-V in an FPGA

All the source code, processor hardware design intellectual property, and development tools required to build and implement a complete RISC-V processor in a low-cost FPGA are freely available on the internet. This section provides a high-level overview of the open-source RISC-V design and the steps for bringing it up in an FPGA device. The total cost for the hardware to accomplish this task is less than US$200.

The RISC-V FPGA target in this example is the Digilent Arty A7-35T board, available at https://store.digilentinc.com/arty-a7-artix-7-fpga-development-board-for-makers-and-hobbyists/. The Arty A7-35T costs US$129 at the time of writing.

The Arty A7-35T contains a Xilinx Artix-7 XC7A35TICSG324-1L FPGA, which can be programmed to implement a RISC-V processor. The XC7A35TICSG324-1L has the following features:

- 5,200 logic slices.
- 1,600 of the logic slices can implement a 64-bit RAM.
- 41,600 flip-flops. Each logic slice contains eight flip-flops.
- 90 DSP slices supporting high-performance DSP MAC operations.
- 400 kbits of distributed RAM.
- 1,800 kbits of total RAM.

The Artix-7 FPGA architecture uses **lookup tables (LUTs)** to implement combinational logic. Each of the Artix-7 LUTs has six input signals and one output signal, where each signal is one data bit. A single LUT can represent any feedback-free circuit composed of AND, OR, NOT, and XOR gates operating on the six input signals by simply storing the results of each input combination as a bit in a small ROM. With six input bits, the ROM contains 64 (2^6) bits of data addressed by the six input signals. If desired, each LUT can instead be configured as two 32-bit LUTs operating on five shared inputs with 2 output bits. Optionally, the LUT output can be stored in a flip-flop.

A *logic slice* contains four LUTs and eight flip-flops plus additional multiplexer and arithmetic carry logic. Four of the eight flip-flops in a slice can be configured as latches. Each of the 1,600 slices with 64-bit RAM capability can instead implement a 32-bit shift register or two 16-bit shift registers.

The low-level LUTs and other facilities provided by the several thousand logic slices represent the raw materials needed to assemble a complete RISC-V processor plus peripheral devices within a single FPGA. The FPGA programming process interconnects components within the FPGA to form a complex digital device defined in a hardware definition language.

From the perspective of the system designer, it is not necessary to understand the detailed inner workings of the Xilinx FPGA. The designer works at the hardware design language level. A tool such as Vivado, introduced in the solutions to the exercises in *Chapter 2, Digital Logic*, translates the hardware design language (typically, VHDL or Verilog, though the original RISC-V design is implemented in the Chisel and Scala languages) into a compiled format suitable for programming an FPGA device.

The designer's primary concerns regarding the FPGA are to ensure that the system design is capable of fitting within the resource constraints of the FPGA device and that the resulting implementation operates with acceptable performance. In this example, the XC7A35TICSG324-1L FPGA provides more than enough resources to implement the RISC-V processor.

To develop and run programs on the Arty A7-35T RISC-V processor, you also need a low-cost hardware debugger. The Olimex ARM-TINY-USB-H debugger is available for US$47.65 at `https://www.digikey.com/product-detail/en/olimex-ltd/ARM-USB-TINY-H/1188-1013-ND/3471388`. You will also need some jumper wires to connect the debugger to the Arty A7-35T board. These are available for US$3.95 at `https://www.adafruit.com/product/826`. Finally, the Arty A7-35T processor requires a USB cable to connect its USB Micro-B connector to your host computer system. All the software and design data required to implement the RISC-V in Arty is available for free download from the internet.

The processor we will implement in the Arty A7-35T is the *Freedom E310 Arty*, an open-source implementation of an RV32IMAC core with support for interrupt processing. Peripheral devices include 16 **general-purpose I/O (GPIO)** signals and a serial port.

The Freedom E310 processor is provided as source code and is therefore modifiable by users who wish to implement customized versions of the processor. The hardware design languages used in the RISC-V processor hardware code are Chisel and Scala.

Chisel is a domain-specific language targeted at the development of complex digital hardware devices such as SoCs. Chisel runs on top of **Scala**, a modern, general-purpose programming language supporting the functional and object-oriented programming paradigms. Scala is a pure object-oriented language in which every value is an object. It is also a functional language in the sense that every function is a value. Scala compiles to Java bytecode and runs on the standard Java Virtual Machine. Scala programs can directly use any of the thousands of available Java libraries.

RISC-V SUPPORT FOR CUSTOMIZED EXTENSIONS

The RISC-V architecture explicitly supports user-defined variations in the form of custom opcodes, coprocessors, and other modifications, so long as they are compatible with the RISC-V customization rules. Starting from the open-source RISC-V design, you can implement custom modifications that will be guaranteed to remain compatible with future versions of RISC-V standards and extensions.

Chisel and Scala are preferred in the design of some categories of complex digital systems today because of the higher-level nature of these languages compared to traditional hardware design languages such as VHDL and Verilog. While it's true that any circuit you might design in Chisel can also be designed in VHDL, there are some substantial benefits to using Chisel. For example, the compilation process transforms the Chisel/Scala code into a form called **Flexible Intermediate Representation for RTL (FIRRTL)**, where **RTL** stands for **register-transfer level**. RTL is the abstraction level used in synchronous circuit hardware design languages such as VHDL. Using freely available tools, it is possible to perform optimizations on the FIRRTL representation of a circuit that result in a better-performing FPGA implementation than a comparable design in VHDL or Verilog would likely provide.

One way to appreciate the difference between Chisel and VHDL/Verilog is the analogous differentiation between the Python and C programming languages. While you can implement the functional equivalent of any Python program in C, Python programs can express far more high-level functionality in a few lines of code than a similar size program in C.

We can compare Chisel code to the VHDL example we looked at in the *Hardware description languages* section of *Chapter 2, Digital Logic*. Consider the VHDL version of the single-bit full adder presented in that chapter, shown in the following code:

```
-- Load the standard libraries

library IEEE;
```

```vhdl
    use IEEE.STD_LOGIC_1164.ALL;

-- Define the full adder inputs and outputs

entity FULL_ADDER is
  port (
    A     : in   std_logic;
    B     : in   std_logic;
    C_IN  : in   std_logic;
    S     : out  std_logic;
    C_OUT : out  std_logic
  );
end entity FULL_ADDER;

-- Define the behavior of the full adder

architecture BEHAVIORAL of FULL_ADDER is

begin

  S     <= (A XOR B) XOR C_IN;
  C_OUT <= (A AND B) OR ((A XOR B) AND C_IN);

end architecture BEHAVIORAL;
```

The Chisel equivalent of the full adder is shown in the following code block:

```scala
import chisel3._

class FullAdder extends Module {
  val io = IO(new Bundle {
    val a     = Input(UInt(1.W))
    val b     = Input(UInt(1.W))
    val c_in  = Input(UInt(1.W))
    val s     = Output(UInt(1.W))
    val c_out = Output(UInt(1.W))
  })
  io.s := (io.a ^ io.b) ^ io.c_in
  io.c_out := (io.a & io.b) | ((io.a ^ io.b) & io.c_in)
}
```

In the Chisel code, the IO bundle defines the module inputs and outputs. The argument to each Input and Output parameter defines the data type (Uint) and the bit width (1.W, indicating each input and output signal is 1 bit wide).

While this simple example does not demonstrate the full range of benefits of developing complex circuits in Chisel, it shows that at the level of detailed implementation, the code does not look too different from VHDL. We won't delve further into the details of Chisel here. For more information, consult the Chisel repository at https://github.com/freechipsproject/chisel3.

The process of building the RISC-V processor and programming it into the Arty A7-35T board consists of the following steps:

1. Translate the Chisel and Scala code into the FIRRTL form.
2. Translate the FIRRTL into Verilog.
3. Compile the Verilog into an FPGA image.
4. Program the FPGA image onto the Arty A7-35T board.

A step-by-step walkthrough for implementing a RISC-V processor in the Arty A7-35T board is available from the Digi-Key Electronics TechForum at https://forum.digikey.com/t/digilent-arty-a7-with-xilinx-artix-7-implementing-sifive-fe310-risc-v.

Once you have programmed the RISC-V image onto the Arty board, it is possible to connect a software development suite to the board through the debugger interface. From this point, you can develop RISC-V code in assembly language or high-level languages, compile it, and run it on the FPGA RISC-V processor in the same manner you would with a hardware processor.

Summary

This chapter introduced the RISC-V processor architecture and its instruction set. The RISC-V architecture defines a complete user-mode and privileged instruction set specification and several extensions to support general-purpose computing, high-performance computing, and embedded applications requiring minimal code size. RISC-V processors are offered commercially, and free open-source products are available to implement RISC-V in FPGA devices.

Having completed this chapter, you should understand the architecture and features of the RISC-V processor and its optional extensions.

You learned the basics of the RISC-V instruction set and now understand how RISC-V can be tailored to target a variety of application domains, from low-end micropower embedded systems to warehouse-scale cloud server farms. You also learned how to implement a RISC-V processor in a low-cost FPGA board.

The next chapter introduces the concept of processor virtualization, where rather than running code directly on a host processor, an entire virtual environment is implemented to run one or more virtual processors, each with its own operating system and applications, on a single physical processor.

Exercises

1. Visit https://www.sifive.com/software/ and download *Freedom Studio*. Freedom Studio is an Eclipse **integrated development environment** (**IDE**)-based development suite with a complete set of tools for building a RISC-V application and running it on a hardware RISC-V processor or in the emulation environment included with Freedom Studio. Follow the instructions in the *Freedom Studio User Manual* to complete the installation. Start Freedom Studio and create a new Freedom E SDK project. In the project creation dialog, select **qemu-sifive-u54** as the target (this is a single-core 64-bit RISC-V processor in the RV64GC configuration). Select the **hello** example program and click the **Finish** button. This will start a build of the example program and the RISC-V emulator. After the build completes, the **Edit Configuration** dialog box will appear. Click **Debug** to start the program in the emulator debug environment. Single-step through the program and verify that the text Hello, World! appears in the console window.

2. With the project from *Exercise 1* still open, locate the hello.c file in the **src** folder in the **Project** window. Right-click on the file and rename it hello.s. Open hello.s in the editor and delete the entire contents. Insert the assembly language program shown in the *RISC-V assembly language* section in this chapter. Perform a clean, and then rebuild the project (press *Ctrl* + *9* to initiate the clean operation). Select **Debug** under the **Run** menu. Once the debugger starts, open windows to display the hello.s source file, the **Disassembly** window, and the **Registers** window. Expand the **Registers** tree to display the RISC-V processor registers. Single-step through the program and verify that the text Hello, Computer Architect! appears in the console window.

3. Write a RISC-V assembly language program that computes the following expression and prints the result as a hexadecimal number: [(129 − 66) × (445 + 136)] ÷ 3. As part of this program, create a callable function to print 1 byte as 2 hex digits.

Join our community Discord space

Join the book's Discord workspace for a monthly *Ask me Anything* session with the author: https://discord.gg/7h8aNRhRuY

12
Processor Virtualization

This chapter introduces the concepts underlying processor virtualization and explores the many benefits to individual users and large organizations from its effective use. We will discuss the principal virtualization techniques and the open source and commercial tools that implement them.

Virtualization tools enable the emulation of instruction set-accurate representations of various computer architectures and operating systems on general-purpose computers. Virtualization is used widely in the deployment of real-world software applications in cloud environments.

After completing this chapter, you will understand the technology and benefits associated with hardware virtualization and how modern processors support virtualization at the instruction set level. You will have learned the technical features of several open source and commercial tools providing virtualization capabilities and will understand how virtualization is used to build and deploy scalable applications in cloud computing environments.

The following topics will be presented in this chapter:

- Introducing virtualization
- Virtualization challenges
- Virtualizing modern processors
- Virtualization tools
- Virtualization and cloud computing

Technical requirements

The files for this chapter, including the answers to the exercises, are available at https://github.com/PacktPublishing/Modern-Computer-Architecture-and-Organization-Second-Edition.

Introducing virtualization

In the domain of computer architecture, **virtualization** refers to the use of hardware and software to create an emulated version of an environment in which a piece of software runs, as opposed to the *real* environment in which the code normally expects to run.

We have already looked at one form of virtualization in *Chapter 7, Processor and Memory Architectures*, in some depth: virtual memory. Virtual memory uses software, with supporting hardware, to create an environment in which each running application functions as if it has exclusive access to the entire computer, including all the memory it requires at the addresses it expects. This allows the virtual address ranges used by a program to be the same as those in use by other currently running processes.

Systems using virtual memory create multiple sandboxed environments in which each application runs without interference from other applications, except in competition for shared system resources.

In the virtualization context, a **sandbox** is an isolated environment in which code runs without interference from anything outside its boundaries, and which prevents code inside the sandbox from affecting resources external to it. This isolation between applications is rarely absolute, however. For example, even though a process in a virtual memory system cannot access another process's memory, it may do something else, such as delete a file that is needed by a second process, which may cause problems for the other process.

Our primary focus in this chapter will be on virtualization at the processor level, which allows one or more operating systems to run in a virtualized environment on a computer system. This virtual environment operates at an abstracted level relative to the system's physical hardware.

The next section will briefly describe the various categories of virtualization you are likely to encounter.

Types of virtualization

The term *virtualization* is applied in several different computing contexts, especially in larger network environments such as businesses, universities, government organizations, and cloud service providers. The definitions that follow will cover the most common types of virtualization you are likely to come across.

Operating system virtualization

A virtualized operating system runs under the control of a hypervisor. A **hypervisor** is a combination of software and hardware that is capable of instantiating and running virtual machines. A **virtual machine** is an emulation of an entire computer system. The prefix *hyper* in hypervisor refers to the fact that the hypervisor is more privileged than the supervisor mode of the operating systems running in its virtual machines. Another term for hypervisor is **virtual machine monitor**.

There are two general types of hypervisor:

- A **type-1 hypervisor**, sometimes referred to as a *bare metal* hypervisor, is a software environment for managing virtual machines that runs directly on the hardware of a host computer
- A **type-2 hypervisor**, also called a *hosted* hypervisor, runs as an application program that manages virtual machines under a host operating system

> **HYPERVISOR VERSUS VIRTUAL MACHINE MONITOR**
>
> Technically, a virtual machine monitor is not precisely the same thing as a hypervisor, but for our purposes, we will treat the terms as synonymous. A virtual machine monitor is responsible for virtualizing a processor and other computer system components. A hypervisor combines a virtual machine monitor with an underlying operating system, which may be dedicated to hosting virtual machines (a type-1 hypervisor), or it may be a general-purpose operating system (a type-2 hypervisor).

The computer running the hypervisor is referred to as the **host**. Operating systems running within hypervisor-managed virtual environments on a host system are called **guests**.

Regardless of its type, a hypervisor enables guest operating systems and applications running within them to be brought up and executed in virtualized environments. A hypervisor can support multiple virtual machines running on a single processor simultaneously.

The hypervisor is responsible for managing all requests for privileged operations initiated by guest operating systems and the applications running within them. Each request requires a transition from user mode to kernel mode and back to user mode. As in non-virtualized operating systems, all I/O requests from applications on guest operating systems involve privilege-level transitions.

Since operating system virtualization in a type-2 hypervisor involves running an operating system under the hypervisor in the host operating system, a natural question is, what happens if you run another copy of the hypervisor within the operating system of the virtual machine? The answer is that this approach is supported in some, but not all, combinations of hypervisor, host OS, and guest OS. This configuration is referred to as **nested virtualization**.

The next thing you might wonder about nested virtualization is why anyone would want to do such a thing. Here is one scenario where nested virtualization is useful: assume your business's primary web presence is implemented as a virtualized operating system image containing a variety of installed and custom software components. If your cloud service provider goes offline for some reason, you will need to bring the application up at an alternative provider quickly.

Google Compute Engine (https://cloud.google.com/compute), for example, provides an execution environment implemented as a virtual machine. Compute Engine allows you to install a hypervisor in this virtual machine and bring up your application virtual machine within it, putting your web presence back online with minimal installation and configuration.

Application virtualization

Instead of creating a virtual environment to encapsulate an entire operating system, it is possible to virtualize at the level of a single application. Application virtualization abstracts the operating system from the application code and provides a degree of sandboxing.

This type of virtualization allows programs to run in an environment that differs from the intended application target environment. For example, *Wine* (https://www.winehq.org/) is an application compatibility layer that allows programs written for Microsoft Windows to run under POSIX-compliant operating systems, typically Linux variants.

The **Portable Operating System Interface (POSIX)** is a set of IEEE standards providing application programming compatibility between operating systems. Wine translates Windows library and system calls to equivalent POSIX calls in an efficient manner.

Application virtualization replaces portions of the runtime environment with a virtualization layer and performs tasks such as intercepting disk I/O calls and redirecting them to a sandboxed, virtualized disk environment.

Application virtualization can encapsulate a complex software installation process, consisting of hundreds of files installed in various directories, as well as numerous Windows registry modifications, in an equivalent virtualized environment contained within a single executable file.

Simply copying the executable to a target system and running it brings up the application as if the entire installation process had taken place on the target.

Network virtualization

Network virtualization is the connection of software-based emulations of network components, such as switches, routers, firewalls, and telecommunication networks, in a manner that represents a physical configuration of these components. This allows operating systems and the applications running on them to interact with and communicate over the virtual network in the same manner they would on a physical implementation of the same network architecture.

A single physical network can be subdivided into multiple **virtual local area networks (VLANs)**, each of which appears to be a complete, isolated network to all systems connected on the same VLAN.

Multiple computer systems at the same physical location can be connected to different VLANs, effectively placing them on separate networks. Conversely, computers at distant geographic separations can be placed on the same VLAN, making it appear as if they are interconnected within a small local network.

Storage virtualization

Storage virtualization is the abstraction of physical data storage from the logical storage structure used by operating systems and applications. A storage virtualization system manages the process of translating logical data requests to physical data transfers.

Logical data requests are addressed as block locations within a disk partition. Following the logical-to-physical translation, data transfers may ultimately interact with a storage device that has an organization completely different from the logical disk structure.

The process of accessing physical data given a logical address is much like the virtual-to-physical address translation process in virtual memory systems. The logical disk I/O request includes information such as a device identifier and a logical block number. This request is translated to a physical device identifier and block number. The requested read or write operation then takes place on the physical disk.

Storage virtualization in data centers often includes several enhancements that increase the reliability and performance of data storage systems. Some of these improvements are:

- **Centralized management** enables monitoring and control of a large collection of storage devices, possibly of different sizes and from different vendors.

Because all virtualized storage appears the same to client applications, any vendor-specific variations in storage devices are hidden from users.

- **Replication** provides transparent data backup and disaster recovery capabilities for mission-critical data. When performing real-time replication, writes to the storage array are immediately copied to one or more remote replicas.
- **Data migration** allows administrators to move data to a different physical location or switch to a replica while concurrent data I/O operations continue without interruption. Because the storage virtualization management system has full control over disk I/O, it can switch the target of any logical read or write operation to a different physical device at any time.

The next section will introduce some of the most common methods of processor virtualization in use today.

Categories of processor virtualization

The ideal mode of operation for a processor virtualization environment is **full virtualization**. In full virtualization, binary code in operating systems and applications runs in the virtual environment with no modifications whatsoever. Guest operating system code performing privileged operations executes under the illusion that it has complete and sole access to all machine resources and interfaces. The hypervisor manages interactions between guest operating systems and host resources and takes any steps needed to deconflict access to I/O devices and other system resources for each virtual machine under its control.

Our focus in this chapter is processor virtualization, which enables the execution of complete operating systems and applications running on them in a virtualized environment.

Historically, there have been several different methods used for the implementation of virtualization at the processor level. We'll take a brief look at each of them, beginning with an approach first implemented on systems such as the IBM VM/370, which was introduced in 1972. VM/370 was the first operating system specifically designed to support the execution of virtual machines.

Trap-and-emulate virtualization

In a 1974 article entitled *Formal Requirements for Virtualizable Third Generation Architectures*, Gerald J. Popek and Robert P. Goldberg described the three properties a hypervisor must implement to virtualize a computer system efficiently and fully, including the processor, memory, storage, and peripheral devices:

- **Equivalence**: Programs (including the guest operating system) running in a hypervisor must exhibit essentially the same behavior as when they run directly on machine hardware, excluding the effects of timing
- **Resource control:** The hypervisor must have complete control over all the resources used by the virtual machine
- **Efficiency**: A high percentage of instructions executed by the virtual machine must run directly on the physical processor without hypervisor intervention

For a hypervisor to satisfy these criteria, the hardware and operating system of the computer on which it is running must grant the hypervisor the power to fully control the virtual machines it manages.

The code within a guest operating system assumes it is running directly on the physical processor hardware and has full control of all the features accessible via system hardware. Guest operating system code executing at the kernel privilege level must be able to execute privileged instructions and access regions of memory reserved for the operating system.

In a hypervisor implementing the trap-and-emulate virtualization method, portions of the hypervisor run with kernel privilege, while all guest operating systems (and, of course, the applications running within them) operate at the user privilege level. Kernel code within the guest operating systems executes normally until a privileged instruction attempts to execute or a memory-access instruction attempts to read or write memory outside the user-space address range available to the guest operating system. When the guest attempts any of these operations, a trap occurs.

> **EXCEPTION TYPES: FAULTS, TRAPS, AND ABORTS**
>
> The terms *fault*, *trap*, and *abort* are used to describe similar exception events. The primary differences between each of these exception types are as follows:
>
>
>
> A **fault** is an exception that ends by restarting the instruction that caused the exception. For example, a page fault occurs when a program attempts to access a valid memory location that is currently inaccessible. After the page fault handler completes, the triggering instruction is restarted, and execution continues from that point.
>
> A **trap** is an exception that ends by continuing the execution with the instruction following the triggering instruction. For example, execution resumes after the exception triggered by a debugger breakpoint by continuing with the next instruction.
>
> An **abort** represents a serious error condition that may be unrecoverable. Problems such as errors accessing memory may cause aborts.

The fundamental trick (if you want to think of it that way) to enable trap-and-emulate virtualization is in the handling of the exceptions generated by privilege violations. While it is starting up, the hypervisor routes the host operating system exception handlers into its own code. Exception handlers within the hypervisor process these exceptions before the host operating system has a chance to handle them.

The hypervisor exception handler examines the source of each exception to determine if it was generated by a guest operating system under the hypervisor's control. If the exception originated from a guest the hypervisor manages, the hypervisor handles the exception, emulating the requested operation, and returns execution control directly to the guest. If the exception did not come from a guest belonging to the hypervisor, the hypervisor passes the exception to the host operating system for processing in the normal manner.

For trap-and-emulate virtualization to work in a comprehensive and reliable manner, the host processor must support the criteria defined by Popek and Goldberg. The most critical of these requirements is that any guest instruction attempting to access privileged resources must generate a trap. This is necessary because the host system has only one set of privileged resources (we're assuming a single-core system here for simplicity) and the host and guest operating systems cannot both exercise control of those resources.

As an example of the types of privileged information controlled by the hypervisor, consider the page tables used to manage virtual memory.

The host operating system maintains a collection of page tables that oversee the entirety of the system's physical memory. Each guest operating system has its own set of page tables that it believes it is using to manage physical and virtual memory on the system it controls. These two sets of page tables contain substantially different data, even though both ultimately interact with the same physical memory regions.

Through the trapping mechanism, the hypervisor intercepts all guest operating system attempts to interact with page tables and directs those interactions to a guest-specific memory region containing page tables used only by the guest operating system. The hypervisor then manages the necessary translation between addresses used by instructions executing in the guest operating system and the host system's physical memory.

The greatest barrier to the widespread use of virtualization in the late 1990s and early 2000s was the fact that the general-purpose processors in common use at the time (x86 variants) did not support the Popek and Goldberg virtualization criteria.

The x86 instruction sets contained several instructions that allowed unprivileged code to interact with privileged data without generating a trap. Many of these instructions merely permitted unprivileged code to read selected privileged registers. While this may seem harmless, it caused a severe problem for virtualization because there is only one copy of each of those registers in the entire machine, and each guest OS may need to maintain different values in those registers.

Later versions of the x86, beginning in 2006, added hardware features (**Intel virtualization technology (VT-x)**, and **AMD virtualization (AMD-V)**) that enabled full virtualization under the Popek and Goldberg criteria.

The virtualization requirements defined by Popek and Goldberg assumed that the use of the trap-and-emulate technique, widely viewed as the only practical virtualization method in the 1970s, was the only feasible method for processor virtualization. In the following sections, we will see how it is possible to perform effective and efficient virtualization on a computer system that does not fully comply with the Popek and Goldberg criteria.

Paravirtualization

Because most, if not all, of the instructions that require special handling in the virtualized environment reside in the guest operating system and its device drivers, one method for rendering the guest virtualizable is to modify the operating system and its drivers to explicitly interface with the hypervisor in a non-trapping manner. This is called **paravirtualization**.

This approach can result in substantially better guest OS performance than a system running under a trap-and-emulate hypervisor because the paravirtualized hypervisor interface is composed of optimized code rather than a series of trap handler invocations. In the trap-and-emulate method, the hypervisor must process every trap in a generic handler that begins by determining whether the trap comes from a guest OS it controls before further processing to determine the desired operation and emulate its effects.

The primary drawback of paravirtualization is the need to modify the guest operating system and its drivers to implement the hypervisor interface. There has been limited interest among the maintainers of major operating system distributions in fully supporting a paravirtualization interface.

Binary translation

One way to deal with problematic instructions within processor architectures that lack full support for virtualization is to scan the binary code prior to execution to detect the presence of nonvirtualizable instructions. When such instructions are found, the code is translated into virtualization-friendly instructions that produce identical effects.

This has proven to be a popular approach for virtualization in the x86 architecture. The combination of trap-and-emulate and binary translation of nonvirtualizable instructions permits reasonable guest OS performance. This technique limits the processing required to deal with nonvirtualizable instructions to a tolerable level.

Binary translation can be performed on a static or dynamic basis. Static binary translation recompiles a set of executable images into a form ready for execution in the virtual environment. This translation takes some time, but it is a one-time process creating a set of system and user images that will continue to work until new image versions are installed, necessitating a recompilation procedure for the new images.

Dynamic binary translation scans sections of code during program execution to locate problematic instructions. When such instructions are encountered, they are replaced with virtualizable instruction sequences. Dynamic binary translation avoids the recompilation step required by static binary translation, but it results in reduced performance due to the ongoing scanning and translation process as code runs. Each code segment only needs to be scanned and translated once each time a program runs and it is then cached—so, for example, code within a loop will not be rescanned on each iteration.

Hardware emulation

All the virtualization techniques that we have discussed to this point have assumed the guest OS is expecting to run on a processor with the same instruction set architecture as the host processor. There are many situations in which it is desirable to run an operating system and application code on a host processor with a completely different instruction set architecture (ISA) from the guest OS.

When emulating processor hardware, each instruction executing in an emulated guest system must be translated to an equivalent instruction or sequence of instructions in the host ISA. As with binary translation, this process can take place in a static or dynamic manner:

- **Static translation** can produce an efficient executable image capable of running in the target processor ISA. There is some risk in static translation because it may not be straightforward to identify all code paths in the executable file, particularly if branch target addresses are computed in code rather than statically defined. This risk also applies to the static binary translation technique described in the previous section.

- **Dynamic translation** avoids potential errors that may occur in static translation, but performance can be quite poor. This is because dynamic translation with hardware emulation involves translating every instruction from one architecture to another.

This contrasts with dynamic binary translation for the same ISA, which, although it must scan every instruction, typically only needs to perform translation for a small percentage of executed instructions.

One example of hardware emulation tools is the open source QEMU (https://www.qemu.org/) machine emulator and virtualizer. QEMU runs operating systems for a wide variety of processor architectures on an impressive list of differing architectures, with reasonably good performance.

The Freedom Studio tool suite for the RISC-V processor includes a QEMU implementation of the RV64GC instruction set architecture. We used this virtualized environment to run the code in the exercises for *Chapter 11, The RISC-V Architecture and Instruction Set*.

In the next section, we will discuss the challenges and benefits related to virtualization in the processor architectures discussed in the preceding chapters.

Virtualization challenges

In simple terms, the goal of processor virtualization is to run an operating system within a hypervisor, which itself either runs on the bare metal of a computer system or executes as an application under the control of another operating system.

In this section, we will focus on the hosted (type 2) hypervisor because this mode of operation presents a few additional challenges that a bare-metal hypervisor may not face because the type-1 hypervisor has been optimized to support virtualization.

In a type-2 hypervisor, the host operating system supports kernel and user modes, as does the guest operating system (in the guest's perception). As the guest operating system and the applications running within it request system services, the hypervisor must intercept each request and translate it into a suitable call to the host operating system.

In a nonvirtualized system, peripheral devices such as the keyboard and mouse interact directly with the host operating system. In a virtualized environment, the hypervisor must manage the interfaces to these devices whenever the user requests interactions with the guest OS.

The degree of difficulty involved in implementing these capabilities depends on the instruction set of the host computer. Even if an instruction set was not designed to facilitate virtualization, it may or may not be possible for that architecture to support virtualization in a straightforward manner. The ease of virtualization on a particular processor ISA is a function of the way the processor handles unsafe instructions.

Unsafe instructions

The name of the trap-and-emulate virtualization method refers to the ability of the hypervisor to take control of exception processing that would normally be dealt with by kernel mode handlers in the host operating system. This allows the hypervisor to process privilege violations and system calls from guest operating systems and the applications that run within them.

Each time an application running on a guest operating system requests a system function, for example opening a file, the hypervisor intercepts the request, adjusts the parameters of the request to align with the virtual machine configuration (perhaps by redirecting the file open request from the host filesystem to the guest's virtual disk sandbox), and passes the request on to the host operating system. The process of inspecting and handling exceptions by the hypervisor is the emulation phase of the trap-and-emulate approach.

In the context of virtualization, processor instructions that either rely on or modify privileged system state information are referred to as **unsafe**. For the trap-and-emulate method to function in a comprehensively secure and reliable manner, all unsafe instructions must generate exceptions that trap to the hypervisor. If an unsafe instruction is allowed to execute without trapping, the isolation of the virtual machine is compromised, and virtualization may fail.

As we learned in the *Trap-and-emulate virtualization* section of this chapter, early versions of the x86 architecture were affected by this issue.

Shadow page tables

Protected data structures used in the allocation and management of virtual and physical memory present another challenge to full virtualization. A guest operating system kernel presumes it has full access to the hardware and data structures associated with the system MMU. The hypervisor must translate guest operating system requests for memory allocation and deallocation in a manner that is functionally equivalent to running the guest OS on bare metal.

A particular problem arises in the x86 architecture because virtual memory page table configuration data must be stored within the processor to properly configure the system, but that information becomes inaccessible once it has been stored. To resolve this issue, the hypervisor maintains its own copy of the page table configuration data, referred to as **shadow page tables**.

Because the shadow page tables are not actual page tables managing memory for the host OS, it is necessary for the hypervisor to set access permission restrictions on shadow page table memory regions and intercept the resulting traps when the guest OS attempts to access its page tables.

The hypervisor then emulates the requested operation by interacting with the physical MMU through calls to the host OS.

The use of shadow page tables incurs a significant performance penalty and has been an area of focus for the development of hardware-assisted virtualization enhancements.

Security

There is nothing inherently insecure about using a hypervisor to virtualize one or more guest applications. It is, however, important to understand the added opportunities for malicious actors to attempt to infiltrate a virtualized environment.

A guest virtual machine presents essentially the same collection of vulnerabilities to remote attackers as an identical operating system and set of applications running directly on hardware. The hypervisor provides an additional avenue that an attacker may attempt to exploit in a virtualized environment.

If malicious actors manage to penetrate and take control of the hypervisor, they will have full access to all the guest operating systems, as well as the applications and data accessible from within the guests. The guests are accessible in this scenario because they operate at a lower privilege level than the hypervisor, which has full control over them.

When implementing virtualization in a context that permits public access, such as web hosting, it is vital that credentials enabling access to hypervisors be strictly limited to a small number of personnel. All reasonable protective measures must be rigorously maintained to prevent unauthorized hypervisor access.

In the next section, we will examine some key technical aspects of virtualization as implemented in modern processor families.

Virtualizing modern processors

The hardware architectures of most general-purpose processor families have matured to the point where they fully support the execution of virtualized guest operating systems, at least in their higher-end variants. The following sections briefly introduce the virtualization capabilities provided by modern general-purpose processor families.

x86 processor virtualization

The x86 architecture was not originally designed to support the execution of virtualized operating systems.

As a result, x86 processors, from the earliest days through to the Pentium series, implemented instruction sets containing several unsafe but non-trapping instructions. These instructions caused problems with virtualization by, for example, allowing the guest operating system to access privileged registers that did not contain data corresponding to the state of the virtual machine.

X86 CURRENT PRIVILEGE LEVEL AND UNSAFE INSTRUCTIONS

In the x86 architecture, the lower two bits of the **code segment (CS)** register contain the **current privilege level (CPL)**, identifying the currently active protection ring. The CPL is generally 0 for kernel code and 3 for user applications in a nonvirtualized operating system. In most hypervisor implementations, virtual machines run at CPL 3, causing many unsafe x86 instructions to trap upon execution. Unfortunately for the early adopters of x86 virtualization, not all unsafe x86 instructions in Pentium processors caused traps when executed at CPL 3.

For example, the `sidt` instruction permits unprivileged code to read the 6-byte **interrupt descriptor table register (IDTR)** and store it at a location provided as an operand. There is only one IDTR in a single-core x86 processor.

When a guest operating system executes this instruction, the IDTR contains data associated with the host operating system, which differs from the information the guest operating system expects to retrieve. This will lead to erroneous execution of the guest operating system.

Writing to the physical system's IDTR is only possible for code running at CPL 0. When a guest operating system attempts to write to the IDTR while running at CPL 3, a privilege violation occurs and the hypervisor processes the ensuing trap to emulate the write operation by writing to a shadow register instead. The shadow register is just a location in memory allocated by the hypervisor.

Reads from the IDTR, however, are permitted at CPL 3. User-mode software can read the IDTR and no trap occurs. Without a trap, the hypervisor is unable to intercept the read operation and return data from the shadow register. In short, writes to the IDTR are virtualizable, while reads from the IDTR are not.

Of the hundreds of instructions in the Pentium ISA, 17 were found to be unsafe but non-trapping. In other words, these instructions are nonvirtualizable. For the Pentium x86 architecture, implementing a pure trap-and-emulate virtualization approach is therefore not possible.

Unsafe but non-trapping instructions are used frequently in operating systems and device drivers but are rarely found in application code. The hypervisor must implement a mechanism to detect the presence of unsafe, non-trapping instructions in the code and handle them.

The approach settled on by several popular virtualization engines has been to combine trap-and-emulate virtualization, where possible, with the binary translation of unsafe instructions into functionally equivalent code sequences suitable for the virtualized environment.

Most guest user applications do not attempt to use unsafe instructions at all. This allows them to run at full speed, once the hypervisor has scanned the code to ensure no unsafe instructions are present. Guest kernel code, however, may contain numerous, frequently encountered unsafe instructions. To achieve reasonable performance from binary-translated code, it is necessary to cache the modified code the first time it executes and reuse the cached version on future execution passes.

x86 hardware virtualization

In 2005 and 2006, Intel and AMD released versions of the x86 processors containing hardware extensions supporting virtualization.

These extensions resolved the problems caused by unsafe but non-trapping instructions, enabling full system virtualization under the Popek and Goldberg criteria. The extensions were named AMD-V in AMD processors and VT-x in Intel processors. The virtualization extensions in modern Intel processors are referred to as VT.

The initial implementations of these hardware virtualization technologies removed the requirements for the binary translation of unsafe instructions, but overall virtual machine performance did not improve substantially following the removal of binary translation. This was because page table shadowing was still required. Page table shadowing had been the cause of most of the performance degradation observed during virtual machine execution.

Later versions of hardware virtualization technology removed many of the performance barriers in virtual machine execution, leading to the widespread adoption of x86 virtualization across a variety of domains. Today, multiple tools and frameworks are available for implementing x86 virtualization solutions within a standalone workstation, with options available to scale up to a fully managed data center with potentially thousands of servers, each capable of running several virtual machines simultaneously.

ARM processor virtualization

The ARM architecture supports virtualization in both the 32-bit and 64-bit execution states. Hardware support for virtualization includes the following:

- Full trap-and-emulate virtualization
- A dedicated exception category for hypervisor use
- Additional registers supporting hypervisor exceptions and stack pointers

The ARM architecture provides hardware support for the translation of guest memory access requests to physical system addresses.

Systems running ARM processors offer a comprehensive capability for virtual machine execution using either a type-1 or type-2 hypervisor. 64-bit ARM processor performance is comparable to x64 servers with similar specifications. For many applications, such as large data center deployments, the choice between x64 and ARM as the server processor may revolve around factors unrelated to processor performance, such as system power consumption and cooling requirements.

RISC-V processor virtualization

Unlike the other ISAs discussed in this chapter, the architects of the RISC-V ISA included comprehensive virtualization support as a baseline requirement from the beginning of the design process. The RISC-V hypervisor extension provides a full set of capabilities to support the efficient implementation of type-1 and type-2 hypervisors.

The RISC-V hypervisor extension fully implements the trap-and-emulate virtualization method and provides hardware support for the translation of guest operating system physical addresses to host physical addresses.

RISC-V implements the concept of foreground and background control and status registers, which allows the rapid swapping of supervisor registers in and out of operation as virtual machines transition into and out of the running state.

Each hardware thread in RISC-V runs at one of three privilege levels:

- **User (U)**: This is the same as user privilege in a traditional operating system
- **Supervisor (S)**: This is the same as supervisor or kernel mode in a traditional operating system
- **Machine (M)**: The highest privilege level, with access to all system features

Individual processor designs may implement all three of these modes, or they may implement the M and U mode pair, or M mode alone. Other combinations are not allowed.

In a RISC-V processor supporting the hypervisor extension, an additional configuration bit, the **V bit**, controls the virtualization mode. The V bit is set to 1 for hardware threads executing in a virtualized guest. Both user and supervisor privilege levels can execute with the V bit set to 1. These are named the **virtual user (VU)** and **virtual supervisor (VS)** modes.

In the RISC-V hypervisor context, supervisor mode with V = 0 is renamed the **hypervisor-extended supervisor mode (HS)**. This name indicates HS is the mode in which the hypervisor itself, regardless of whether it is type 1 or type 2, runs. The remaining privilege level, M, only functions in a non-virtualized manner with V = 0.

In both VU and VS modes, RISC-V implements a two-level address translation scheme to convert each guest virtual address first to a guest physical address and then to a supervisor physical address. This procedure efficiently performs the translation from virtual addresses in applications running in guest operating systems to physical addresses in system memory.

The next section provides overviews of several popular tools for processor and operating system virtualization.

Virtualization tools

In this section, we will look at several widely available open source and commercial tools that implement different forms of processor virtualization. This information may be useful as a starting point the next time you initiate a project involving virtualization.

VirtualBox

VirtualBox is a free, open source type-2 hypervisor from Oracle Corporation. Supported host operating systems include Windows and several Linux distributions. One or more guest operating systems on a single host can simultaneously run Windows, macOS, Solaris, Open Solaris, and a variety of Linux distributions.

GUEST OS LICENSING REQUIREMENTS

For organizations and individuals to remain in compliance with copyright laws, operating systems requiring licensing, such as Windows, must be properly licensed even when running as guest operating systems.

Individual virtual machines can be started, stopped, and paused under the control of the interactive VirtualBox management program or from the command line. VirtualBox can capture snapshots of executing virtual machines and save them to disk. Later, a snapshot can resume execution from the precise point at which it was taken.

VirtualBox requires hardware-assisted virtualization provided by platforms with the AMD-V or Intel VT extensions. Multiple mechanisms are provided that enable virtual machines to communicate with the host OS and with each other.

A shared clipboard supports copy-and-paste between host and guest machines and from guest to guest. An internal network can be configured within VirtualBox for guests to interact with each other as if they were connected on an isolated local area network.

VMware Workstation

VMware Workstation, first released in 1999, is a type-2 hypervisor that runs on 64-bit versions of Windows and Linux. VMware products are offered commercially and require the purchase of licenses by some users.

A version of Workstation called VMware Workstation Player is available at no cost with the provision that it only be used for non-commercial purposes.

VMware Workstation supports the execution of potentially multiple copies of Windows, Linux, and MS-DOS operating systems within the host Linux or Windows operating system. Like VirtualBox, Workstation can capture snapshots of the virtual machine state, save that information to disk, and later resume execution from the captured state. Workstation also supports host-to-guest and guest-to-guest communication features such as a shared clipboard and local network emulation.

VMware ESXi

ESXi is a type-1 hypervisor intended for enterprise-class deployments in data centers and cloud server farms. As a type-1 hypervisor, ESXi runs on the bare metal of the host computer system. It has interfaces with the computer system hardware, each guest operating system, and a management interface called the service console.

From the service console, administrators can oversee and manage the operation of a large-scale data center, bringing up virtual machines and assigning them tasks (referred to as **workloads**).

ESXi provides additional features necessary for large-scale deployments such as performance monitoring and fault detection. In the event of hardware failure or to enable system maintenance, virtual machine workloads can be transitioned seamlessly to different host computers.

KVM

The **kernel-based virtual machine (KVM)** is an open source type-2 hypervisor initially released in 2007. KVM supports full virtualization for guest operating systems. When used with x86 or x64 hosts, the system hardware must include the AMD-V or Intel VT virtualization extensions. The KVM hypervisor kernel is included in the main Linux development line.

KVM supports the execution of one or more virtualized instances of Linux and Windows on a host system without any modification of the guest operating systems.

Although originally developed for the 32-bit x86 architecture, KVM has been ported to x64, ARM, and PowerPC. KVM supports paravirtualization for Linux and Windows guests using the Virtio API. In this mode, paravirtualized device drivers are provided for Ethernet, disk I/O, and the graphics display.

Xen

Xen, first released in 2003, is a free and open source type-1 hypervisor. The current version of Xen runs on x86, x64, and ARM processors. Xen supports guest virtual machines running under hardware-supported virtualization (AMD-V or Intel VT) or as paravirtualized operating systems. Xen is implemented in the mainline Linux kernel.

The Xen hypervisor runs one virtual machine at the most privileged level, referred to as domain 0, or dom0. The dom0 virtual machine is typically a Linux variant and has full access to the system hardware. The dom0 machine provides the user interface for managing the hypervisor.

Some of the largest commercial cloud service providers, including Amazon EC2, IBM SoftLayer, and Rackspace Cloud, use Xen as their primary hypervisor platform.

Xen supports live migration, where a virtual machine can be migrated from one host platform to another without downtime.

QEMU

QEMU, an abbreviation for **quick emulator**, is a free and open source emulator that performs hardware virtualization. QEMU can emulate at the level of a single application or an entire computer system. At the application level, QEMU can run individual Linux or macOS applications that were built for a different ISA than the execution environment.

When performing system emulation, QEMU represents a complete computer system, including peripherals.

QEMU supports the execution of multiple guest operating systems on a single host simultaneously. Supported ISAs include x86, MIPS, ARMv7, ARMv8, PowerPC, Sparc, Alpha, OpenRISC, and RISC-V.

QEMU supports the setup and migration of KVM machines, performing hardware emulation in conjunction with the virtual machine running under KVM. Similarly, QEMU can provide hardware emulation for virtual machines running under Xen.

QEMU is unique among virtualization tools in that it is not necessary for it to run at elevated privilege because it entirely emulates the guest system in software. The downside to this approach is the performance degradation resulting from the software emulation process.

The next section will discuss the synergistic effects that flow from the use of virtualization in a cloud computing environment.

Virtualization and cloud computing

The terms *virtualization* and *cloud computing* are often tossed about with vague, sometimes overlapping meanings. Here is an attempt to highlight the difference between them:

- Virtualization is a technology for abstracting software systems from the environment in which they operate
- Cloud computing is a methodology for employing virtualization and other technologies to enable the deployment, monitoring, and control of large-scale data centers

The use of virtualization in cloud computing environments enables the flexible deployment of application workloads across an array of generic computing hardware in a controlled, coherent manner. By implementing applications such as web servers within virtual machines, it is possible to dynamically scale online computing capacity to match varying load conditions.

Commercial cloud service providers generally offer the use of their systems on a pay-per-capacity-used basis. A website that normally receives a small amount of traffic may spike substantially if, for instance, it receives a mention on a national news program. If the site is deployed in a scalable cloud environment, the management software will detect the increased load and bring up additional instances of the website and potentially of the backend database as well.

This increased resource consumption will result in a larger bill from the cloud service provider, which most businesses will happily pay if the result is a website that remains operational and responsive to user input even under a heavy traffic load.

Cloud management environments such as VMware ESXi and Xen provide comprehensive tools for the configuration, deployment, management, and maintenance of large-scale cloud operations. These configurations may be intended for local use by an organization, or they may offer public-facing facilities for online service providers such as Amazon Web Services.

Electrical power consumption

Electrical power consumption is a significant expense for cloud service providers. Each computer in a large-scale server farm consumes power whenever it is running, even if it is not performing any useful work. In a facility containing thousands of computers, it is important to the bottom line that servers consume power only when needed by paying customers.

Virtualization helps substantially with the effective utilization of server systems.

Since a single server can potentially host several guest virtual machines, customer workloads can be allocated efficiently across server hardware in a manner that avoids low utilization of available computers. Servers that are not needed at a given time can be powered off completely, thereby reducing energy consumption, which, in turn reduces costs to the cloud provider and enables more competitive pricing for end users.

This section has provided a brief introduction to the use of virtualization in the context of cloud computing. Most organizations and individuals that establish a presence on the internet make use of virtual servers in a cloud computing environment, whether they know it or not.

Summary

This chapter presented the concepts underlying processor virtualization and explained the many benefits to individual users and large organizations that derive from the effective use of virtualization. We examined the principal virtualization techniques and the open source and commercial tools that implement them.

We also saw the benefits of virtualization in the deployment of real-world software applications in cloud environments.

You should now understand the technology and benefits associated with processor virtualization and how modern processor ISAs support virtualization at the instruction set level. We learned about several open source and commercial tools providing virtualization capabilities. You should now understand how virtualization can be used to build and deploy scalable applications in cloud computing environments.

In the next chapter, we will look at the architecture of some specific application categories, including mobile devices, personal computers, gaming systems, systems that process big data, and neural networks.

Exercises

1. Download and install the current version of VirtualBox. Download, install, and bring up Ubuntu Linux as a virtual machine within VirtualBox. Connect the guest OS to the internet using a bridged network adapter. Configure and enable clipboard sharing and file sharing between the Ubuntu guest and your host operating system.

2. Within the Ubuntu operating system you installed in *Exercise 1*, install VirtualBox and then install and bring up a virtual machine version of FreeDOS, available from `https://www.freedos.org/download/`.

    Verify that DOS commands, such as `echo Hello World!` and `mem`, perform properly in the FreeDOS virtual machine. After completing this exercise, you will have implemented an instance of nested virtualization.

3. Create two separate copies of your Ubuntu guest machine in your host system's VirtualBox environment. Configure both Ubuntu guests to connect to the VirtualBox *internal* network. Set up the two machines with compatible IP addresses. Verify each of the machines can receive a response from the other using the `ping` command. By completing this exercise, you will have configured a virtual network within your virtualized environment.

Join our community Discord space

Join the book's Discord workspace for a monthly *Ask me Anything* session with the author: `https://discord.gg/7h8aNRhRuY`

13
Domain-Specific Computer Architectures

This chapter brings together the topics discussed in previous chapters as we examine a variety of computer system architectures designed to meet unique user requirements. We will gain an understanding of the user-level requirements and performance capabilities associated with several categories of real-world computer systems.

This chapter will cover the following topics:

- Architecting computer systems to meet unique requirements
- Smartphone architecture
- Personal computer architecture
- Warehouse-scale computing architecture
- Neural networks and machine learning architectures

After completing this chapter, you will understand the decision process used in defining computer architectures to support specific needs. You will have learned the key requirements driving the architectural designs of mobile devices, personal computers, cloud server systems, and neural networks and other machine learning architectures.

Technical requirements

The files for this chapter, including answers to the exercises, are available at https://github.com/PacktPublishing/Modern-Computer-Architecture-and-Organization-Second-Edition.

Architecting computer systems to meet unique requirements

Every device containing a digital processor is designed to perform a particular function or collection of functions. This includes general-purpose devices such as personal computers. A comprehensive list of the required and desired features and capabilities for a particular device or computer system provides the raw information to begin designing the architecture of its digital components.

The list that follows identifies some of the considerations a computer architect must weigh in the process of organizing the design of a digital system:

- **The types of processing required:** Does the device need to process audio, video, or other analog information? Is a high-resolution graphics display included in the design? Will extensive floating-point or integer mathematics be required? Will the system support multiple, simultaneously running applications? Are special algorithms, such as neural network processing, going to be used?

- **Memory and storage requirements:** How much RAM will the operating system and anticipated user applications need to perform as intended? How much non-volatile storage will be required?

- **Hard or soft real-time processing:** Is a real-time response to inputs within a time limit mandatory? If real-time performance is not absolutely required, are there desired response times that must be met most, but not necessarily all, of the time?

- **Connectivity requirements:** What kinds of wired connections, such as Ethernet and USB, does the device need to support? How many physical ports for each type of connection are required? What types of wireless connections (cellular network, Wi-Fi, Bluetooth, NFC, GPS, and so on) are needed?

- **Power consumption:** Is the device battery-powered? If so, what is the tolerable level of power consumption for digital system components during periods of high usage, as well as during idle periods? If the system runs on externally provided power, is it more important for it to have high processing performance or low power consumption? For battery-powered systems and externally powered systems, what are the limits of power dissipation before overheating becomes an issue?

- **Physical constraints:** Are there tight constraints on the size of the digital processing components? Is there a limit on the system weight?

- **Environmental limits:** Is the device intended to operate in very hot or cold environments? What level of shock and vibration must the device withstand? Does the device need to operate in extremely humid, dry, or dusty atmospheric conditions? Will the device be exposed to a saltwater environment or to radiation in space?

The following sections examine the top-level architectures of several categories of digital devices and discuss the answers the architects of those systems arrived at in response to questions like those in the preceding list. We'll begin with mobile device architecture, looking specifically at the iPhone 13 Pro Max.

Smartphone architecture

At the architectural level, there are three key features a smartphone must provide to gain wide acceptance: small size (except for the display), long battery life, and very high processing performance upon demand. Obviously, the requirements for long battery life and high processing power are in conflict and must be balanced to achieve an optimal design.

The requirement for small size is generally approached by starting with a screen size (in terms of height and width) large enough to render high-quality video and function as a user-input device (especially as a keyboard), yet small enough to be easily carried in a pocket or purse. To keep the overall device size small in terms of total volume, it must be as thin as possible.

In the quest for thinness, the mechanical design must provide sufficient structural strength to support the screen and resist damage from routine handling, drops on the floor, and other physical assaults, while simultaneously providing adequate space for batteries, digital components, and subsystems such as the cellular radio transceiver.

Because users will have unrestricted access to the external and internal features of their phones, any trade secrets or other intellectual property, such as system firmware, that the manufacturer wishes to prevent from being disclosed must be protected from all types of extraction. Yet, even with these protections in place, it must also be straightforward for end users to securely install firmware updates while blocking the installation of unauthorized firmware images.

We will examine the digital architecture of a current high-end smartphone in the context of these requirements in the next section.

iPhone 13 Pro Max

The iPhone 13 Pro Max was released in September 2021. The iPhone 13 Pro Max was Apple's flagship smartphone at the time of its release and contained some of the most advanced technologies on the market.

Because Apple releases only limited information on the design details of its products, some of the following information comes from teardowns and other types of analysis by iPhone 13 Pro Max reviewers and should therefore be taken with a grain of salt.

The computational architecture of the iPhone 13 Pro Max is centered on the Apple A15 Bionic SoC, an ARMv8 six-core processor constructed with 15 billion CMOS transistors. Two of the cores, with an architecture code-named **Avalanche**, are optimized for high performance and support a maximum clock speed of 3.23 GHz. The remaining four cores, code-named **Blizzard**, are designed for energy-efficient operation at up to 1.82 GHz. All six cores are out-of-order superscalar designs. When executing multiple processes or multiple threads within a single process concurrently, it is possible for all six cores to run in parallel.

Of course, running all six cores simultaneously creates a significant power drain. Most of the time, especially when the user is not interacting with the device, several of the cores are placed in low-power modes to maximize battery life.

The iPhone 13 Pro Max contains up to 8 GB of fourth-generation **low-power double data rate RAM (LP-DDR4x)**. Each LP-DDR4x device supports a 4,266 Mbps data transfer rate. The enhancement indicated by the x in LP-DDR4x reduces the I/O signal voltage from the 1.1 V of the previous DDR generation (LP-DDR4) to 0.6 V in LP-DDR4x, reducing RAM power consumption.

The A15 SoC integrates a five-core GPU designed by Apple. In addition to accelerating traditional GPU tasks such as three-dimensional scene rendering, the GPU contains several enhancements supporting machine learning and other data-parallel tasks suitable for implementation on GPU hardware.

The 3D rendering process implements an algorithm tailored to resource-constrained systems like smartphones called **tile-based deferred rendering (TBDR)**. TBDR attempts to identify objects within the field of view that are not visible (in other words, those that are obscured by other objects) as early in the rendering process as possible, thereby avoiding the work of completing their rendering. This rendering process divides the image into sections (the tiles) and performs TBDR on multiple tiles in parallel to achieve maximum performance.

The A15 contains a neural network processor called the **Apple Neural Engine**. This processor contains 16 cores capable of a total of 15.8 trillion operations per second. This subsystem appears to be used for tasks such as identifying and tracking objects in the live video feed received from the phone's cameras.

The A15 contains a motion coprocessor. This is a separate ARM processor dedicated to collecting and processing data from the phone's gyroscope, accelerometer, compass, and barometric sensors. The processed output of this data includes an estimated category of the user's current activity, such as walking, running, sleeping, or driving. Sensor data collection and processing continues at a low power level even while the remainder of the phone is in sleep mode.

The A15, fully embracing the term *system on chip*, also contains a high-performance **solid-state drive (SSD)** controller that manages access to up to 1 TB of internal drive storage. The interface between the A15 SoC and flash memory is PCI Express.

The following diagram displays the major components of the iPhone 13 Pro Max:

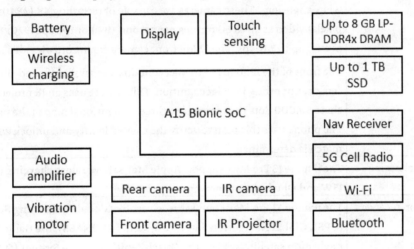

Figure 13.1: iPhone 13 Pro Max components

The iPhone 13 Pro Max contains several high-performance subsystems, each described briefly in the following table:

Subsystem	Description
Batteries	The iPhone 13 Pro Max contains a battery with 3095 **milliamp-hours (mAh)** of energy.
Display	The display is a 6.1 inch diagonal flat panel with 2532 x 1170 pixel resolution. The display technology is **organic light-emitting diode (OLED)**, where *organic* refers to the use of organic compounds in the luminescent material.

Touch sensing	Capacitive sensors are integrated into the display to detect touch interactions. These sensors detect changes in capacitance resulting from the proximity of a conductive object, such as a human finger, to the sensors in the display. After filtering and processing the raw sensor measurements, the accurate locations of multiple simultaneous touchpoints can be determined. In addition, sensors measure the pressure applied during touch interactions. This allows software to react differently to hard and soft presses on the screen.
Dual cameras, IR projector, IR camera	The three rear cameras each produce 12 **megapixels (MP)** images. One camera has a standard lens, one has a telephoto lens, and one has an ultra-wide-angle lens. These cameras are capable of recording 4K (3840x2160 pixels) video at up to 60 **frames per second (fps)** or 1080p (1920x1080 pixels) up to 240 fps. The 12 MP front camera can record 4K video at 60 fps. The front of the iPhone 13 Pro Max contains a separate **infrared (IR)** lidar sensor supporting facial recognition. This feature uses an IR projector to shine 30,000 dots that generate a three-dimensional map of the user's face. The phone uses this map to verify the user's identity and unlock when a match is determined.
Wireless charging	The iPhone 13 Pro Max support Apple **MagSafe wireless charging** at up to 15W and **Qi wireless charging** at up to 7.5W.
Navigation receivers	The iPhone 13 Pro Max contains receivers for **Global Positioning System (GPS), Global Navigation Satellite System (GLONASS)**, the Galileo navigation satellite system, the **Quasi-Zenith Satellite System (QZSS),** and the BeiDou navigation satellite system.
Cellular radio	The iPhone 13 Pro Max contains a **5th generation (5G)** cellular radio modem.
Wi-Fi and Bluetooth	The iPhone 13 Pro Max includes a Wi-Fi interface supporting Wi-Fi 6 (802.11ax) with 2x2 MIMO. 2x2 MIMO provides two transmitter antennas and two receiver antennas to minimize signal dropouts. The Bluetooth interface supports version 5.0 of the Bluetooth standard.
Audio amplifier, vibration motor	The iPhone 13 Pro Max audio amplifier is designed for extremely low power consumption when idle and provides high efficiency and superior sound quality when in operation. Vibration is produced by a device called the **Taptic Engine**, a linear oscillator capable of generating a variety of types of interactive feedback to the user.

Table 13.1: iPhone 13 Pro Max subsystems

The iPhone 13 Pro Max brought together the most advanced, small, lightweight mobile electronic technologies available at the time of its design and assembled them into a sleek, attractive package that served as Apple's flagship smartphone product.

Next, we will look at the architecture of a high-performance personal computer.

Personal computer architecture

The next system we'll examine is a gaming PC with a processor that, at the time of writing (in late 2021), leads the pack in terms of raw performance. We will look in detail at the system processor, the GPU, and the computer's major subsystems.

Alienware Aurora Ryzen Edition R10 gaming desktop

The Alienware Aurora Ryzen Edition R10 desktop PC is designed to provide maximum performance for gaming applications. To achieve peak speed, the system architecture is built around the fastest main processor, GPU, memory, and disk subsystems available at prices that at least some serious gamers and other performance-focused users are willing to tolerate. However, the number of customers for the high-end configuration options described in this section is likely to be limited by its cost, which is over US $4,000.

The Aurora Ryzen Edition R10 is available with a variety of AMD Ryzen processors at varying performance levels and price points. The current highest-performing processor for this platform is the AMD Ryzen 9 5950X, which launched in November 2020.

The Ryzen 9 5950X implements the x64 ISA in a superscalar, out-of-order architecture with speculative execution and register renaming. Based on AMD-provided data, the Zen 3 microarchitecture of the 5950X has up to 19% higher **instructions per clock** (**IPC**) than the previous generation (Zen 2) AMD microarchitecture.

The Ryzen 9 5950X processor boasts the following features:

- 16 cores
- 2 threads per processor (for a total of 32 simultaneous threads)
- Base clock speed of 3.4 GHz with a peak frequency of 4.9 GHz when overclocking
- 32 KB L1 instruction cache with 8-way associativity for each core
- 32 KB L1 data cache with 8-way associativity for each core
- 8 MB L2 cache
- 64 MB L3 cache

- 20 PCIe 4.0 lanes
- Total dissipated power of 105 watts

At the time of its release, Ryzen 9 5950X was arguably the highest performing x86 processor available for the gaming and performance enthusiast market.

Ryzen 9 5950X branch prediction

The Zen 3 architecture includes a sophisticated branch prediction unit that caches information describing the branches taken and uses this data to increase the accuracy of future predictions. This analysis covers not only individual branches, but also correlates among recent branches in nearby code to further increase prediction accuracy. Increased prediction accuracy reduces the performance degradation from pipeline bubbles and minimizes the unnecessary work involved in speculative execution along branches that end up not being taken.

The branch prediction unit employs a form of machine learning called the **perceptron**. Perceptrons are simplified models of biological neurons that form the basis for many applications of artificial neural networks. Refer to the *Deep learning* section in *Chapter 6, Specialized Computing Domains*, for a brief introduction to artificial neural networks.

In the 5950X, perceptrons learn to predict the branching behavior of individual instructions based on recent branching behavior by the same instruction and by other instructions. Essentially, by tracking the behavior of recent branches (in terms of branches taken and not taken), it is possible to develop correlations involving the branch instruction under consideration that lead to increased prediction accuracy.

Nvidia GeForce RTX 3090 GPU

The Aurora Ryzen Edition R10 offers as an option the Nvidia GeForce RTX 3090 GPU. In addition to the generally high level of graphical performance you would expect from a top-end gaming GPU, this card provides substantial hardware support for raytracing and includes dedicated cores to accelerate machine learning applications.

In traditional GPUs, visual objects are described as collections of polygons. To render a scene, the location and spatial orientation of each polygon must first be determined, then those polygons visible in the scene are drawn at the appropriate location in the image.

Raytracing uses an alternative, more sophisticated approach. A ray-traced image is drawn by tracing the path of light emitted from one or more illumination sources in the virtual world. As the light rays encounter objects, effects such as reflection, refraction, scattering, and shadows occur.

Ray-traced images generally appear much more visually realistic than polygon-rendered scenes; however, raytracing incurs a much higher computational cost.

Today, most popular, visually rich, highly dynamic games take advantage of raytracing at least to some degree. For game developers, it is not an all-or-nothing decision to use raytracing. It is possible to render portions of scenes in the traditional polygon-based mode while employing raytracing to render the objects and surfaces in the scene that benefit the most from its advantages. For example, a scene may contain background imagery displayed as polygons, while a nearby glass window renders reflections of objects from the glass surface along with the view seen through the glass, courtesy of raytracing.

At the time of its release, the RTX 3090 was the highest-performing GPU available for running deep learning models with TensorFlow. **TensorFlow**, developed by Google's Machine Intelligence Research organization, is a popular open-source software platform for machine learning applications. TensorFlow is widely used in research involving deep neural networks.

The RTX 3090 leverages its machine learning capability to increase the apparent resolution of rendered images without the computational expense of rendering at the higher resolution. It does this by intelligently applying antialiasing and sharpening effects to the image. The technology learns image characteristics during the rendering of tens of thousands of images and uses this information to improve the quality of subsequently rendered scenes. This technology can, for example, make a scene rendered at 1080p resolution (1,920 x 1,080 pixels) appear as if it is being rendered at 1440p (1,920 x 1,440 pixels).

In addition to its raytracing and machine learning technologies, the RTX 3090 has the following features:

- **10496 NVIDIA CUDA® Cores**: The CUDA cores provide a parallel computing platform suitable for general computational applications such as linear algebra.
- **328 tensor cores**: The tensor cores perform the tensor and matrix operations at the center of deep learning algorithms.
- **A PCIe 4.0 x16 interface**: This interface communicates with the main processor.
- **24 GB of GDDR6X memory**: GDDR6X improves upon the prior generation of GDDR6 technology by providing an increased data transfer rate (up to 21 Gbit/sec per pin versus a maximum of 16 Gbit/sec per pin for GDDR6).
- **Nvidia Scalable Link Interface (SLI)**: The SLI links two to four identical GPUs within a system to share the processing workload. A special bridge connector must be used to interconnect the collaborating GPUs.

- **Three DisplayPort 1.4a video outputs:** The DisplayPort interfaces support 8K (7,680 x 4,320 pixels) resolution at 60 Hz.
- **HDMI 2.1 port:** The HDMI output supports 4K (3,840 x 2,160 pixels) resolution at 60 Hz.

The next section summarizes the subsystems with the Alienware Aurora Ryzen Edition R10.

Aurora subsystems

The major subsystems of the Alienware Aurora Ryzen Edition R10 are described briefly in the following table:

Subsystem	Description
Motherboard	The motherboard supports PCIe 4.0, doubling the bandwidth between the processor and graphics card over PCIe 3.0. Four slots are provided for DDR4 memory modules. Four PCIe slots are provided, though the double-width Nvidia GPU consumes two of them.
Chipset	The AMD B550A chipset supports processor and memory overclocking and PCIe 4.0.
Cooling	An Alienware liquid cooling system is provided to cool the processor, which is critically needed when overclocking.
Memory	The system includes up to 32 GB of dual-channel DDR4 XMP operating at 3200MHz. The **Extreme Memory Profiles (XMP)** configuration capability permits simultaneously changing several memory performance-related settings by simply selecting among different profiles. This function is usually used to select between a standard memory clocking configuration and an overclocked configuration.
Storage	The Aurora Ryzen Edition R10 includes a 1 TB NVMe M.2 solid state drive. **Non-volatile memory express (NVMe)** is an interface standard for connecting solid state drives to PCIe 4.0. The M.2 standard defines a small form factor for expansion cards such as SSDs.
Front panel	Three USB 3.2 Gen 1 Type A ports, a USB 3.2 Gen 1 Type C port, and audio input and output jacks are provided.
Rear panel	The system includes six USB 2.0 ports, 4 USB 3.2 Gen 1 Type A ports, a USB 3.2 Gen 1 Type C port, an Ethernet port, and digital and analog audio input and output jacks.

Table 13.2: Alienware Aurora Ryzen Edition R10 subsystems

The Alienware Aurora Ryzen Edition R10 gaming desktop integrates the most advanced technology available at the time of its introduction in terms of the raw speed of its processor, memory, GPU, and storage, as well as its use of machine learning to improve instruction execution performance.

The next section will take us from the level of the personal computer system discussed in this section and widen our view to explore the implementation challenges and design solutions employed in large-scale computing environments consisting of thousands of integrated, cooperating computer systems.

Warehouse-scale computing architecture

Providers of large-scale computing capabilities and networking services to the public and to sprawling organizations such as governments, research universities, and major corporations often aggregate computing capabilities in large buildings, each containing perhaps thousands of computers.

To make the most effective use of these capabilities, it is not sufficient to consider the collection of computers in a **warehouse-scale computer (WSC)** as simply a large number of individual computers. Instead, in consideration of the immense quantity of processing, networking, and storage capability provided by a warehouse-scale computing environment, it is more appropriate to think of the entire data center as a single, massively parallel computing system.

Early electronic computers were huge systems, occupying large rooms. Since then, computer architectures have evolved to today's fingernail-size processor chips possessing vastly more computing power than those early systems. We can imagine that today's warehouse-sized computing environments are a prelude to computer systems a few decades in the future that might be the size of a pizza box, or a smartphone, or a fingernail, packing as much processing power as today's WSCs, if not far more.

Since the internet rose to prominence in the mid-1990s, a transition has been in progress, shifting application processing from programs installed on personal computers over to centralized server systems that perform algorithmic computing, store and retrieve massive data content, and enable direct communication among internet users.

These server-side applications employ a thin application layer on the client side, often provided by a web browser. All the data retrieval, computational processing, and preparation of information for display takes place in the server. The client application merely receives instructions and data regarding the text, graphics, and user interface controls to present to the user. The browser-based application then awaits user input and sends requests for action back to the server.

Online services provided by internet companies such as Google, Amazon, and Microsoft rely on the power and versatility of very large data center computing architectures to provide services to millions of users. One of these WSCs might run a small number of very large applications providing services to thousands of users simultaneously.

Service providers strive to provide exceptional reliability, often promising 99.99% uptime, corresponding to approximately 1 hour of downtime per year.

The following subsections introduce the hardware and software components of a typical WSC and discuss how these pieces work together to provide fast, efficient, and highly reliable internet services to large numbers of users. This section concludes with a walkthrough of the steps involved in building and deploying a simple web application in a commercial cloud environment.

WSC hardware

Building, operating, and maintaining a WSC is an expensive proposition. While providing the necessary quality of service (in terms of metrics such as response speed, data throughput, and reliability), WSC operators strive to minimize the total cost of owning and operating these systems.

To achieve very high reliability, WSC designers might take one of two approaches in implementing the underlying computing hardware:

- **Invest in hardware that has exceptional reliability**: This approach relies on costly components with low failure rates. However, even if each individual computer system provides excellent reliability, by the time several thousand copies of the system are in operation simultaneously, occasional failures will occur at a statistically predictable frequency. This approach is very expensive and, ultimately, it doesn't solve the problem because failures continue to occur.

- **Employ lower-cost hardware with average reliability and design the system to tolerate individual component failures at the highest expected rates**: This approach permits much lower hardware costs compared to high-reliability components, though it requires a sophisticated software infrastructure capable of detecting hardware failures and rapidly compensating with redundant systems in a manner that maintains the promised quality of service.

Most providers of standard internet services, such as search engines and email services, employ low-cost generic computing hardware and perform **failover** by transitioning workloads to redundant online systems when failures occur.

To make this discussion concrete, we will examine the workloads a WSC must support to function as an internet search engine. WSC workloads supporting internet searches must possess the following attributes:

- **Fast response to search requests**: The server-side turnaround for an internet search request must be a small fraction of a second. If users are routinely forced to endure a noticeable delay, they are likely to switch to a competing search engine for future requests.
- **State information related to each search need not be retained at the server, even for sequential interactions with the same user**: In other words, the processing of each search request is a complete interaction. After the search completes, the server forgets all about it. A subsequent search request from the same user to the same service does not leverage any stored information from the first request.

Given these attributes, each service request can be treated as an isolated event, independent of all other requests, past, present, and future. The independence of each request means it can be processed as a thread of execution in parallel with other search requests coming from other users or even from the same user. This workload model is an ideal candidate for acceleration through hardware parallelism.

The processing of internet searches is less a compute-intensive task than it is data intensive. As a simple example, when performing a search where the search term consists of a single word, the web service receives the request from the user, then it extracts the search term and consults its index to determine the most relevant pages for the given term.

The internet contains, at a minimum, hundreds of billions of pages, most of which users expect to be able to locate via searches. This is an oversimplification, though, because a large share of the pages accessible via the internet are not indexable by search engines. However, even limiting the search to the accessible pages, it is simply not possible for a single server, even one with a large number of processor cores and the maximum installable amount of local memory and disk storage, to respond to internet searches in a reasonable time period for a large user base. There is just too much data and too many user requests. Instead, the search function must be split among many (hundreds, possibly thousands) of separate servers, each containing a subset of the entire index of web pages known to the search engine.

Each index server receives a stream of lookup requests filtered to those relevant to the portion of the index it manages. An index server generates a set of results based on matches to the search term and returns that set for higher-level processing. In more complex searches, separate searches for multiple search terms may need to be processed by different index servers. The results of those searches must then be filtered and merged during higher-level processing.

As the index servers generate results based on search terms, these subsets are fed to a system that processes the information into a form to be transmitted to the user. For standard searches, users expect to receive a list of pages ranked in order of relevance to their query. For each page returned, a search engine generally provides the URL of the target page along with a section of text surrounding the search term within the page's content to provide some context.

The time required to generate these results depends more on the speed of database lookups associated with the page index and the extraction of page content from storage than it does on the raw processing power of the servers involved in the task. For this reason, many WSCs providing web search and similar services use servers containing inexpensive motherboards, processors, memory components, and disks.

Rack-based servers

WSC servers are typically assembled in racks with each server consuming one 1U slot. A 1U server slot has a front panel opening 19" wide and 1.75" high. One rack might contain as many as 40 servers, consuming 70" of vertical space.

Each server is a fairly complete computer system containing a moderately powerful processor, RAM, a local disk drive, and a 1 Gbit/sec or faster Ethernet interface. Since the capabilities and capacities of consumer-grade processors, DRAM, and disks are continuing to grow, we won't attempt to identify the performance parameters of a specific system configuration.

Although each server contains a processor with integrated graphics and some USB ports, most servers do not have a display, keyboard, or mouse directly connected, except perhaps during their initial configuration. Rack-mounted servers generally operate in a so-called **headless** mode, in which all interaction with the system takes place over its network connection.

The following diagram shows a rack containing 16 servers:

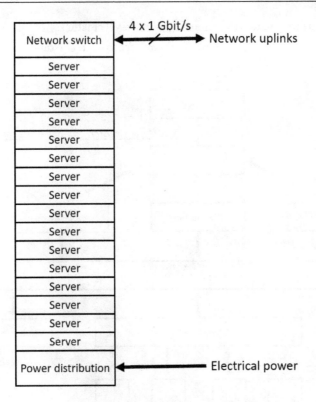

Figure 13.2: A rack containing 16 servers

Each server connects to the rack network switch with a 1 Gbit/s Ethernet cable. The rack in this example connects to the higher-level WSC network environment with four 1 Gbit/s Ethernet cables. Servers within the rack communicate with each other through the rack switch at the full 1 Gbit/s Ethernet data rate. Since there are only four 1 Gbit/s external connections leading from the rack, all 16 servers obviously cannot communicate at full speed with systems external to the rack. In this example, the rack connectivity is **oversubscribed** by a factor of 4. This means the external network capacity is one quarter of the peak communication speed of the servers within the rack.

Racks are organized into clusters that share a second-level cluster switch. The following diagram represents a configuration in which four racks connect to each cluster-level switch that, in turn, connects to the WSC-wide network:

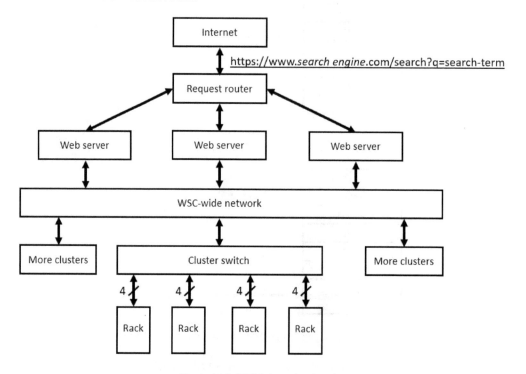

Figure 13.3: WSC internal network

In the WSC configuration of *Figure 13.3*, a user request arrives over the internet and is initially processed by a routing device that directs the request to an available web server. The server receiving the request is responsible for overseeing the search process and sending the response back to the user.

Multiple web servers are online at all times to provide load sharing and redundancy in case of failure. *Figure 13.3* shows three web servers, but a busy WSC may have many more servers in operation simultaneously. The web server parses the search request and forwards queries to the appropriate index servers in the rack clusters of the WSC. Based on the terms being searched, the web server directs index lookup requests to one or more index servers for processing.

To perform efficiently and reliably, the WSC must maintain multiple copies of each subset of the index database, spread across multiple clusters, to provide load sharing and redundancy in case of failures at the server, rack, or cluster level.

Index lookups are processed by the index servers and relevant target page text is collected from document servers. The complete set of search results is assembled and passed back to the responsible web server. The web server then prepares the complete response and transmits it to the user.

The configuration of a real-world WSC will contain additional complexity beyond what is shown in *Figure 13.2* and *Figure 13.3*. Even so, these simplified representations permit us to appreciate some of the important capabilities of a WSC implementing an internet search engine workload.

In addition to responding to user search requests, the search engine must regularly update its database to remain relevant to the current state of web pages across the internet. Search engines update their knowledge of web pages using applications called **web crawlers**. A web crawler begins with a web page address as its starting point, reads the targeted page, and parses its content. The crawler stores the page text in the search engine document database and extracts any links it contains. For each link it finds, the crawler repeats the page reading, parsing, and link-following process. In this manner, the search engine builds and updates its indexed database of internet content.

This subsection summarized a conceptual WSC design configuration, which is based on racks filled with commodity computing components. The next section examines the measures the WSC must take to detect component failures and compensate without compromising the overall quality of service.

Hardware fault management

As we've seen, WSCs contain thousands of computer systems. We can expect hardware failures will occur on a regular basis, even if more costly components have been selected to provide a higher, but not perfect, level of reliability.

As an inherent part of the multilevel dispatch, processing, and return of results implied in *Figure 13.3*, each server sending a request to a system at a lower level of the diagram must monitor the responsiveness and correctness of the system assigned to process the request. If the response is unacceptably delayed, or if it fails to pass validity checks, the lower-level system must be reported as unresponsive or misbehaving.

When such an error is detected, the requesting system immediately re-sends the request to a redundant server for processing. Some response failures may be due to transient events such as a momentary processing overload. If the lower-level server recovers and continues operating properly, no response is required.

If a server remains persistently unresponsive or erroneous, a maintenance request must be issued to troubleshoot and repair the offending system. When a system is identified as unavailable, WSC management (both the automated and human portions) may choose to bring up a system to replicate the failed server from a pool of backup systems and direct the replacement system to begin servicing requests from users.

Electrical power consumption

One of the major cost drivers of a WSC is electrical power consumption. The primary consumers of electricity in a WSC are the servers and networking devices that perform data processing for end users, as well as the air conditioning system that keeps those systems cool.

To keep the WSC electricity bill to a minimum, it is critical to only turn on computers and other power-hungry devices when there is something useful for them to do. The traffic load to a search engine varies widely over time and may spike in response to events in the news and on social media. A WSC must maintain enough servers to support the maximum traffic level it is designed to handle. When the total workload is below the maximum, any servers that do not have work to do must be powered down.

A lightly loaded server consumes a significant amount of electrical power. For best efficiency, the WSC management system should completely turn off servers and other devices when they are not needed. As the traffic load increases, servers and associated network devices can be powered up and brought online quickly to maintain the required quality of service.

The WSC as a multilevel information cache

We examined the multilevel cache architecture employed in modern processors in *Chapter 8, Performance-Enhancing Techniques*. To achieve optimum performance, a web service such as a search engine must employ a caching strategy that, in effect, adds levels to those that already exist within the processor.

To achieve the best response time, an index server should maintain a substantial subset of its index data in an in-memory database. By selecting content for in-memory storage based on historic usage patterns, as well as recent search trends, a high percentage of incoming searches can be satisfied without the need to access disk storage.

To make the best use of an in-memory database, the presence of a large quantity of DRAM in each server is clearly beneficial. The selection of the optimum amount of DRAM to install in each index server is dependent upon attributes such as the relative cost of additional DRAM per server in comparison to the cost of additional servers containing less memory, as well as the performance characteristics of more servers with less memory relative to fewer servers with more memory.

We won't delve any further into such analysis, other than to note that such evaluations are a core element of WSC design optimization.

If we consider DRAM to be the first level of WSC-level caching, the next level is the local disk located in each server. For misses of the in-memory database, the next place to search is the server's disk. If the result is not found in the local disk, the next search level takes place in other servers located in the same rack. Communications between servers in the same rack can run at full network speed (1 Gbit/s in our example configuration).

The next level of search extends to racks within the same cluster. Bandwidth between racks is limited by the oversubscription of the links between racks and the cluster switch, which limits the performance of these connections. The final level of search within the WSC stretches across clusters, which likely has further constraints on bandwidth.

A large part of the challenge of building an effective search engine infrastructure is the development of a high-performance software architecture. This architecture must satisfy a high percentage of search requests by the fastest, most localized lookups achievable by the search engine index servers and document servers. This means a high percentage of search lookups must be completed via in-memory searches in the index servers.

Deploying a cloud application

In this section, we walk through the steps to develop and deploy a simple web application on the Microsoft Azure cloud platform. This demonstrates how developers take full advantage of large and complex WSC service providers using readily available development tools. This example uses software tools and a cloud service that are available at no cost:

1. Visit https://azure.microsoft.com/en-us/free/ and create a free Azure account. Azure is a cloud computing service provided by Microsoft.
2. Visit https://nodejs.org/en/ and install Node.js. Node.js is a runtime environment for web server applications written in the JavaScript language.
3. Visit https://code.visualstudio.com/ and install Visual Studio Code. Visual Studio Code (abbreviated VS Code) is a multi-language source code editor.
4. Visit https://marketplace.visualstudio.com/items?itemName=ms-azuretools.vscode-azureappservice and install the Azure App Service for VS Code. Azure App Service is an extension to VS Code that assists with creating and deploying web applications in Azure.

The following instructions are demonstrated on a Windows host operating system, but similar (or identical in many cases) commands will work on Linux hosts.

After installing the tools, open a Windows Command Prompt and enter the following commands to create a simple web application and run it on your computer:

```
C:\Projects>npx express-generator webapp --view pug
C:\Projects>cd webapp
C:\Projects\webapp>npm install
C:\Projects\webapp>npm start
```

Open a web browser and navigate to http://localhost:3000. You will see a web page that looks like *Figure 13.4*:

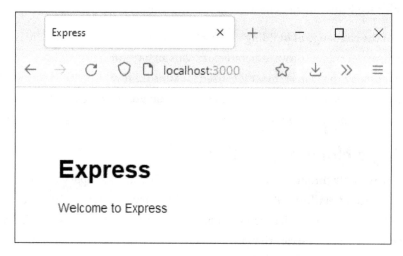

Figure 13.4: Simple Node.js application display

While this application does not do anything objectively useful, the code produced by the preceding steps provides a solid basis for building a sophisticated, scalable web application.

In the following steps, you will deploy this application to the Azure cloud environment using a free Azure account.

1. Start VS Code in the **webapp** directory:

   ```
   C:\Projects\webapp>code
   ```

2. Select the Azure logo as shown in the bottom left corner of *Figure 13.5*:

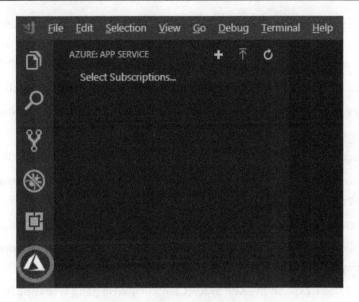

Figure 13.5: Adding an application setting

3. Click **Sign in to Azure...** and complete the process to create a new free account and sign in.
4. Click the cloud icon to the right of **APP SERVICE** as shown in *Figure 13.6*. If the icon is not visible, move the cursor to this area and it will appear. When prompted, select the **webapp** folder:

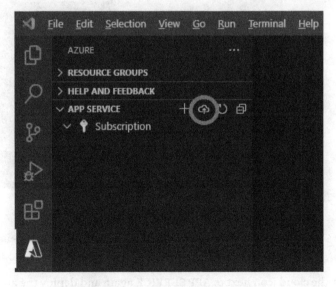

Figure 13.6: Deploying to the cloud

5. Click **+ Create new Web App... Advanced**. You will be prompted to create a unique name for your application. For example, the name **webapp** was already in use, but **webapp228** was available. You will need to select a different name.

6. Click **+ Create a new resource group** and give it a name. For example, **webapp228-rg**.

7. You will be prompted to select a runtime stack. In this case, select **Node 16 LTS**.

8. Select the **Windows** operating system.

9. Select the geographic location for deployment. For example, **West US 2**.

10. Select **+ Create new App Service plan**. Give it a name such as **webapp228-plan**. Select the **Free (F1)** pricing tier.

11. When prompted with **+ Create new Application Insights**, select **Skip for now**.

12. Application provisioning in Azure will begin and will take some time. When prompted **Always deploy the workspace "webapp" to "webapp228"?** select **Yes**.

13. In VS Code, expand the **APP SERVICE** node then expand **webapp228**. Right-click **Application Settings**. Select **Add New Setting...** as shown in *Figure 13.7*.

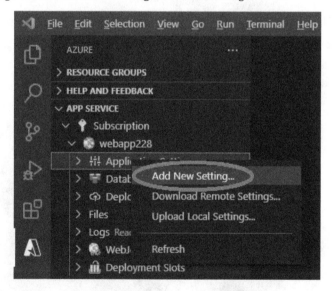

Figure 13.7: Adding an application setting

14. Enter SCM_DO_BUILD_DURING_DEPLOYMENT as the setting key and true as the setting value. This step will generate a **web.config** file automatically. This file is required for deployment.

15. Select the cloud icon next to **APP SERVICE** again and deploy the application.

16. After deployment completes, click **Browse Website** to open the website in your browser. In this example, the URL for the website is https://webapp228.azurewebsites.net/. This is a publicly accessible website, available to anyone on the internet.

These steps demonstrated the procedure for building and deploying a simple web application to a full-featured cloud environment. By deploying to the Azure environment, our application takes full advantage of Azure cloud platform capabilities that provide performance, scalability, and security.

The next section looks at the high-performance architectures employed in dedicated neural network processors.

Neural networks and machine learning architectures

We briefly studied the architecture of neural networks in *Chapter 6, Specialized Computing Domains*. This section examines the inner workings of a high-performance, dedicated neural net processor.

Intel Nervana neural network processor

In 2019, Intel announced the release of a pair of new processors, one optimized for the task of training sophisticated neural networks and the other for using trained networks to conduct inference, which is the process of generating neural network outputs given a set of input data.

The Nervana **neural network processor for training (NNP-T)** is essentially a miniature supercomputer tailored to the computational tasks required in the neural network training process. The NNP-T1000 is available in the following two configurations:

- The NNP-T1300 is a dual-slot PCIe card suitable for installation in a standard PC. It communicates with the host via PCIe 4.0 x16. It is possible to connect multiple NNP-T1300 cards within the same computer system or across computers using cables.
- The NNP-T1400 is a mezzanine card suitable for use as a processing module in an **Open Compute Project (OCP) accelerator module (OAM)**. A **mezzanine card** is a circuit card that plugs into another plug-in circuit card such as a PCIe card. OAM is a design specification for hardware architectures that implement artificial intelligence systems requiring high module-to-module communication bandwidth. Development of the OAM standard has been led by Facebook, Microsoft, and Baidu. Up to 1,024 NNP-T1000 modules can be combined to form a massive NNP architecture with extremely high-speed serial connections among the modules.

The NNP-T1300 fits in a standard PC and is something an individual developer might use. A configuration of multiple NNP-T1400 processors, on the other hand, quickly becomes very costly and begins to resemble a supercomputer in terms of performance.

The primary application domains for powerful NNP architectures such as Nervana include **natural language processing (NLP)** and machine vision. NLP performs tasks such as processing sequences of words and attempting to extract the meaning behind them and generating natural language for computer interaction with humans. When you call a company's customer support line and a computer asks you to talk to it, you are interacting with an NLP system.

Machine vision is a key enabling technology for autonomous vehicles. Automotive machine vision systems process video camera feeds to identify and classify road features, road signs, and obstacles, such as vehicles and pedestrians. This processing must produce results in real time to be useful while driving a vehicle.

Building a neural network to perform a human-scale task, such as reading a body of text and interpreting its meaning or driving a car in heavy traffic, requires an extensive training process. Neural network training involves sequential steps of presenting the network with a set of inputs together with the response the network is expected to produce given that input. This information, consisting of pairs of input datasets and known correct outputs, is called the **training set**. Each time the network sees a new input set and is given the output it is expected to produce from that input, it adjusts its internal connections and weight values slightly to improve its ability to generate correct outputs. For complex neural networks, such as those targeted by the Nervana NNP, the training set might consist of millions of input/output dataset pairs.

The processing required by NNP training algorithms boils down to mostly matrix and vector manipulations. The multiplication of large matrices is one of the most common and most compute-intensive tasks in neural network training. These matrices may contain hundreds or even thousands of rows and columns. The fundamental operation in matrix multiplication is the multiply–accumulate, or MAC, operation we learned about in *Chapter 6, Specialized Computing Domains*.

Complex neural networks contain an enormous number of weight parameters. During training, the processor must repetitively access these values to compute the signal strengths associated with each neuron in the model and make training adjustments to the weights. To achieve maximum performance for a given amount of memory and internal communication bandwidth, it is desirable to employ the smallest usable data type to store each numeric value. In most applications of numeric processing, the 32-bit IEEE single-precision, floating-point format is the smallest data type used. When possible, it can help to use an even smaller floating-point format.

The Nervana architecture employs a specialized floating-point format for storing network signals. The **bfloat16** format is based on the IEEE-754 32-bit single-precision, floating-point format, except the mantissa is truncated from 24 bits to 8 bits. The *Floating-point mathematics* section in *Chapter 9, Specialized Processor Extensions*, discussed the IEEE-754 32-bit and 64-bit floating-point data formats in some detail.

The reasons for using the bfloat16 format instead of the IEEE-754 half-precision 16-bit floating-point format for neural network processing are as follows:

- The IEEE-754 16-bit format has a sign bit, 5 exponent bits, and 11 mantissa bits, one of which is implied. Compared to the IEEE-754 single-precision (32-bit), floating-point format, this half-precision format loses three bits in the exponent, reducing the range of numeric values it can represent to one-eighth the range of 32-bit floating point.
- The bfloat16 format retains all eight exponent bits of the IEEE-754 single-precision format, allowing it to cover the full numeric range of the IEEE-754 32-bit format, though with substantially reduced precision.

Based on research findings and customer feedback, Intel suggests the bfloat16 format is most appropriate for deep learning applications because the greater exponent range is more critical than the benefit of a more precise mantissa. In fact, Intel suggests the quantization effect resulting from the reduced mantissa size does not significantly affect the inference accuracy of bfloat16-based network implementations in comparison to IEEE-754 single-precision implementations.

The fundamental data type used in ANN processing is the **tensor**, which is represented as a multidimensional array. A vector is a one-dimensional tensor, and a matrix is a two-dimensional tensor. Higher-dimension tensors can be defined as well. In the Nervana architecture, a tensor is a multidimensional array of bfloat16 values. The tensor is the fundamental data type of the Nervana architecture – the NNP-T operates on tensors at the instruction set level.

The most compute-intensive operation performed by deep learning algorithms is the multiplication of tensors. Accelerating these multiplications is the primary goal of dedicated ANN processing hardware such as the Nervana architecture. Accelerating tensor operations requires not just high-performance mathematical processing; it is also critical to transfer operand data to the core for processing in an efficient manner and move output results to their destinations just as efficiently. This requires a careful balance of numeric processing capability, memory read/write speed, and communication speed.

Processing in the NNP-T architecture takes place in **tensor processor clusters (TPCs)**, each of which contains two **multiply–accumulate (MAC)** processing units and 2.5 MB of high-bandwidth memory. **High-bandwidth memory (HBM)** differs from DDR SDRAM by stacking several DRAM die within a package and providing a much wider (1,024 bits in comparison to 64 bits for DDR) data transfer size. Each MAC processing unit contains a 32 x 32 array of MAC processors operating in parallel.

An NNP-T processor contains up to 24 TPCs, running in parallel, with high-speed serial interfaces interconnecting them in a fabric configuration. The Nervana devices provide high-speed serial connections to additional Nervana boards in the same system and to Nervana devices in other computers.

A single NNP-T processor can perform 119 **trillion operations per second (TOPS)**. The following table shows a comparison between the two processors:

Feature	NNP-T1300	NNP-T1400
Device form factor	Double width card, PCIe 4.0 x16	OAM 1.0
Processor cores	22 TPCs	24 TPCs
Processor clock speed	950 MHz	1,100 MHz
Static RAM	55 MB SRAM with ECC	60 MB SRAM with ECC
High bandwidth memory	32 GB second generation **high bandwidth memory (HBM2)** with ECC	32 GB HBM2 with ECC
Memory bandwidth	2.4 Gbit/s (300 MB/s)	2.4 Gbit/s (300 MB/s)
Serial **inter-chip link (ICL)**	16 x 112 Gbit/s (448 GB/s)	16 x 112 Gbit/s (448 GB/s)

Table 13.3: Features of the NNP T-1000 processor configurations

The Nervana **neural network processor for inference (NNP-I)** performs the inference phase of neural network processing. Inference consists of providing inputs to pretrained neural networks, processing those inputs, and collecting the outputs from the network. Depending on the application, the inference process may involve repetitive evaluations of a single, very large network on time-varying input data, or it may involve applying many different neural network models to the same set of input data at each input update.

The NNP-I is available in two form factors:

- A PCIe card containing two NNP I-1000 devices. This card is capable of 170 TOPS and dissipates up to 75 W.
- An M.2 card containing a single NNP I-1000 device. This card is capable of 50 TOPS and dissipates just 12 W.

The Nervana architecture is an advanced, supercomputer-like processing environment optimized for training neural networks and performing inferencing on real-world data using pretrained networks.

Summary

This chapter presented several computer system architectures tailored to particular user needs and identified some key features associated with each of them. We looked at application categories including smartphones, gaming-focused personal computers, warehouse-scale computing, and neural networks. These examples provided a connection between the more theoretical discussions of computer and systems architectures and components presented in earlier chapters and the real-world implementations of modern, high-performance computing systems.

Having completed this chapter, you should understand the decision processes used in defining computer architectures to support specific user needs. You have gained insight into the key requirements driving smart mobile device architectures, high-performance personal computing architectures, warehouse-scale cloud-computing architectures, and advanced machine learning architectures.

The next chapter presents the categories of cybersecurity threats modern computer systems face and introduces computing architectures suitable for applications that require an exceptional assurance of security, such as national security systems and financial transaction processing.

Exercises

1. Draw a block diagram of the computing architecture for a system to measure and report weather data 24 hours a day at 5-minute intervals using SMS text messages. The system is battery powered and relies on solar cells to recharge the battery during daylight hours. Assume the weather instrumentation consumes minimal average power, only requiring full power momentarily during each measurement cycle.

2. For the system of *Exercise 1*, identify a suitable commercially available processor and list the reasons that processor is a good choice for this application. Factors to consider include cost, processing speed, tolerance of harsh environments, power consumption, and integrated features such as RAM and communication interfaces.

Join our community Discord space

Join the book's Discord workspace for a monthly *Ask me Anything* session with the author: https://discord.gg/7h8aNRhRuY

14

Cybersecurity and Confidential Computing Architectures

This chapter introduces computing architectures suitable for applications that require an exceptional assurance of security. Critical application areas like national security systems and financial transaction processing demand this high level of protection. These systems must be resilient against a broad range of cybersecurity threats, including malicious code, covert channel attacks, and attacks enabled by physical access to the computing hardware. Topics addressed in this chapter include cybersecurity threats, encryption, digital signatures, and secure hardware and software design.

After completing this chapter, you will be able to identify the categories of cybersecurity threats a system will face and understand the security features of modern computer hardware. You will understand how to avoid security gaps in system architectures and how a secure computer architecture can help enforce security in software applications.

The following topics will be presented in this chapter:

- Cybersecurity threats
- Features of secure hardware
- Confidential computing
- Designing for security at the architectural level
- Ensuring security in system and application software

Technical requirements

The files for this chapter, including solutions to the exercises, are available at https://github.com/PacktPublishing/Modern-Computer-Architecture-and-Organization-Second-Edition.

Cybersecurity threats

The first step in ensuring the cybersecurity of a computer system is to understand the threats it is likely to encounter and must defend against. We can place these threats into a few broad categories and identify the key features of each category. With this information, we can design a computer system architecture with attributes that will provide an appropriate level of protection against those threats.

Unfortunately, developing a completely secure computer system is not a straightforward process. This is because new vulnerabilities are identified regularly in existing operating systems, software libraries, user applications, and web applications. It is not unusual for flaws to be identified in widely used cryptographic components such as encryption algorithms and authentication protocols. As new software products are developed, they often contain entirely new vulnerabilities that will be discovered sooner or later. Most importantly, experience shows the greatest security weaknesses related to the computer systems we use are created by the humans who operate them.

Given this reality, we can only strive to reduce the risk of a damaging cyberattack to an acceptably low level, rather than eliminate it entirely. In the process of hardening a system against attack, we must keep the cost of protective measures to an acceptable level, and the system cannot be overly cumbersome for users to operate. The protections must also avoid reducing the reliability of the system in terms of performing its intended function.

Cybersecurity threat categories

As a first step to understanding the landscape of cybersecurity threats, we will examine the categories of people and organizations who plan and carry out cyberattacks. Based on historical patterns of cyberattacks against individuals, governments, businesses, and other groups, these are the primary threat categories:

- **National governments**: Many countries, including some that may be considered "hostile" and others one might normally think are "friendly," have been credibly identified as conducting aggressive cyber actions against other governments as well as businesses and other organizations in different countries, and even against specific individuals. Military cyber organizations plan and carry out cyberattacks against enemy nations to achieve tactical and strategic goals.

- **Terrorists:** Terrorist organizations have expressed the desire to conduct cyberattacks with potentially devastating effects against governments and industrial systems with potential outcomes ranging from widespread power blackouts to the destruction of large-scale infrastructure, such as dams, refineries, and pipelines. While the potential for such attacks has been repeatedly demonstrated, actual occurrences of destructive attacks have, so far, been rare.
- **Industrial spies:** In many industries, particularly those involving high technology, secret information about software components and hardware systems can be a prime target for hackers. If the attackers can gain access to the computer network where the desired information resides, they may be able to copy it and use it for their own purposes, or simply sell it to the highest bidder.
- **Criminal groups:** Groups of criminal hackers carry out attacks such as penetrating computer systems containing private information like credit card data with the goal of selling the information to other criminals or using it directly to make illegal purchases. Ransomware is another category of cybercrime. In a ransomware attack, the hackers load software onto the victim's computer that encrypts valuable data files. The attacker demands payment from the victim before providing a key that will purportedly decrypt the user's data. At the lower end of the technological scale, mass-emailing scams continue to proliferate. In these efforts, attackers attempt to lure recipients into engaging the criminals in a conversation that can lead to expensive disappointment for the unwary.
- **Hacktivists:** Some cyberattackers claim to be driven by just motives in their attacks against perceived immoral enemies such as disfavored corporations or governments. These individuals are labeled **hacktivists**, a combination of the words *hacker* and *activist*. Hacktivists perform actions such as bombarding the website of their target with so much traffic that the site becomes unavailable for regular users. Hacktivists may also attempt to steal sensitive data belonging to their target and then publish the data to embarrass or shame the organization that was the source of the information.
- **White-hat hackers:** Traditionally, hackers might investigate a target and then plan and execute a cyberattack against it for the pure intellectual thrill. In other cases, cybersecurity researchers will investigate a target, possibly a website or a digital device such as a smartphone, and identify one or more vulnerabilities. The researcher then shares the information with the owner of the site or the manufacturer of the device with the goal of encouraging improved security. After allowing sufficient time for the issues to be resolved, the researcher then makes the information about the vulnerability public. This process is referred to as **white-hat hacking**, a reference to early American Western films in which the protagonists wore white hats and the villains wore black hats.

- **Insiders (witting and unwitting):** The most common precipitating factor for cyber intrusions into protected networks is an action taken by an authorized user of a computer system on that network. In many cases, the action is something as simple as clicking on a link in an email that references a topic of interest to the recipient. This can lead to the installation of malware that spreads quickly throughout the supposedly protected network. In other cases, perhaps due to an employee's discontent with the working environment, a user intentionally takes steps to enable intrusion into the network by malicious actors. In either case, the network is compromised, and the attackers have a foothold that enables further exploitation of the supposedly secure network. Once attackers have achieved this level of access, they can take any action they wish, such as extracting confidential company information or installing ransomware on critical computer systems.

As of 2021, ransomware is the most widespread malware threat to individuals, businesses, and other organizations.

Cyberattack techniques

Regardless of a cyberattacker's motivating factors, these actors employ a few common types of attacks against computer systems, whether the target is a web server, an industrial control system, or a PC owned by an individual. These are some of the attack types that are most likely to occur:

- **Phishing: Phishing** describes email-based attempts to convince the recipient to take some action that will further the attacker's goals. The email may request the recipient to click a link, or possibly download and open a file, or sometimes just respond to the sender with an answer indicating interest in the topic of the original email. In most cases, the goal of the phishing effort is to convince the recipient to take some action that results in the installation of malware, giving the malware sender control over the recipient's computer.
- **Botnets:** A **botnet** (short for *robot network*) is a collection of computers that have been infected by malware and are under the control of a single attacker. This individual is known as a **bot herder**. The computers within a botnet can be commanded to perform nefarious tasks such as sending out phishing emails in an attempt to recruit more members for the botnet or conducting **denial-of-service (DoS)** attacks by attempting to overload websites with thousands of simultaneous requests for service from the botnet member computers.
- **Password attacks:** In most computer systems, the passwords users enter to log in to the system are not stored in the same format they are typed in at the keyboard. Instead, the passwords are passed through a cryptographic hashing algorithm that converts the password to an unrecognizable string of binary data.

The important features of a cryptographic hash algorithm are that entering the same password will always produce the same hashed value, and knowledge of the hashed value does not give the attacker any easy way to recover the corresponding password. Even though possession of a hashed password value does not allow direct discovery of the password, it is still possible for attackers to determine a user's password given its hashed value. The standard approach for discovering a user's password given a known hash is to simply try all possible passwords until a matching hash value is found. This technique is referred to as a **brute-force password attack**. The best defense against this type of attack is to use long passwords (15 characters or longer) with a mixture of upper- and lower-case characters along with some symbols, such as * and &. Passwords that are short (less than 10 characters) or can be found in a dictionary are trivially easy to crack, as well as unfortunately common passwords such as `123456`, `qwerty`, and `password1`.

- **Vulnerability exploitation:** White hat security researchers and black hat criminal hackers put great effort into discovering cybersecurity vulnerabilities in existing operating systems, applications, websites, and embedded devices. White hat hackers pursue the goal of improved cybersecurity for everyone by first informing the developer of the vulnerable system or application and then, after allowing time for the problem to be fixed, publicizing the vulnerability for everyone to see. The white hats are not entirely driven by altruistic motives – discovering a critical vulnerability in a major software system can lead to publicity, reputation enhancement, and respect from peer researchers. Black hat hackers use any vulnerability they discover to commit criminal activity against computers that are susceptible to the attack. Often, it is only after systems have been attacked using a previously unknown vulnerability that the system developers and independent white hat researchers dissect the offending malware and discover the vulnerability in the targeted system.

Vulnerabilities are categorized by the length of time (in days) that system administrators and users have to prepare defenses against a newly discovered vulnerability before that vulnerability comes under attack. In the worst-case scenario for the owners and users of the victim computer system, attacks can begin with no warning at all. In this case, the vulnerability is referred to as a **zero-day** vulnerability, which means there was no warning in advance of the attack.

Types of malware

Software installed on a computer system without the owner's permission and that attempts to perform actions the computer owner does not want is called malware.

Malware, an abbreviation of **malicious software**, is a category of code that performs undesired actions and often interferes with the normal operation of the computer. Some of the most common types of malware are:

- **Spyware: Spyware** is software that collects personal information about the computer user and other valuable data and transmits it to the attacker. This may include information such as usernames and passwords for online accounts and the addresses of websites visited. Spyware may use this information to display pop-up advertisements targeted to the perceived interests of the user. The tactics used by spyware are similar to the methods commonly used in advertisements by legitimate businesses, so classifying a piece of apparent spyware as malware may not be a straightforward process.

- **Ransomware**: As indicated earlier in this chapter, ransomware most commonly encrypts the user's data files and then demands payment in return for providing a key to decrypt the data. In another form of ransomware, the attacker steals personal information from an individual or confidential company information and demands payment to prevent the public release of the information. Ransomware attacks have frequently targeted businesses and organizations such as hospitals. These criminals usually demand payment in the form of cryptocurrency, which they perceive to be untraceable. While in many cases payment of the ransom results in the successful decryption of the encrypted data or the withholding of the private information from public release, there are no guarantees that paying the ransom will result in a favorable outcome for the victim.

- **Viruses**: Like a biological virus, a software virus replicates by infecting new hosts with copies of itself. In addition to functioning as spyware, ransomware, or some other form of cyberattack, a virus contains code that attempts to gain access to other computer systems accessible over the network and install copies of itself onto them. A virus inserts itself into an existing computer program by making modifications to the program to incorporate the virus code. The modified program then becomes a virus host and begins attempting to infect other computers. A program that has been modified to incorporate the virus is said to be virus-infected. The requirement for replication via a host software application distinguishes a computer virus from a worm, which does not require a host application for this purpose.

- **Worms**: Similar to a virus, a computer worm is a standalone program that attempts to replicate by installing copies of itself onto other computers accessible over the network. A worm does not require an infected host application in the manner of a computer virus. Like a virus, a worm may contain code to act as spyware, ransomware, or some other type of attack.

- **Man in the middle**: In a **man in the middle (MITM)** attack, the attacker attempts to position software in a communication path between two communicating applications or computers. If successful, an MITM attack permits the attacker to capture sensitive information such as usernames and passwords and allows the attacker to modify information as it passes between the communicating nodes.

- **Denial of service**: In a **denial of service (DoS)** attack, a target system such as a web server is bombarded with an excessive level of spurious requests. This attack may achieve its goal by making the targeted system unusable by legitimate users for an extended period. DoS attacks are popular among hacktivists, who employ the technique to cause problems and draw negative attention to businesses and other organizations that are targets of their ire.

- **SQL injection**: Web applications often use **Structured Query Language (SQL)**, a database programming language, to communicate between the user interface and databases containing information such as user credentials and information on products for sale through a website. If the developers of the web application are not sufficiently careful in employing secure coding practices, it may be possible for users of the application to craft inputs that are interpreted as executable code by the database SQL interpreter. If successful, this enables the attacker to extract and modify data in the database, and it sometimes leads to a more devastating attack against the network on which the database resides.

- **Key logger**: A **key logger** is a type of malware that collects the sequence of keys a computer user presses on the keyboard and forwards this information to the attacker. A key logger can record important private information such as usernames and passwords for bank accounts and credit card accounts. More generally, key logger-type malware can perform functions such as capturing screen shots from the victim's computer and even recording using the video camera and microphone attached to an infected computer or smartphone. Key logger functionality is often a component within a larger piece of malware that contains spyware, ransomware, or virus capabilities.

- **Infrastructure attacks**: Many types of large-scale infrastructure such as power plants, electrical transmission systems, dams, refineries, and pipelines rely on computerized control systems for their operation. Traditionally, the specialized computers used for these purposes have possessed only rudimentary security features. There is great concern among cybersecurity professionals that sophisticated malware attacks against these systems could lead to drastic consequences such as widespread power outages or explosions at refineries.

Some types of malware combine several of these capabilities and perform actions such as conducting an extensive analysis of the newly infected system before it decides if it will proceed with an attack. This analysis may attempt to determine factors such as the country where the infected system is located, the company the system belongs to, and the particular applications installed on the computer.

Some malware arrives in a mostly-encrypted form and only performs decryption as individual segments of code prepare to execute. A goal of code encryption is to make it as difficult as possible for cybersecurity researchers to examine the code and understand what it does. Code encryption also may make it possible for malicious code to slip through the automated defenses provided by antivirus software.

Post-exploitation actions

Once an attacker has gained remote access to a targeted system, a variety of options become available to further exploit that system as well as other computers on the same network. One of the first steps an attacker normally takes after gaining access to a victim system is to install software that enables continued access even if the currently logged-in user logs out, or if the system reboots. Many types of attacks take this step automatically as part of their initial entry into the victim computer. In other cases, an attacker conducting remote probing of a target system will install this code once access has been achieved.

The attacker's first entry into a victim system often takes place at the privilege level of the user that performed the action enabling the access, for example, clicking a link that downloaded and installed a piece of malware. Many types of malicious activity require only the limited privilege level of an ordinary user. Some examples are ransomware that encrypts the user's data files or botnet malware that bombards a targeted web server with requests using the limited privileges available to an ordinary user.

Some types of attacks, such as those attempting to gain access to protected information like a database containing customer information, require administrator-level privileges. Once access into a computer system at the unprivileged user level has been achieved and software has been installed to enable ongoing communication between the attacker and the victim system, the next step may be to attempt privilege escalation.

Privilege escalation is the process an attacker uses to achieve a higher privilege level within the target system, thereby enabling access to system resources such as files and databases that are off-limits to regular users.

Privilege escalation can be achieved by a variety of means. Often, known flaws exist in operating systems, driver software, or applications that run at higher privilege levels. If the system is not updated regularly to install fixes for known vulnerabilities, attackers can exploit these flaws in a manner that enables them to raise their privilege to a higher level, granting them access to protected system resources.

Having established persistent connectivity into a victim system and gained administrative privileges, the attacker has full control of the computer. Any data present on the computer is available for extraction and any software the attacker desires can be installed. In hacker terminology, a computer that is under the full administrative control of a remote attacker is considered to be *owned* by the attacker. If the attacker achieved this level of control without any notice by legitimate users or by software such as antivirus tools, the attack is said to be a *perfect own*, shortened to *pwn* (pronounced "pone") in hacker slang.

One constraint the attacker must usually keep in mind is a need to avoid making the fact the computer is infected with malware overly obvious to its legitimate users. Most users are not too concerned if the computer is a bit more sluggish than usual, but seeing a command window pop up onscreen followed by the display of commands typed by the attacker would probably alert anyone that the system has been breached. Sophisticated attackers strive to avoid making moves that expose their presence to the victims of their attacks.

This section provided an overview of some of the more common cybersecurity threat categories and techniques. The methods and malware types listed here are not exhaustive and new methods for attacking digital systems and devices are under development continuously. For the computer system architect, it is not sufficient to understand the types of cyberattacks that have been common in the past. It is vital to understand the full range of attacks that are even theoretically possible, even if some of those attacks have not been observed in past attacks. The next section will examine some key features that computing devices must implement to provide a high level of assurance the system and its interfaces to other networks, users, and devices will remain secure for the lifetime of the system.

Features of secure hardware

When initiating the design of a new computer or digital device, or when revisiting the design of an existing system, it is vital that the computer architect considers security as a top-level requirement. Even the most basic decisions in the development process, such as selecting a processor model, are likely to have profound effects on the security of the resulting design. As a first step in this process, it is necessary to understand what important types of data and other technology-related information must be protected from disclosure to unauthorized individuals.

Identify what needs to be protected

Some types of information present on computers and networks that generally need to be protected against unauthorized disclosure are:

- Personal information such as passwords, social security numbers, financial data, and medical history.
- Confidential information belonging to a business including customer lists, product design data, and strategic plans.
- Proprietary technology such as the digital circuit design within a smartphone.
- Government information such as national defense information and intelligence data collected by law enforcement.

In designing a particular system, the computer architect must remain constantly aware of the types of information that need to be protected. These concerns relate to the digital data contained within a system as well as hardware features that may expose vulnerabilities that grant attackers access into the device.

The system security analysis must include an assessment of the degree to which the system can be physically accessed by potential attackers. For items delivered to end users, such as personal computers and smartphones, users can obviously take any steps they desire with the hardware, including disassembling it and examining the components under microscopes.

For hardware intended to be operated in a controlled environment, such as a cloud server farm, protection against outright physical attacks may be less of a concern. However, given the potential for attacks by malicious insiders, some thought must be given to at least detecting the occurrence of unacceptable tampering, even if such tampering may be the result of actions by well-intentioned employees.

A particular category of intellectual property worthy of protection is the firmware and software code installed in devices delivered to users. If the manufacturer desires to maintain this code as a trade secret, the development process must include steps to secure the code and prevent its exposure by even the most skilled and capable attackers.

Anticipate all types of attacks

In designing the security features of a digital system, it is important to factor in not just the types of attacks that have been observed in the past.

The architect must also cast a wide net to identify attack categories that may have never been seen but are at least theoretically possible. Such thinking may lead to technologies that seem to be of little threat in the near term (such as, perhaps, quantum computing) but nevertheless can be effectively mitigated with an acceptable expenditure of resources.

For computer systems that do not require access to the internet or to other external networks, it is common to construct the computing environment in a manner that ensures isolation from connection to outside networks. This configuration is referred to as **air-gapped**, which means there is a significant physical distance between the computer hardware and any potential connection to external networks.

While constructing an air-gapped computing environment theoretically provides a substantial security enhancement, in real-world usage the benefit of this architecture has proven to be limited. For any computer system to provide substantial value, it is usually necessary to regularly transfer information into the system in the form of software updates and updates to data used on the computer. For an air-gapped system, these updates typically arrive on optical disks or on portable hard drives. Despite the best intentions and security efforts of the system operators, this data transfer process offers a substantial opportunity for malware to travel into the protected system and, once infected, enables the malware to attempt to transfer data out of the system and back to the attackers using the same data transfer mechanisms.

Sophisticated hackers are continually working to develop attack methods against air-gapped computer systems and other high-security computing environments. The designers of secure computing systems must consider all the potential avenues these attackers may use in their efforts to gain access and exploit the system. Creativity is vital to ensuring a secure design.

All possible forms of information access and data leakage are fair game for a dedicated attacker. Even if a computer system is intended to never connect to an externally accessible network, other potentially exotic attacks may be viable for a determined adversary. Cyberattacks that enable transfer of data between processes that are not intended to have permission to communicate are called **covert channel attacks**. Some surprising types of attacks that have demonstrated at least some degree of success are listed here:

- **Row hammer**: The modern DRAM devices discussed in earlier chapters have been shown to be vulnerable to a type of attack called **row hammering**. The "row" in row hammer refers to the rows of bit cells in a DRAM device. Due to the tiny size of each bit cell and its proximity to neighboring cells in the same row and in adjacent rows, it becomes possible to change the state ("flip the bits") in cells in adjacent rows under certain conditions.

To cause this effect, code performs repeated accesses to a targeted DRAM row at a high rate. To be successful, the code performing this attack must ensure its memory access requests result in cache misses, thus ensuring the DRAM internal circuitry is activated. Row hammer attacks have demonstrated the ability to achieve privilege escalation in computers based on the x86 architecture.

- **Power consumption fluctuations**: Even if attackers succeed at installing malware on an air-gapped computer system, extraction of data collected by the malware remains a challenge. If it is not feasible to exfiltrate information via the disks used to move data to and from the air-gapped system, an alternative method must be identified. Researchers have demonstrated that malware running on a target computer system can create sufficient variations in the building's electrical power consumption that it may be possible to monitor those fluctuations on the power lines outside the building. By carefully coding digital data into the fluctuations, the attackers can collect data from the air-gapped computer.

- **Thermal fluctuations**: Modern computers have fans and sometimes liquid cooling systems to manage the system temperature and keep it within prescribed limits. When computers are located close together, the temperature of one computer may affect the temperature readings of a nearby computer. If malware is present in two closely spaced computers, one on an air-gapped network and one connected to an external network, it has been demonstrated that causing intentional temperature fluctuations in one computer by inducing processor loading can induce measurable temperature changes in the adjacent computer. Using this technique, malware on the air-gapped computer can transmit digital data to the network-connected computer. Data can be transferred in the opposite direction in the same manner. While extremely slow, in terms of bit transfer rate, this technique permits the transfer of critical data such as encryption keys that can be used to enable further attacks.

- **Electromagnetic emissions**: Any time electrical current flows through a conductor such as a USB cable or across a trace on a printed circuit board an electromagnetic wave is transmitted into the surrounding environment. If a threat actor can place a receiving antenna within range of this signal, it may be possible for the attacker to collect the information that was originally transmitted. Of course, in an electrically noisy environment, with perhaps many computers operating simultaneously, it may be exceptionally difficult to extract the signal radiated from any single system. However, if malware is present on the targeted computer, the malware may take actions to intentionally generate fluctuations in radiated emissions in a pattern that can be detected and decoded by a suitable reception system. Stranger things have happened.

This section has presented some examples of the more exotic types of attacks that a mission-critical computer system may experience. While by no means complete, this list of attacks provides some examples of the types of cyberattacks a system architect must consider when designing a secure system.

Features of secure system design

Given these examples of possible attacks, we can list some important features a secure computer system hardware design must implement to provide a high level of security assurance.

Secure key storage

Any cryptographic keys used by the system to secure data must be stored in a manner that prohibits their extraction by any conceivable means. This typically means the keys must be stored within a device such as a processor chip in a manner that prevents their extraction by any software technique. It also means that attempts to extract the keys by methods such as disassembling the integrated circuit or using sophisticated tools such as a scanning electron microscope must be unsuccessful.

Encryption of data at rest

Any data stored in the system must be protected when the system is powered off. This means that even if the hardware is disassembled and the contents of individual memory devices are extracted by an attacker, the data within them remains secure. The most common method of achieving this level of security is to encrypt the data using an encryption key that is available for use by the computer but fully protected against disclosure outside of its intended use. One way to store this key is in special registers within a processor that cannot be accessed even by malware running at elevated privilege. Many modern processors, even tiny embedded processors, have begun to provide carefully designed cryptographic capabilities for this purpose.

Encryption of data in transit

Any data transferred from a processor or communication interface can be accessed by an attacker with physical access to the communication path. Whether the data is transferring through a trace on a circuit board or over a global network, any information that must remain secure needs to be protected over the entire path from source to destination. Once again, the common approach for implementing this protection is the use of encryption. Securing data in transit between two endpoints is more challenging than encrypting and decrypting data in local storage because it is necessary to perform the encryption and decryption at two different locations.

Conceptually, the most straightforward method to achieve this is to give the encryption key to the systems at both ends of the communication path. However, it is challenging to transport a secret key to both systems if there is not already a secure communication path between them. The standard approach used today to set up secure communication paths between systems that don't already have a shared secret key is to use public key encryption to transfer a secret key from one end of the communication path to the other. The secret key is then used to encrypt and decrypt the data at each end of the communication path. While it would be possible to just use the public key encryption process on all the data shared between the systems, it turns out that using public key encryption is far more computationally intensive than using a shared secret key.

Cryptographically secure key generation

When a fresh secret key is needed, perhaps when setting up a secure communication path between two computer systems, it is vital that the key is completely unpredictable to any outside attackers. Any newly generated secret key should appear entirely random and completely unrelated to any prior or subsequent keys created by the same system or by other systems. Traditionally, the easiest way to create a random-looking number on a computer was to use the pseudorandom number generation capabilities available in many programming language libraries. It turned out that the number sequences produced by many of these algorithms did not produce random-like outputs and these algorithmic flaws could be exploited in ways that made it far easier to crack encryption algorithms than an initial analysis would make it appear. Modern cryptographic random number generators use specialized hardware to generate *truly* random numbers, enabling the creation of cryptographic keys that are as secure as possible.

Secure boot procedure

In a secure system, all code that executes at an elevated privilege level must be verified as authentic before being permitted to run. This includes all code that executes as part of the boot process as well as the operating system kernel and drivers. The standard approach for achieving this goal is to attach a digital signature to each piece of code. The digital signature contains an encrypted hash that is computed over the entire block of executable code, whether it resides as a file on disk or is stored in flash memory within the processor. The key used to decrypt the digital signature hash must be stored securely within the processor hardware. With this level of protection, any attempt to replace the legitimate code with maliciously modified code will fail because the attacker lacks the key necessary to prepare a valid digital signature for the modified code.

Tamper-resistant hardware design

The features of secure computer architectures described in this section all rely to some degree on processor hardware that is able to securely store secret information like cryptographic keys. To ensure long-term protection of the sensitive information processed by the computer, the hardware architecture must remain secure against any conceivable type of attack. Some examples of methods that an attacker might use to try to recover sensitive information given physical access to a computing device are:

- **Physical access into the device**: By performing actions such as carefully grinding or chemically eroding the casing of the integrated circuit, the attacker can gain access to internal circuit components. Having achieved this access, it may be possible to probe circuit components electrically and extract information from them.
- **Monitoring electromagnetic emissions**: After gaining internal access to a device containing sensitive information, it may be possible to use a microscopic antenna to monitor activity within particular components or information traveling through interconnections.
- **Microscopic examination**: It may be possible to use exotic laboratory instrumentation such as a *scanning voltage microscope* to measure the voltage distribution across the surface of a circuit such as a memory array.

While the deployment of some of the attack techniques described in this section against ordinary computer systems may seem to be extremely unlikely, you may be surprised at how accessible these techniques can be. Although most of us do not own or have access to a scanning voltage microscope, there are companies that own these devices and are willing to perform scans for customers for a surprisingly reasonable price. This is one reason why it is important for architects of computer systems that will contain extremely sensitive or valuable information to take all potential attack types into account during system design.

In the next section, we will discuss confidential computing, where strong security measures are employed to ensure data is protected through its processing lifecycle.

Confidential computing

Confidential computing is a recent development that aims to use cryptography and hardware-level security features to ensure data is always protected. Data can be in one of three states: at rest, in transit, or in use. **Data at rest** is typically located in files on a storage device. **Data in transit** refers to data traveling over some type of communication medium. **Data in use** is being actively operated on by a processor and resides in the processor's main memory.

Confidential computing aims to ensure a comprehensive level of protection for data in all three of these possible states. Traditional security mechanisms focus on one state at a time, such as encrypting data on disk or while transferring information to and from a website. These approaches neglect the necessity to provide the same level of protection to data in use.

Securing data in use requires support from processor hardware to isolate applications from each other and to ensure the protection of sensitive code and data. As one example of hardware supporting confidential computing, the Intel **Secure Guard Extensions (SGX)** enable robust application isolation and protection of data in use.

Intel claims that SGX protects application data even when the BIOS, operating system, and the application itself have been compromised and the attacker has full control of the platform.

The SGX technology creates isolated memory regions called enclaves. Each enclave contains non-addressable memory regions that hold an application's code and data in encrypted memory pages.

Applications built with SGX technology consist of two parts: an untrusted part and a trusted part. The untrusted part of the application creates the trusted part, which contains the secure enclave. Enclaves are considered trusted because they cannot be modified after they have been built. If an enclave is modified, the change will be detected and it will not be permitted to execute. Code running within the trusted part of the application accesses data within the enclave as clear (unencrypted) text. Any code that is outside the enclave, such as the BIOS, operating system, or even the untrusted part of the same application, is denied access to the enclave memory region. Even code outside the enclave that is executing at the kernel privilege level is unable to gain access to the secure data within the enclave.

Memory pages within a secure enclave can be evicted to secondary storage using traditional page swapping algorithms. The encryption of code and data within the enclave pages protects the information they contain when secure pages are in the untrusted swap file.

The SGX technology supports the concept of software attestation. Using a **software attestation** procedure, code that is remotely accessing the functionality offered by a secure enclave can validate it is communicating with the specific enclave it intends to work with and not with an imposter. The attestation procedure relies on the exchange of cryptographic digital signatures to reliably verify the identity of the enclave.

Confidential computing is particularly applicable in the context of remote computing. **Remote computing** describes the use of computing resources that are owned and operated by an untrusted party that is not the application owner.

This description applies, for example, to a company that uses a cloud service provider to operate its enterprise computing capabilities. A remote application can work with sensitive data in the untrusted computing environment and maintain security throughout all phases of operation.

An example of remote computing is a commerce web server running in a commercial cloud environment. The web application collects and retains sensitive information such as credit card details for the site's customers. Using SGX technology and other standard cryptographic techniques, it is possible to encrypt each customer' sensitive data within the user's computer and transport the data into a secure enclave for order processing. To update the user's record within the application database, the sensitive information is encrypted within the secure enclave before passing it out for storage in the database. At every instant after the user types the sensitive information into their browser, the information is secured cryptographically and protected against malicious actors who may have achieved a presence within the untrusted computing environment.

Intel SGX technology was released in 2015 and is present in most modern Intel processors. To make use of this technology, a computer system must provide BIOS support that takes the steps required to enable it. There is broad support for SGX across motherboards and computers for sale today, but it is not automatically available in all systems. If you decide your application requires SGX, you will need to ensure the processor, motherboard, BIOS, and operating system of the computer you plan to use all provide support for this technology.

Unfortunately, as with many earlier attempts to improve computer security through hardware advances, security researchers have identified vulnerabilities in the SGX technology. In fact, researchers have demonstrated the ability to extract cryptographic keys and other valuable information from applications running within SGX secure enclaves.

The technique attackers use against SGX rely on the eager (or speculative) execution features of modern processors. As we discussed *Chapter 8, Performance-Enhancing Techniques*, eager execution is an optimization technique in which a processor begins executing code along both paths leading from a branch instruction until it becomes clear which path from the branch the code will take. Eager execution results in data being stored in cache memory that will ultimately be discarded. This cached data is the source of the vulnerability.

The specific attack technique against SGX is called **load value injection (LVI)**. An LVI attack can not only read values from the supposedly secure enclave – it can also inject data values into the enclave. Intel has identified steps that software developers can take to mitigate LVI; however, the mitigations may impose a substantial performance impact on software execution. Intel has also been working to provide updates to processor microcode that will resolve this vulnerability.

Designing for security at the architectural level

The design process for a highly secure computer system must factor in a wide range of security requirements from the very start of the process. All aspects of system design, such as processor selection and printed circuit board characteristics, must be evaluated in the light of how those components may enhance the system's overall security, or the ways they might degrade security.

In addition to ensuring security at the lowest level of integrated circuits and printed circuit boards, it is also important to apply security design principles in a layered manner. For example, after selecting suitably secure digital components and designing a security-focused circuit layout for a digital device, it may be necessary to design a tamper-resistant case to enclose the circuit board. This case might include embedded wires intended to detect if an attacker attempts to cut or drill holes through the case and thereby gain access to the internal components. This approach is often used in financially critical end-user devices such as point-of-sale credit card readers.

The best system security comes from applying a secure design methodology at all layers of system design, both in hardware and software. The following sections will review some design principles that will help ensure a secure system design.

Avoid security through obscurity

One tempting approach that has been used over the years in digital system development is to try to make it as unobvious as possible what the different components and connections within a digital device are and what they do. One step in this effort might involve grinding off the nomenclature from some of the integrated circuits within a device, thereby making it more difficult for someone analyzing the device to identify the components used to construct it.

Another way of obscuring the functionality available within a system is to run some of the wiring on a printed circuit board in an apparently nonsensical manner. The goal here is to make it as challenging as possible for someone to examine the device and reverse engineer its design.

The term **reverse engineering** has long been used to describe the analytical process of understanding how a device or software program was constructed without access to any of the documentation that went into the development process. While there are legitimate reasons for performing reverse engineering (such as when repairing a system for which no documentation exists), our focus is on more malicious applications of this process. Malicious reverse engineering occurs when someone attempts to gain access to protected information such as trade secrets or copyrighted software with nefarious intent.

Historically, the greatest flaw in the use of intentionally obscure designs in valuable digital systems has been the degree to which the system designers have underestimated the capabilities and resourcefulness of reverse engineers. It seems the designers who relied on the obscurity-insertion process assumed that because they themselves would never endure the tedium required to reverse engineer an intentionally baroque, overly complicated design, no one else would try to do it either.

There have been many cases in which reverse engineers have been able to perform tasks such as identifying the individual signals associated with a debugging interface hidden within a complex circuit board design. With knowledge of those connections, they were able to connect a hardware debugging system and extract all of the proprietary code and other information contained within the device.

It is common among experts in the development of a particular system to assume that only they could possibly understand the design and behavior of such a complicated device, and their understanding is only possible because they have access to system documentation that is unavailable to an outside attacker. This assumption is often invalid for two reasons:

- First, many reverse engineers are highly intelligent and detail-oriented, enabling them to methodically map out a system's features and behavior in a surprisingly complete manner.
- Second, it may not be reasonable to assume the system documentation will remain secure and hidden from attackers over the long term. Cyberattacks and industrial espionage against high technology companies are widespread. There is a significant possibility that determined attackers will gain access to some or all of the system documentation at some point. This possibility must be considered when evaluating the use of obscurity injection rather than using proven security mechanisms to protect critical information.

After dispensing with the approach of using obscure design processes to hinder attackers attempting to compromise a system, we must turn to effective, proven secure design approaches that can be expected to perform reliably now and in the future.

Comprehensive secure design

Using the approaches discussed earlier in this chapter, a highly secure computer system design must be approached from its fundamental components through to the final details to ensure the highest level of security. Where feasible, it should be possible to mathematically prove the system is secure by design. While it will not be possible to prove security in all aspects of a system architecture (because with even the most basic software application it is usually infeasible to demonstrate security mathematically) it may be possible to do so with critical aspects of a system design, such as the user login process.

By identifying all access points through which a cyberattacker may attempt to exploit a system and demonstrating a comprehensive security implementation at each of those points, system designers can demonstrate a level of confidence in the security of the system against attacks. This analysis must incorporate all the potential exploitation methods described earlier in this chapter, including those methods that are perceived to be most unlikely, if a very high level of system security is required.

In addition to incorporating security concerns across all aspects of the system hardware and software design, it is important to ensure that users of the system operate with the minimum level of privilege they require to perform their job duties. This is the topic of the next section.

The principle of least privilege

While some users require privileged access to manage and maintain a secure computer system, many other users require only the minimal privilege of a general user to do their jobs. The *principle of least privilege* summarizes the idea that each user should have no higher privilege on the computer system than is required for their job responsibilities. For example, a user that needs to examine and update information in a corporate database should have the privileges required to perform those tasks, but should not have additional privileges such as those required for performing system administration tasks that are not part of the job description.

The principle of least privilege ensures that users have the authorizations and access rights they require to do their jobs, but nothing beyond that.

When a user is transferred to a different job function, or is terminated, it is particularly important for a security-conscious organization to update the employee's access rights immediately upon the change in duties to ensure the employee (who may be disgruntled, if disciplined or terminated) no longer has access to information that is no longer part of their job description.

The effective use of the principle of least privilege requires more than a careful design process that grants minimal privileges to users and applications during system development. It is important for system administrators and operators to ensure on an ongoing basis that any newly introduced applications and users are given only the minimum privilege levels required to perform authorized functions.

Zero trust architecture

The traditional approach to secure computer system design described in the previous sections relies on a layered security model, with the goal that even if a particular security feature fails, the remaining layers will be sufficient to maintain overall security.

Given the wide range of successful attacks we see regularly publicized that feature total failure of security in systems that had been trusted to protect critical personal, business, and government data, this approach clearly has some limitations.

In the traditional security model, the outer perimeter of the network is considered the security boundary. Inside this protective layer, communication on the internal network is considered trusted and many types of access are available to computer systems and users.

In the zero trust architectural model, any communication that reaches a computer system containing protected information must be treated as potentially hostile. The request will only be processed after the source has been authenticated and the validity of the requested action has been confirmed.

The United States **National Security Agency (NSA)** has identified a set of guiding principles to apply when developing a zero trust computer architecture:

- **Never trust, always verify**: Every computer system, user, network device, or other source of data on the network must always be treated as untrusted. Every recipient of user input or other consumer of data must explicitly authenticate every piece of input data received. In addition, when granting privileges to users or applications the principle of least privilege dictates that only the minimum required privileges are granted.

- **Assume breach**: Assume at all times that hostile actors have penetrated the network and operate accordingly. This means the default response to any input or service request is to ignore or deny the request and only respond after rigorously validating the source and confirming the requested action is allowable. System administrators and automated tools must continuously monitor all aspects of the network configuration and identify when unauthorized access attempts or configuration changes occur and react promptly to stop any malicious activity and restore to an approved operating configuration. All relevant user, application, and network activity must be logged and inspected to rapidly detect any deviation from approved operations.

- **Verify explicitly**: Each access attempt to protected resources by a user or application requires a separate verification operation. The verification process must incorporate multiple attributes of the requesting user or application to reliably authenticate the source of the request. For example, the system can require the use of two-factor authentication, where a user inserts a cryptographically secure access card into a reading device and then enters a PIN. The card and the PIN represent two independent attributes that enable authentication as a trusted user.

To fully implement a zero trust architecture requires the implementation of a decision engine that assesses access decisions in the context of all available information about the requester and the destination of the request. In addition to authenticating the requester and verifying that the least privilege is applied to the request, the decision engine must address any additional information that may indicate an increased risk associated with the request. Only after validating all available information and determining that the risk associated with granting the request does not exceed a predefined threshold is the user's request permitted to proceed.

The next section will examine some of the ways in which security weaknesses in software running on otherwise secure hardware can introduce vulnerabilities into a computer system.

Ensuring security in system and application software

When developing a secure system design at the hardware level, it is important to follow through with a secure and verified design approach at all levels of the software architecture. We will next examine some of the ways that software code can introduce vulnerabilities into otherwise secure computer systems.

Common software weaknesses

This section lists several categories of software weaknesses that have traditionally caused severe security issues in operating systems, applications, and web servers. These vulnerabilities sometimes occur because software developers make assumptions about user behavior that turn out to be invalid. At other times, developers are simply unaware that particular software patterns lead to insecure designs.

Some of the techniques listed below are more likely to occur in specific programming languages, but software developers should be aware that it is possible to create insecure code in any programming language.

This is a list of some of the most common software weaknesses that have been observed in recent years.

Buffer overflow

A buffer overflow occurs when an input data set is larger than the memory allocated for input and the excess data overwrites memory that may contain other important data. This has traditionally been an issue with software written in the C and C++ programming languages. In a typical scenario, the code prompts a user to enter a text string that is expected to be short, such as a username.

The developer may have allocated a buffer of perhaps 80 characters to receive the input, assuming no username would be even close to that long. If the code receiving the input accepts more than 80 characters of input and stores it into memory, the data at addresses beyond the 80-character buffer will be overwritten. In traditional C language implementations, the input buffer is likely to be located within the processor stack. By inserting a carefully constructed input string, an attacker may be able to insert code containing processor instructions of the attacker's choosing and also overwrite the return address for the current function, which is stored on the stack. By overwriting the return address with the address of the malicious code, the attacker can execute any instructions they desire beginning when the input function returns. The initial goal of the attacker is usually an attempt to achieve remote access to a command prompt (also called a **command shell**) on the victim system. For this reason, the injected code that executes this type of exploit is called **shellcode**.

Cross-site scripting

Cross-site scripting is a vulnerability specific to web applications. In a cross-site scripting attack, the attacker finds a way of injecting malicious executable scripts into web pages that are then delivered from a reputable site to its users. These scripts execute in the context of the victim's browser and can potentially perform actions such as forwarding the victim's login credentials (such as an authorization cookie) to the attacker. This allows the attacker to impersonate the victim on the reputable website and gain access to the victim's personal information. Several mechanisms are available for software developers to validate and sanitize input from website users that will effectively prevent cross-site scripting. Unfortunately, not all web developers take full advantage of these security features, so this vulnerability continues to exist on many websites.

SQL injection

As discussed earlier in this chapter, many web applications use a database to store user information such as usernames, hashed passwords, and site content like user posts and uploaded images. Many sites use SQL to manipulate information within the database and insert new data in response to user inputs. A potential vulnerability arises if a user intentionally provides input to the site that may be interpreted as SQL code, and the site fails to sanitize its input in a manner that prevents the execution of the user-provided code.

As an example, this server code retrieves the username typed into a box on a web page after a user clicks the Submit button:

```
txtUserName = getRequestString("UserName");
```

In the next processing step, a naïve implementation of server code might create a command in SQL syntax that will retrieve the data record associated with the provided username from the database:

```
txtSqlCmd = 'SELECT * FROM Users where UserName = "' + txtUserName + '"';
```

For example, if the user entered `Alice` as the username, the text string containing the resulting SQL command would be:

```
SELECT * FROM Users where UserName = "Alice"
```

The server then passes this command to the database command interpreter and, if a valid username was provided, the user's data record will be returned.

The problem with this approach is the use of the text string entered by the user directly within the SQL command string. A malicious user can use SQL syntax to alter the behavior of the database access operation. For example, instead of entering a valid username the attacker might enter the following text string:

```
" or ""="
```

The SQL command that results when this string is provided as the username is:

```
SELECT * FROM Users where UserName = "" or ""=""
```

This command tells the database to return all records where the username is an empty string (which is probably none of them) or where an empty string is equal to an empty string. Since an empty string is always equal to an empty string, the condition will be true for all records and the database will return all the user records in the database. If the server code then presents the results of the SQL command to the user's browser, the attacker will have succeeded in extracting the entire user database for the website.

While this might seem like an obscure method for attacking a website, you should understand that many websites, often owned by famous and richly valued companies, have fallen prey to devastating attacks very similar to the one described here.

Path traversal

A path traversal vulnerability occurs when a network application, typically a web server, unintentionally grants some degree of access to its directory structure to users of the application. Normally, the site operators intend for users to access subdirectories beneath a main application directory to retrieve data organized in those directories. The vulnerability may arise if the logic in the web server allows users to go *up* one or more directory levels using this technique.

In both the Windows and Linux operating systems, a directory path element consisting of two adjacent periods means to go up a directory level. As an example, the following URL shows how to attempt to retrieve the file containing encrypted passwords on a Linux system running a standard web server:

```
http://www.example.com/../../../../etc/shadow
```

If successful, visiting this URL in a browser enables the attacker to retrieve the hashed versions of all passwords on the system. The attacker can then use brute force password cracking techniques to attempt to recover the users' passwords.

This section has listed just a few of the most common software vulnerabilities that have been exploited historically to cause severe harm to companies and individuals with private information stored on computer systems.

The **Common Weakness Enumeration (CWE)** database at https://cwe.mitre.org/index.html contains a list of software and hardware weaknesses derived from inputs provided by users around the world. In particular, the Top 25 Most Dangerous Software Weaknesses are listed at https://cwe.mitre.org/data/definitions/1337.html. This list provides a broad overview of the currently observed software weaknesses causing serious security issues.

Source code security scans

One way to quickly gain information about security weaknesses present in an existing body of software is to use an automated source code security scanning tool to assess the codebase and categorize the issues identified within it by severity.

Depending on the programming language or languages used to write your code, you may be able to find free tools to perform this type of scan. For any free code scanner you are considering using, be sure to examine reviews from users and ensure you are comfortable that it is a legitimately useful tool. For some programming languages, it may be necessary to purchase a security scanning tool, which may be quite expensive.

Automated security scanning tools can identify many categories of problems with source code, including the weaknesses listed above as well as others identified in the CWE database. These tools also identify other issues in the code such as the presence of deprecated features and the use of performance-compromising constructs.

Rather than attempt to list these tools here, I suggest you perform web searches for automated security scanning tools targeted at the languages you use for your critical applications.

Summary

This chapter introduced computing architectures suitable for uses that require an exceptional assurance of security. Critical application areas such as national security systems and financial transaction processing demand this high level of protection. These systems must be resilient against a broad range of cybersecurity threats including malicious code, covert channel attacks, and attacks enabled by physical access to the computing hardware. Topics addressed in this chapter included cybersecurity threats, encryption, digital signatures, and secure hardware and software design.

Having completed this chapter, you are able to identify many of the categories of cybersecurity threats a system will face and understand the security features of modern computer hardware. You understand some best practices for avoiding security gaps in system architectures and know how a secure computer architecture can help enforce security in software applications.

In the next chapter, we will introduce the concepts associated with blockchain, a public, cryptographically secured ledger recording a sequence of transactions. The chapter provides an overview of the bitcoin mining process and discusses the hardware architectures of bitcoin mining computer systems.

Exercises

1. Where supported, set up two-factor authentication for all your internet-accessible accounts containing data that you care about. This includes bank accounts, email accounts, social media, code repositories (if you are a software developer), medical services, and anything else you value. Ensure at all stages that you are using only information and software applications from trusted sources.

2. Create strong passwords for all your internet-accessible accounts containing information of value that cannot be protected by two-factor authentication. A strong password is long (15 characters or more) and includes uppercase, lowercase, numerical, and special characters (for example: ! " # $ % & ' ( ) * +). To keep track of these complicated passwords, install and use a reputable password safe application. Use care when selecting a password safe and consider its source.

3. Update the operating system and other applications and services (such as Java) on all computers and other devices under your control. This will ensure the security updates included in those updates start working to protect you soon after they become available. Set up a plan to continue regularly installing updates as they are released to ensure you are protected in the future.

Join our community Discord space

Join the book's Discord workspace for a monthly *Ask me Anything* session with the author: https://discord.gg/7h8aNRhRuY

15

Blockchain and Bitcoin Mining Architectures

This chapter begins with a brief introduction to the concepts associated with blockchain, a public, cryptographically secured ledger recording a sequence of transactions. We continue with an overview of the process of bitcoin mining, which appends transactions to the bitcoin blockchain and rewards those who complete this task with payment in the form of bitcoin. Bitcoin processing requires high-performance computing hardware, which is presented in terms of a current-generation bitcoin mining computer architecture. The chapter concludes with a brief introduction to some cryptocurrency alternatives to bitcoin.

After completing this chapter, you will understand the concepts of blockchain and how this technology is used. You will have learned the steps in the bitcoin mining process and will understand the key features of bitcoin mining computer architectures and the attributes of some currently popular cryptocurrencies.

The following topics will be presented in this chapter:

- Introduction to blockchain and bitcoin
- The bitcoin mining process
- Bitcoin mining computer architectures
- Alternative types of cryptocurrency

Technical requirements

The files for this chapter, including answers to the exercises, are available at https://github.com/PacktPublishing/Modern-Computer-Architecture-and-Organization-Second-Edition.

Introduction to blockchain and bitcoin

The concept of **bitcoin** first became public in a paper by Satoshi Nakamoto in 2008 entitled *Bitcoin: A Peer-to-Peer Electronic Cash System*. The author's name appears to have been a pseudonym and the identity of the author (or authors) of the paper is not publicly known. The paper laid out the mathematical and cryptographic underpinning of a system for performing decentralized financial transactions.

In a centralized financial system, the operation of the system relies on entities such as governments and banks to monitor and control system activities and to regulate what users of the system are allowed to do.

The bitcoin concept has no centralized regulator and relies entirely on networked peers to competitively interact in a manner that maintains stable system operation. Anyone can join the network as a peer and immediately gain all the privileges available to network participants.

One important feature of the bitcoin design is that it is not necessary for individuals using the currency to trust that everyone interacting with the bitcoin ecosystem will act in an honest or honorable manner. If honest cooperating peers control the majority of the computing power available to the bitcoin network, users can be confident in the trustworthiness of the system.

Of course, this assurance of trustworthiness relies on an assumption that the only viable path for attackers to compromise the bitcoin ledger or the transactions it contains is via compromising the consensus-generated integrity of the system. If a software flaw in the bitcoin code exposes a vulnerability that attackers can exploit, or if an encryption algorithm the system relies upon is found to be weak, there may be other ways the system could be compromised.

Bitcoin stores transaction information in the bitcoin **blockchain**, a distributed ledger containing cryptographically secure records of all transactions that have occurred with bitcoin since its inception. Each network peer can request and receive a complete copy of the blockchain at any time. As part of the startup process for a newly joined network peer, the new system must download and validate all blockchain transactions, beginning with the very first bitcoin transaction that took place and continuing to the end of the blockchain and the most recently added block. This is a mandatory step to ensure the new peer has validated all transactions up to the current state of the blockchain.

Bitcoin users rely on software applications called digital wallets to hold their funds. A **digital wallet** tracks the balance of bitcoin held by its owner and facilitates transfers to and from other bitcoin users.

The wallet stores the secret key used by the owner to access bitcoin funds that belong to the owner. If a hacker gains access to a bitcoin wallet secret key, the hacker can transfer the funds held in the wallet anywhere they wish.

When a bitcoin user initiates a transaction to transfer some quantity of bitcoin funds to or from another user, an entry must be added to the blockchain and then verified by peers on the network, in addition to being approved by the transaction sender.

Although the blockchain itself is publicly available, the information contained within it does not identify the sender and receiver involved in a particular transaction in any way other than by exposing a digital key that is associated with a particular user's digital wallet. This number does not connect to the user's identity other than by means such as the use of the same identifier to transact with another person who happens to know the wallet owner's identity. To avoid this reduction in anonymity, a user can create a different wallet identifier for each new transaction, if desired. This partial anonymity is the reason cybercriminals prefer to use bitcoin for purposes such as ransomware payments.

To convert a quantity of bitcoin to or from a more traditional form of currency, such as US dollars, a financial institution authorized to operate in that currency can perform the transaction for a fee. Alternatively, a user can transfer a quantity of bitcoin to another user in exchange for a mutually agreeable quantity of cash.

Each bitcoin transaction incurs a small (but optional) fee, which is paid to the first bitcoin miner that successfully performs the work required to append a collection of transactions from different users, including the user's transaction, to the blockchain ledger. Network nodes that perform these computations are called **bitcoin mining systems** and the people who own and operate the computer systems performing this work are called **bitcoin miners**. The voluntary inclusion of a fee with each transaction increases the likelihood that miners will include the transaction in the next block, thereby clearing it more quickly.

Adding a block to the blockchain is an intentionally computation-intensive task. Mining nodes compete to be the first to successfully perform the computation required to append a block to the blockchain. The first peer to demonstrate a correct solution for a new block is awarded a fee associated with the block as well as receiving all fees offered for transactions contained within the block. We will discuss bitcoin mining in more detail later in this chapter.

As the name implies, the blockchain is a chain of blocks. Each block contains cryptographically secure descriptions of multiple bitcoin transactions between users of the network.

Beginning with the first block placed at the start of the chain when it was created, each later block contains a link to the immediately previous block on the chain. Cryptographic techniques are used to ensure all the blocks and the transactions within them remain unmodified and that the block-to-block links have not been tampered with.

Figure 15.1 presents a simplified representation of the end of the blockchain after a new block, identified as Block *X*, has been appended to the chain. Each block contains cryptographically secure references to the details of each transaction contained within it.

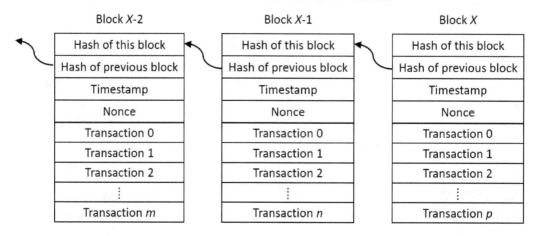

Figure 15.1: Simplified view of the blockchain

Each block is numbered, indicating its sequence in the chain. At the time of this writing, there are over 700,000 blocks in the chain. A new block is created approximately every 10 minutes containing transactions that had been initiated during the preceding minutes and gathered to form the block's transaction list. This means it typically takes at least 10 minutes for a transaction to "clear," using the analogy of depositing a check into a bank account. The number of transactions included in each block varies over time depending on the number of transactions initiated by users.

Data within each block is cryptographically hashed along with the hash of the previous block in the chain. Miners then compete to identify a 32-bit value that can be placed within the block (called a **nonce**) that will produce a block hash value that is numerically equal to or below a hash target value provided by the bitcoin network software. The process of adding blocks to the chain involves first finding a nonce that satisfies the network hash target, then publishing the new block to the network, and finally receiving confirmation from multiple peer nodes that the new block is indeed valid.

The bitcoin hash target varies over time with the goal of maintaining the stability of transaction processing. This means the amount of work a miner must do to earn a return varies over time and depends on factors such as the amount of mining processing power currently active on the network.

As a peer-to-peer network, newly mined blocks must be validated by network peers to confirm each new block contains a nonce that produces a block hash below the target value and that the information contained in the block is otherwise correct. Once a consensus has been reached among peers that the block is valid and that it was the winner in terms of being the first block with a valid nonce, the block is appended to the blockchain.

The blockchain architecture has proven to be robust in the presence of security threats. Attempts to insert invalid blocks into the blockchain are easily detected by hash verification, and any invalid blocks are discarded.

By design, only 21 million bitcoins can ever be created. To date, over 18 million are in circulation, leaving less than 3 million still to be mined. It is estimated it will take over 120 years to exhaust the remaining unmined bitcoin.

The bitcoin core software takes several steps to maintain the stability of the network and to achieve the goal of limiting the total number of bitcoins to 21 million:

- The algorithm tries to maintain a block creation interval of one block every 10 minutes by varying the network hash target. If each block took exactly 10 minutes to mine, there would be 2,016 blocks mined in every two-week interval (which is equal to 6 blocks per hour times 24 hours/day times 14 days in 2 weeks). Since the mining time per block varies with the processing power available to the network, at intervals of every 2,016 blocks, the bitcoin core software updates the hash target to a calculated value that would have caused the previous 2,016 blocks to take 2 weeks of mining time. This hash target is then used for mining the next 2,016 blocks.

- On January 3, 2009, Satoshi Nakamoto mined the first block of the bitcoin blockchain. This block, numbered block 0 in the chain, is referred to as the *bitcoin genesis block*. The block reward was 50 bitcoins for the miner. The same reward was provided to the miners of the first 210,000 blocks. The block reward was then cut in half, to 25 bitcoins for the next 210,000 blocks. The reward drops by half after each subsequent set of 210,000 blocks. If each block took exactly 10 minutes to mine, it would take 4 years to mine 210,000 blocks. In 2021, the block reward is 6.25 bitcoins, having fallen from 12.5 bitcoins on May 11, 2020. This halving process ensures the total number of bitcoins will be limited to 21,000,000.

>
> **BLOCKCHAIN TECHNOLOGY**
>
> Bitcoin uses a blockchain to maintain a cryptographically secure ledger of bitcoin transactions, but this is not the only application of blockchain technology. Blockchain provides a more general capability that can be used in any application where it is necessary to securely track a series of transactions over time. For example, a library could use a blockchain to record book lending and return events over time.

In the very early days of bitcoin mining (May 2010), miner Laszlo Hanyecz famously purchased two pizzas for 10,000 bitcoins. This appears to have been the first use of bitcoin to purchase physical goods. This event has come to be seen as a key moment in the development of bitcoin and its use of a blockchain-based distributed ledger. 10,000 bitcoins would be worth about $500 million at the time of writing.

It is estimated that 4 million bitcoins have been permanently lost by their owners. Bitcoin can be lost if the owner loses the secret key to the wallet containing the bitcoin, for example, by deleting all copies of the key or by disposing of the only hard drive containing the key. Bitcoin that has been lost still belongs to its owner, but no one can recover and use it.

We will next examine the secure hash algorithm that gives bitcoin its cryptographic protections and forms the core of the bitcoin mining process.

The SHA-256 hash algorithm

The fundamental operation underlying the computations used in bitcoin mining and by many other cryptocurrencies is a secure hash. Bitcoin uses SHA-256 as its secure hash algorithm. SHA-256 is a published standard cryptographic hash algorithm that has been codified by the US Government in **Federal Information Processing Standards (FIPS)** Publication 180-4.

SHA-256 works on data blocks that are a multiple of 512 bits in length. The algorithm defines a procedure for appending padding bits to the data to reach the required length.

The output of an SHA-256 computation is a 256-bit hash value, which is most commonly represented as 64 hexadecimal characters. The most important features of the relationship between the input data block and the output of the SHA-256 hash of that block are:

- The hash output is always 256 bits independent of the size of the input data block. The input data can be less than or much more than 256 bits in length.
- Computing the SHA-256 hash of a particular data block always produces the same result.

- Changing any part of the data block, even a single bit, will generally result in a completely different SHA-256 hash compared to the hash of the original data block.
- While it is theoretically possible to make changes to a data block in a manner that results in the same SHA-256 block hash as the unmodified original version, it is practically impossible to create two different data blocks that produce the same SHA-256 hash result.

A **hash collision** occurs when two different data blocks produce the same hash result. In the realm of cryptographic hash functions, the possibility of collisions presents a security threat. An effective and secure cryptographic hash algorithm must ensure that the likelihood of hash collisions remains extremely low. For SHA-256, we can consider the likelihood of a hash collision to be astronomically low.

The reason it is so difficult to identify two different data blocks that produce the same SHA-256 hash is the amount of work required to identify the contents of the second block such that its hash matches the hash of the first block. A straightforward approach for identifying a second data block that produces the same hash as the first block is to use a brute force procedure.

A brute force algorithm to search for a data block matching a given (target) hash value can treat the second data block as a series of bits we interpret as a very long integer, perhaps 256 bits in length. This is an example of the steps to search for a data block that produces the given hash using brute force:

1. Input the target hash to be matched.
2. Set the data block to all zero bits.
3. Compute the SHA-256 hash of the data block.
4. Does the block's hash match the target hash? If yes, exit and display the matching data block. If no, continue with *step 5*.
5. Increment the integer contained in the data block.
6. Go to *step 3*.

While the brute force hash collision search algorithm described here is simple and will eventually find a matching hash (if one exists), it is unrealistic to expect it will ever return a useful result for the SHA-256 hash of a large enough (256 bits or more) data block.

Our brute force search algorithm must execute the loop (from *steps* 3 through 6) 2^{256} times to have a near certain chance of producing a hash collision. How long would that take? The answer involves all the computing power on Earth (now and in the future) and all the years remaining until the Sun burns out, and, very likely, still not being anywhere close to finding a data block with a matching hash value.

In other words, given the current state of computing capabilities and reasonable predictions about the future growth in those capabilities, SHA-256 is very likely safe against hash collision vulnerabilities for some time to come, unless, of course, someone identifies an exploitable vulnerability in the SHA-256 algorithm. This is always a risk in the realm of cryptographic algorithms.

We will next look at the steps involved in computing the SHA-256 hash of a data block.

Computing SHA-256

SHA-256 works on a data block that can have any length from 1 to $2^{64}-1$ bits. The data block, also called a message, is treated as a linear string of bits in this discussion.

The message is first padded to a length that is a multiple of 512 bits using the following procedure. Message padding is performed even if the original message length happens to be a multiple of 512 bits.

1. Append a 1 bit to the message
2. Append the minimum number of zero bits to the message such that the length of the message in bits is 64 less than a multiple of 512
3. Append a 64-bit unsigned integer to the message containing the length in bits of the original message

The building blocks of the SHA-256 algorithms are simple logical and mathematical operations: AND, OR, XOR, NOT, integer addition, right shift, right rotate, and concatenation of bits. These operations are performed on 32-bit words. When performing addition, the processor's carry flag is ignored.

The SHA-256 algorithm defines several more complex operations that mix the simple building blocks in the previous paragraph with predefined constants to scramble the data content of the input block and produce a random-appearing 256-bit value as the hashed output.

The SHA-256 algorithm processes each 512-bit section of input data sequentially. After padding the message, the 512 bits within each section are separated into 64 words of 32 bits each. The processing of each section is essentially a scrambling procedure that repeatedly mixes the bits in the words together in a series of logical operations. The algorithm executes a 64-pass loop performing a series of intensive Boolean and mathematical operations on the data words in each section.

To summarize, computing an SHA-256 hash for even a small data block requires a substantial sequence of computation. By design, there are no shortcuts that would allow skipping any of the computational steps.

As an example of SHA-256 input and output, a data block consisting of just the ASCII characters abc will, after padding and executing the computational manipulations of SHA-256, produce the following 64 hexadecimal digits as the hash output:

ba7816bf8f01cfea414140de5dae2223b00361a396177a9cb410ff61f20015ad

In the next section, we will discuss some of the key attributes of the source code that runs the bitcoin network. This code is referred to as bitcoin core.

Bitcoin core software

Anyone with a computer and access to the internet can set up a bitcoin node. To gain full access to the capabilities offered by the bitcoin network, a computer owner needs to set up a full node. A bitcoin **full node** performs the operations required to validate transactions and blocks. This process involves accepting transactions and blocks from other full nodes, validating each transaction and block, and forwarding the blocks and the validation results to other full nodes.

The bitcoin code is open source and can be downloaded from https://bitcoincore.org/en/download/. The code runs on Windows, macOS X, and Linux computer systems that satisfy nominal memory and disk space requirements and have a broadband internet connection. It is possible to build the bitcoin executable applications from source code or download and install the executable files directly.

Once the bitcoin software is installed and running, the application will download the entire bitcoin blockchain from peer nodes, beginning with the genesis block through the most recently added blocks. The full blockchain consists of over 400 GB of data as of 2021. It is possible to have the bitcoin application delete early blocks after they have been analyzed to avoid consuming excessive disk space. The initial download and analysis of the early blocks is essential to validate the entire history of the blockchain and to ensure the current state of recent blocks is valid and accurate.

If you choose to restrict the disk usage by the bitcoin software to a limited amount, such as 2 GB, your node will have to request copies of earlier blocks from other peers when attempting to validate transactions that draw funds from transactions located in those earlier blocks. Each time the code receives data from an (untrusted) network peer, it will perform hash verification and digital signature verification to ensure all elements within the data are valid and trustworthy.

A full node can also act as a client node. A **client node** allows a bitcoin user to initiate bitcoin transactions with other users and to respond to transactions initiated by other users. Most bitcoin users operate a client node, which might be an app running on a smartphone.

Next, we will look in more depth at the processing requirements of bitcoin mining and learn how the price of electricity determines whether a miner can make a profit.

The bitcoin mining process

The computational complexity of the SHA-256 algorithm relates directly to the feasibility of bitcoin mining as a profitable endeavor. The only way to determine the SHA-256 hash of a particular data block is to perform all the steps of the SHA-256 algorithm over all the bits in the block.

A key feature of the bitcoin mining process is that it is intentionally very difficult to find a valid nonce that produces a block hash below the current target network hash target. In fact, it is likely to take an enormous number of guesses of different nonce values before a target-satisfying value is found. Due to the lack of any predictable relationship between the block data content and the SHA-256 hash of that block, there is no more efficient method to determine a suitable nonce value than simply hashing the data block repeatedly with varying nonces until a hash turns up that satisfies the target criteria.

The process of identifying a nonce value that satisfies the target hash requirement is called **proof of work**. To perform the proof of work required for bitcoin mining, the miner must provide suitable hardware and electrical power, and allow time for the algorithm to run.

It is straightforward for a modern PC to try all possible values for the 32-bit nonce within a few seconds. Unfortunately for miners, this computation rarely identifies a hash satisfying the network hash target. After exhausting all possible 32-bit values for the nonce, the miner must modify the tentative block it is working with before it can begin trying all possible nonce values on the new block content. There are a few ways that miners can cause the contents of a tentative block to change in a manner accepted by the bitcoin network:

- **Alter the timestamp within the block**: Although each block in the chain contains a timestamp that represents the time it was created, the timestamp itself is not used for any critical purpose such as determining the order in which blocks were created. If the miner exhausts all possible values for the nonce without success, it is possible to update the timestamp in the block header and again attempt all possible values for the nonce. Changing the timestamp, in effect, increases the size of the nonce search space. Small modifications to the timestamp are tolerable but large changes are not allowed.
- **Update the block with new transactions**: It is likely that transactions continued to occur on the bitcoin network while a previous unsuccessful round of searching for a satisfactory nonce value was in progress. By requesting a new tentative block containing recently added transactions, the miner has increased the size of the nonce search space.

- **Modify data in the transaction header**: The first transaction in a block's transaction list is special in that it represents the payment to the miner for adding a block to the blockchain. It is possible for the miner to insert additional data into this transaction. When used for the purpose of increasing the nonce search space, this added data is referred to as extraNonce. The usual procedure for using extraNonce is to test the block hashes for all possible values for the nonce, then, treating extraNonce as an integer, increment extraNonce and try all possible values of nonce again.

The processes of incrementing the nonce and performing other manipulations to increase the effective nonce search space generally do not consume a large proportion of bitcoin mining processor time. The vast majority of the work involved in the search process occurs during the repeated execution of the SHA-256 hash algorithm as different nonce values are tried in an attempt to find a block hash that is equal to or lower than the network hash target.

In the early days of bitcoin mining (around 2010), a miner could use a personal computer with reasonably good performance specifications to earn a few dollars a day from mining. At the time, the difficulty defined by the network hash target was low enough that standard PC hardware could locate nonce values satisfying the hash target with a reasonable likelihood of success.

As more miners joined the network and the performance specifications of their computing hardware improved, the total amount of computing power working to complete the proof of work to add each block to the chain continued to grow.

The bitcoin hash target adjustment algorithm modifies the network hash target after every 2,016 blocks to maintain an average of one newly added block every 10 minutes. This meant that as the total bitcoin network computing power continued to grow, any individual miner, even one using a very powerful GPU, would have very little chance of being the first to find a solution for any block. A miner would very likely receive no return at all, even after mining for years.

To enable individuals to continue to participate in bitcoin mining with at least some return on their efforts, the concept of bitcoin mining pools arose. This is the topic of the next section.

Bitcoin mining pools

A **bitcoin mining pool** is a group of miners who combine their computing power to increase the chances of receiving block rewards after successfully completing the proof of work for blocks added to the bitcoin blockchain. In joining a mining pool, a miner agrees to contribute computing power to the pool and receive a share of the rewards from mining by the pool members.

In essence, the members of the mining pool split up the large problem of completing the proof of work into a set of smaller problems and pass those problems to the individual pool members to work on. If one of the pool members correctly solves the proof of work, the pool adds the block to the chain and splits the reward among the members.

The organizers of a mining pool must set up a data center to manage the pool's interactions with its miner members and with the bitcoin network. This requires computing hardware and staff members to set up operations and manage the system on a day-to-day basis. Bitcoin mining pool operators charge a fee, typically 1-3% of the mining returns, for this service.

Joining a bitcoin mining pool allows a miner to receive some (typically small) regular returns in exchange for providing computing power to the pool. The more computing power (in terms of the number of SHA-256 hashes evaluated over time) a miner contributes, the greater will be the miner's reward when the pool succeeds at adding a block to the chain.

The likelihood of a particular bitcoin mining pool's success within a given period can be quantified in terms of the hashrate of the mining pool relative to the hashrate of the entire bitcoin network. The bitcoin **hashrate** is the number of SHA-256 hashes executing per second, where each hash operation is an attempt to complete a block proof of work.

During 2021, the estimated total bitcoin network hashrate varied between about 80 million and 180 million terahashes per second. One **terahash** is one trillion hashes, or 10^{12} hashes. Another term for one million terahashes is **exahash**, equivalent to 10^{18} hashes. Using this unit makes conversation a bit easier when describing the 2021 hashrate, which varies between 80 and 180 exahashes per second. This range of hashrates can also be expressed as 80-180 EH/s.

The fraction of the total network hashrate controlled by a mining pool determines the frequency at which the pool can expect to mine blocks. As we've seen, a new block is mined on average every 10 minutes. The following equation shows how often a mining pool can expect to succeed at mining a block based on its share of the total network hashrate:

$$T_B = \frac{10 \ minutes}{\frac{H_P}{H_N}}$$

Equation 15.1

In *Equation 15.1*, T_B represents the average time between blocks mined by the pool in minutes, H_P represents the hashrate of the pool, and H_N represents the total network hashrate.

This equation is valid because the process of performing the block proof of work is essentially statistical in nature, where every processing element on the network is performing a series of guesses and each guess made by each participant has the same very small but equal chance of being successful.

Using this formula, if a pool controls 0.1% (or one one-thousandth) of the total network hashrate, the pool will mine a new block on average every 10,000 minutes, which is about once a week. At the current bitcoin price of $45,000 and block reward of 6.25 bitcoins per block, the block reward will be $281,250 plus any transaction fees offered by bitcoin users.

TRANSACTION FEES

Transaction fees are voluntary payments offered by bitcoin users as an incentive to prioritize their transactions for placement into blocks during mining. Each time a user initiates a new bitcoin transaction, the user has the option to allocate a portion of the transaction funds toward the transaction fee.

Miners can select which transactions they include in each block they work on, which means they generally prefer transactions offering higher fees. During times of high transaction rates, offering a low transaction fee can result in a transaction waiting for a longer time before it gets added to the blockchain.

Eventually, as the number of coins remaining to be mined dwindles toward zero, transaction fees will become the only remaining incentive for miners to continue processing transactions and working on adding blocks to the chain. The weekly median (meaning half were lower and half were higher) bitcoin transaction fee has varied from $0.27 to $26.96 during 2021.

After successfully submitting a new block to the chain, the mining pool manager must divide the block reward among the pool participants. To keep track of the proportion of proof of work that was performed by each miner in the pool, the pool management software sets a hash target for its miners that is substantially higher (easier to satisfy) than the network hash target. This means members of the pool will return many hash solutions to the pool manager that satisfy the pool's target level but don't satisfy the network target. By tracking how many of these pool target-satisfying hashes each miner returns, the pool manager can determine how much hashing was performed by each pool participant. The block reward is then distributed among the pool members in proportion to their contribution to the overall hashing effort.

At a typical network hashrate for 2021 of 140 EH/s, the pool would need to run 140,000 terahashes per second, or 140,000 TH/s, to control 0.1% of the network hashrate. This certainly sounds like a lot of hashes. To get some idea of the computational power required to generate this hashrate, let's first consider the use of standard PC processors for performing the hashing operation. We'll look at this in terms of a solo miner who is not participating in a mining pool.

Mining with a CPU

If you don't want to pay the fee required of members of a bitcoin mining pool, you may decide to do mining by yourself using one or more computing devices you own or control. This is called **solo mining**. We can estimate the proceeds from solo mining based on the hashing capability of the hardware dedicated for this purpose.

One of the top-performing CPUs currently available is the AMD Ryzen Threadripper 3970X. The 3970X has 32 cores and 64 simultaneous threads with a CPU clock frequency that ranges from 3.7 to 4.5 GHz. The large number of simultaneous threads enables parallel computation of hashes for several nonce values. Based on benchmark tests, the 3970X can compute about 19,900 bitcoin hashes per second.

Consider a PC containing one AMD Ryzen Threadripper 3970X processor. We can insert the hashrate of this processor into Equation 15.1 and estimate the interval between successfully mined blocks. We will assume a nominal 2021 network hashrate of 140 EH/s. The result of this computation is shown in *Equation 15.2*:

$$T_B = \frac{10\ minutes}{\left[\frac{19.9 \times 10^3}{140 \times 10^{18}}\right]} = 7.04 \times 10^{16}\ minutes$$

Equation 15.2

From this equation, we see the average interval between successfully mined blocks on a single 3970X is 7.04×10^{16} minutes, which is about 133 billion years. Obviously, this is not a viable configuration for someone trying to get any return whatsoever from bitcoin mining.

Soon after bitcoin was first introduced, a version of mining code was released as open source that took advantage of the parallel processing capability of GPU hardware. We'll look at that next.

Mining with a GPU

The core computation of bitcoin mining is the SHA-256 hash algorithm.

The task of testing a very large number of nonce values is ideal for a parallel processing computer architecture because each test is independent of all the others. Because of this inherent parallelism, it was natural for the bitcoin mining software to migrate into the GPU context. When running on a GPU, mining code takes full advantage of the large number of processing elements to execute the bitcoin hash algorithm at a much higher hashrate than a CPU with even a large number of cores can approach.

A high-end GPU is capable of a much higher SHA-256 hashrate than a microprocessor. During the early years of bitcoin, high-end GPUs were used for bitcoin mining by many miners. Up until around 2014, miners could profitably exploit the increasing computational power of GPUs to perform around 1 gigahashes per second, or 1 GH/s, which is equal to 10^9 hashes per second.

If we plug this hashrate into *Equation 15.1*, we find the average interval between block successes in 2021 shown below:

$$T_B = \frac{10 \; minutes}{\left[\frac{1 \times 10^9}{140 \times 10^{18}}\right]} = 1.40 \times 10^{12} \; minutes$$

Equation 15.3

The result of this computation is an average of 1.40×10^{12} minutes, which is over 2.6 million years between blocks. While this time interval between block successes is much better than using a powerful CPU alone, it is still not something any reasonable individual miner would attempt in 2021.

The use of GPUs for bitcoin mining became unprofitable as specialized ASIC-based devices were brought to market for the sole purpose of performing mining at far higher hashrates than general-purpose processors or GPUs could offer. The computing systems that employ these ASICs are the subject of the next section.

Bitcoin mining computer architectures

The SHA-256 algorithm is a well-defined sequence of steps that consists of simple Boolean operations performed repetitively on 32-bit data items. The algorithm does not require very much memory when operating on a small block (bitcoin block headers are always 80 bytes). This type of problem is ideal for performance optimization using a digital hardware design dedicated to this application.

The input to the mining algorithm is a candidate block header. The bitcoin block header contains the following data items:

- **Bitcoin version number (4 bytes)**: This field identifies the version of the bitcoin core software. The miner selects a version number that is compatible with the miner's bitcoin software.
- **Previous block hash (32 bytes)**: The hash of the previous block in the blockchain. This value is retrieved from the bitcoin network as the hash of the current latest block in the blockchain.
- **Merkle root (32 bytes)**: This hash value protects all the transactions in the candidate block. The term **Merkle root** describes a tree-like data structure that begins with hashes of individual bitcoin transactions and combines those hashes in a way that permits efficient and secure verification of the integrity of each individual transaction in the tree.
- **Time (4 bytes)**: The timestamp of the block in seconds since January 1, 1970, **Coordinated Universal Time (UTC)**. A valid block timestamp must lie within a three-hour window around the current time as determined by the bitcoin network. This window of validity permits some adjustment of the block time to increase the hash search space. Because of this flexibility, the timestamps of blocks on the blockchain cannot be assumed to represent the exact time of block creation.
- **Bits (4 bytes)**: This field defines the network hash target difficulty. This is a floating-point value with a 24-bit mantissa and 8-bit exponent in a unique bitcoin-specified format. This value is supplied by the bitcoin network.
- **Nonce (4 bytes)**: A field the miner varies while attempting to generate varying hashes.

These fields combine to form the 80-byte header for a candidate block. After setting all six parameters to valid values, the miner computes the hash of the header and compares it to the network hash target. If the computed hash value is equal to or less than the network hash target defined in the bits field of the header, the block is valid, and the miner can submit it to the network for verification and addition to the blockchain.

DOUBLE SHA-256

The bitcoin block hash algorithm actually performs SHA-256 twice to compute the hash of the block header. It first computes the hash of the 80-byte block header, and then computes the hash of the hash computed in the first step.

This computation can be expressed in the form SHA-256(SHA-256(header)).

Following almost every guess of a nonce value, the hash output does not result in a valid block. The miner then varies the nonce and, perhaps, the timestamp (within limits) and other parts of the block repeatedly while attempting to compute a hash that results in a valid block. Most of the work in this process involves repetition of the steps in the SHA-256 algorithm.

A first step in developing specialized hardware designs for bitcoin mining brings in **field-programmable gate arrays (FPGAs)**. This is the subject of the next section.

Mining with FPGAs

FPGAs, introduced in *Chapter 2, Digital Logic*, provide a means to create a hardware circuit optimized for a particular task by connecting a collection of generic digital components such as logic gates, flip-flops, and registers to a circuit. Using a hardware description language, a developer can define the logical execution sequence of the SHA-256 algorithm and the HDL compiler will translate this specification into a circuit design that can be downloaded into an FPGA chip.

To be useful for bitcoin mining, it is necessary to include some additional logic in the design to manage the input of data to be hashed and then collect output from the hash algorithm.

In a naïve implementation, the mining software might provide input to the FPGA algorithm that consists of a proposed block header, including the current nonce guess. After the FPGA receives its input data, it performs the SHA-256 hash algorithm and returns the hash to the mining software as its output.

Figure 15.2 shows how this simple mining approach could be implemented:

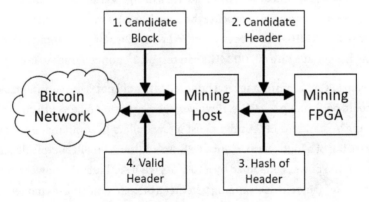

Figure 15.2: Simple FPGA mining configuration

The sequence of operations for this mining configuration consists of the following steps:

1. The mining host (this could be a standard PC or a server system) requests and receives a candidate block from the bitcoin network to work on.
2. The mining host transfers the candidate block header (including the current nonce guess) to the mining FPGA for hashing.
3. The FPGA performs the hash operation and returns the hash value to the mining host.
4. The mining host evaluates the hash. If it satisfies the network difficulty, the host forwards the block to the bitcoin network for inclusion in the blockchain.

While this approach executes all the steps required for bitcoin mining using a high-speed FPGA, the configuration described here is unlikely to provide a sufficient hashrate to interest any serious miner. This is because the overhead needed for transferring data to and from the FPGA for every hash evaluation is likely to drastically degrade the average hashrate.

An improved design would have the FPGA perform the hash operation over a span of nonce values and only stop hashing when it completes the entire span without success, or it finds a satisfactory hash value. The range of nonce values to be tested could be coded into the FPGA firmware, and the range could extend to all 2^{32} possible nonce values. This configuration resembles the configuration of *Figure 15.2* and would allow the FPGA to operate at very close to its maximum achievable hashrate over time.

Various FPGA designs have been developed that perform bitcoin hashing using an approach like the one described here. The maximum hashrate for a design varies drastically depending on the capabilities of the FPGA chip used to execute the algorithm. Hashrates up to several hundred million hashes per second (MH/s) have been achieved with FPGA implementations of SHA-256. An FPGA executing hashes at a rate of 500 MH/s can test all 2^{32} nonce values in less than 10 seconds.

In 2014, a per-chip FPGA hashrate of several hundred MH/s did not seem very impressive compared to GPUs that could perform up to 1 GH/s. However, there are other factors affecting the economics of bitcoin mining. A standard FPGA might cost a few dollars in comparison to the several hundred-dollar price tag of a high-performance GPU. In addition, an FPGA typically consumes only a small fraction of the power consumed by a GPU. As we'll see later, the power consumption of a mining system is a key factor in determining whether a bitcoin miner makes money or loses money.

While the idea of designing a bitcoin mining computer containing several FPGAs might seem like a winning approach in comparison to CPU and GPU mining, the use of ASIC devices instead of FPGAs results in even higher performance and can cost less if the devices can be sold in sufficient quantities. ASIC mining is the subject of the next section.

Mining with ASICs

An Application-Specific Integrated Circuit (ASIC) is a custom-designed chip that implements a specified function or set of functions. In comparison to general-purpose CPUs, GPUs, and FPGAs, the key distinguishing features of ASIC devices are:

- An ASIC contains only the circuitry it needs to perform its intended function. There is no additional circuitry taking up die space and consuming power. This allows an individual ASIC die to be smaller and consume less power than a general-purpose FPGA circuit. This makes the ASIC less expensive to produce and less expensive to operate, if produced in a large enough quantity.
- Producing the first copy of an ASIC device is very expensive. A tremendous amount of engineering and production line development are required to prepare a circuit design and construct the first batch of chips. If a serious error is found in the design after production has started, revising the design to fix the problem is also very expensive.
- Once the production line has been set up and the resulting circuits are shown to work properly, producing the chip in high volumes is very inexpensive in terms of cost-per-chip.

As interest from miners and investors grew in the years following the introduction of bitcoin, the economics of mining hardware grew to provide a level of demand that made the cost and effort to produce mining ASIC devices worthwhile.

Bitmain (https://www.bitmain.com/) is the largest producer of bitcoin mining hardware, which includes ASIC mining chips as well as complete mining computer systems based on these chips. In addition to developing mining hardware, Bitmain mines for its own benefit and runs at least two mining pools: BTC.com (https://pool.btc.com/) and AntPool (https://v3.antpool.com/home).

Bitmain has developed several bitcoin mining ASICs, beginning with the BM1380, which was released in November 2011. This chip could perform up to 2.8 GH/s at its highest operating voltage of 1.10 V.

Bitmain also produced the Antminer S1 mining computer, which contained 64 of the BM1380 chips. This system could perform 180 GH/s while consuming 360 W of electrical power.

This is a much better hashrate than could be achieved with even several GPUs or with 64 FPGA chips. Let's see how long the average block-to-block solution time is for an Antminer S1:

$$T_B = \frac{10\ minutes}{\left[\frac{180 \times 10^9}{140 \times 10^{18}}\right]} = 7.78 \times 10^9\ minutes$$

Equation 15.4

The average time between block successes for the Antminer S1 is 7.78×10^9 minutes, about 14,800 years. This is much better than a fast GPU, but still is not a reasonable configuration for a solo miner in 2021.

Following the BM1380, Bitmain released a series of upgraded mining ASICs named BM1382 (April 2014), BM1384 (September 2014), BM1385 (August 2015), and BM1387 (May 2017). Bitmain has publicly released less detail about newer ASIC versions that followed these. The general trend in each new iteration of the ASIC is to increase the hashrate and decrease power consumption per computed hash.

In 2021, one of the fastest available Bitmain miners is the Antminer S19 Pro, with a hashrate of 110 TH/s. The S19 Pro consumes 3,250 Watts and sells for around $15,000. This system contains three hashing boards, each containing 114 BM1398 chips.

All 114 chips on each board are connected in series, which is referred to as a daisy chain configuration. The S19 Pro has a control board, containing a processor and firmware that communicates with and controls the operation of each of the hash boards and the hash chips on it. The communication protocol used with the chips is a serial data format that is identical to that used by standard serial ports on computers and other digital devices.

The control board sends commands through the serial interface to the first BM1398 ASIC in the daisy chain and this device passes the same command along to the next BM1398 in the chain, and so on.

Each BM1398 has a set of hardware address lines that give it a unique identity on the daisy chain. The ASIC uses these address lines to determine the portion of the nonce search space assigned to it.

Each BM1398 can place messages on the bus, which pass through the daisy chain and are received by the control board. The primary message type generated by a hashing chip is a notification that it has identified a nonce that satisfies the hash target requirement.

Figure 15.3 shows the high-level configuration of the components in the Antminer S19 Pro:

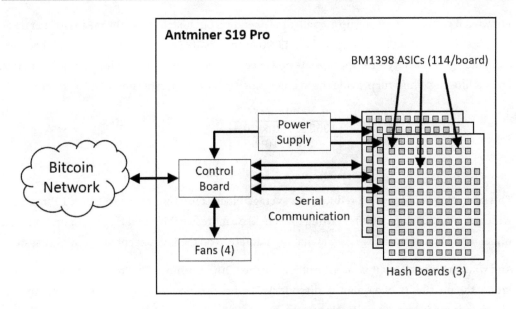

Figure 15.3: Antminer S19 Pro hardware configuration

At the 110 TH/s hashrate, *Equation 15.5* provides the average block solution interval for the Antminer S19 Pro:

$$T_B = \frac{10 \text{ } minutes}{\left[\frac{110 \times 10^{12}}{140 \times 10^{18}}\right]} = 1.27 \times 10^7 \text{ } minutes$$

Equation 15.5

With an average block solution interval of 1.27×10^7 minutes, the Antminer S19 Pro is expected to succeed at adding a block to the blockchain every 24.2 years at a representative 2021 network hashrate. Once again, this is not a useful operating configuration for a solo miner, which is the reason all serious small-time miners join mining pools to produce a reliable, if small, income.

But what if you are not setting up a small-time operation but instead have the means to operate at an industrial scale? Let's say you purchase 1,000 Antminer S19 Pro machines and install them in a computer-grade facility with appropriate power conditioning, air conditioning, humidity control, and air filtering. This will provide the average block solution interval shown in *Equation 15.6*:

$$T_B = \frac{10 \text{ minutes}}{\left[\frac{1000 \times 110 \times 10^{12}}{140 \times 10^{18}}\right]} = 1.27 \times 10^4 \text{ minutes}$$

Equation 15.6

This operating configuration reduces the average block solution interval to 1.27×10^4 minutes, which is about 8.8 days. This is starting to seem like a more feasible operating configuration for a miner interested in receiving a somewhat regular return on the investment in the mining system.

And what an investment it is, at $15 million for the 1,000 Antminer S19 Pro systems alone (if we ignore the possibility of a quantity discount). The next section will examine the economics of bitcoin mining in terms of initial investment, ongoing expenses, and expected return.

Bitcoin mining economics

In addition to network connectivity, successful bitcoin mining at industrial scale requires four main components:

- A suitable facility
- Mining hardware
- Electrical power
- Time

Our discussion to this point has focused on the processing requirements to solve the hash puzzle required to validate a bitcoin block so that it can be added to the blockchain. As bitcoin mining hardware has become more specialized and powerful, the total quantity of electrical power consumed by mining operations has steadily grown. In fact, worldwide bitcoin mining consumes more electrical power on average than Argentina, according to a February 2021 analysis by Cambridge University.

The Antminer S19 Pro discussed in the previous section consumes 3,250 Watts, which equates to 78 kilowatt-hours per day. Depending on the price of electricity in the area where the mining operation is located, the cost of electricity may erase most or all the profit that would otherwise accrue from successfully solving block hashes.

The trend in bitcoin mining hardware has been to increase the hashing power of each new generation of mining system while simultaneously reducing the amount of electrical power required to compute each hash.

This has led to specifications for mining systems that quantify the power consumed per hash operation. One Joule (abbreviated J) is equal to one Watt of power over a period of one second. In other words, 1 W = 1 J/s. The Antminer S19 Pro power efficiency is 3,250 J/s divided by 110 TH/s, which works out to 29.5 J/Th. The Antminer S19 Pro is, besides providing a very high 110 TH/s hashrate, one of the most power-efficient bitcoin mining systems available.

Despite the trend toward increased hash efficiency, the total electrical power consumption by the bitcoin network continues to grow. The massive consumption of electrical power by the network is seen by some as wasteful and as a contribution to environmental and climate concerns.

In addition to a desire to reduce the negative impact of high energy consumption by bitcoin processing, small-scale miners have been unhappy at being crowded out of the ability to mine bitcoin for a reasonable profit by large-scale industrial mining operations located in countries around the world.

Bitcoin is not the only cryptocurrency in use—far from it. There are over 10,000 different cryptocurrencies in active use in 2021. Several have grown in popularity and overall market value to become somewhat competitive with bitcoin. Several of these currencies have specific design features to combat the perceived negative trends associated with bitcoin. We discuss a few of these alternative cryptocurrencies in the next section.

Alternative types of cryptocurrency

Bitcoin mining began as a profitable diversion for computer enthusiasts that made use of spare computing cycles on their processors. As the use of expensive, noisy, power-hungry mining systems has grown to an industrial scale, the ability to make even a minimal profit with a homemade bitcoin mining system has disappeared.

This is one reason behind the development of numerous cryptocurrencies as alternatives to bitcoin, which are commonly referred to as **altcoins**. Some altcoins are designed to make it more difficult and more expensive to design an ASIC to solve the challenge required to mine the coin. Some are designed to avoid the compute-intensive proof of work that forms the basis of bitcoin. By avoiding intentionally compute-intensive operations, these coins substantially reduce the amount of electrical energy required to mine and transact with the coin.

For any cryptocurrency to become widely accepted and used, new users must be confident that any funds they entrust to the currency will retain value over time and that negative outcomes such as finding one's digital wallet has been emptied by a thief are unlikely to occur. These are tough standards for a new cryptocurrency to satisfy. Even so, several altcoins based on blockchain technology have reached a level of widespread acceptance and use. Some of the premier altcoins in 2021 and their key features are listed here:

- **Ethereum**: Ethereum is a decentralized software platform like bitcoin. The Ethereum platform provides support for smart contracts and distributed applications, which can be used to perform functions such as transferring payments from buyer to seller. **Smart contracts** are programs stored on the Ethereum blockchain that execute when predefined conditions have been met. Smart contracts are intended to facilitate the execution of legal agreements in an open and verifiable manner. The Ethereum cryptocurrency is called Ether, which is used to pay for the computing resources and transactions on the Ethereum network. Rather than using the computationally expensive proof of work concept employed by bitcoin, Ethereum uses a process called proof of stake. **Proof of stake** is based on the number of Ether coins a miner holds, and this stake enables the miner to validate transactions on the network.

- **Litecoin**: Litecoin was launched in 2011 as a fork of bitcoin and has much in common with bitcoin. Litecoin was designed with a focus on the rapid approval of transactions even when high transaction volumes are taking place. The block time for Litecoin was reduced to 150 seconds in comparison to the bitcoin block time of 10 minutes. Up to 84 million Litecoins can be created, greater than the total number of bitcoins.

- **Dogecoin**: The Dogecoin cryptocurrency was created in 2013, apparently as a joke, intended as a statement on the ongoing speculation in cryptocurrencies such as bitcoin. Dogecoin features a photo of a Shiba Inu dog as its logo. In May 2021, the market capitalization of Dogecoin was $85 billion. Dogecoin is a fork from the Litecoin code base and differs from many other cryptocurrencies in having no fixed limit on its currency supply. Instead, it has a stable "inflation rate" of 5 million new Dogecoins created per year. The block time for Dogecoin is 60 seconds.

- **Bitcoin Cash**: Bitcoin Cash is a fork of bitcoin. Bitcoin Cash was created following a disagreement between factions of bitcoin supporters regarding a proposed upgrade that would permit larger block sizes for collections of transactions. A larger block size permits more transactions to be included in a single block on the blockchain, which means the time users must wait for a transaction to complete can be shorter. Bitcoin Cash transaction fees also tend to be lower. As with the other altcoins listed here, Bitcoin Cash uses a separate blockchain from the bitcoin blockchain.

These are just a few of the more prominent altcoins in current use. Despite the thousands of alternative cryptocurrencies in existence, bitcoin remains the dominant cryptocurrency in terms of usage and the value of transactions. In November 2021, the market capitalization of bitcoin was over $1 trillion.

Summary

This chapter began with a brief introduction to the concepts associated with the blockchain, a public, cryptographically secured ledger recording a sequence of financial transactions. The discussion continued with an overview of the process of bitcoin mining, which appends transactions to the blockchain sequence and rewards those who complete this task with payment in the form of bitcoin. Bitcoin processing requires high-performance computing hardware, which is often specifically designed for this task. The chapter concluded with a description of the features of these hardware architectures.

Having completed this chapter, you understand what the bitcoin blockchain is and how it is used. You learned the steps in the bitcoin mining process and understand the key features of dedicated bitcoin mining computer architectures.

The next chapter will discuss the capabilities of self-navigating vehicle processing architectures, including the types of sensors and data a self-driving vehicle receives as input while driving and the types of processing required for vehicle control in real-world driving situations.

Exercises

1. Visit the blockchain explorer at https://bitaps.com and locate the list of last blocks on that page. Click on a block number and you will be presented with a display containing the hexadecimal listing of the block header along with its SHA-256 hash. Copy both items and write a program to determine if the hash provided is the correct hash of the header. Remember to perform SHA-256 twice to compute the header hash.

2. Set up a full bitcoin peer node and connect it to the bitcoin network. Download the bitcoin core software from https://bitcoin.org/en/download. It is best to have a fast internet connection and at least 200 GB of free disk space.

Join our community Discord space

Join the book's Discord workspace for a monthly *Ask me Anything* session with the author:
https://discord.gg/7h8aNRhRuY

16

Self-Driving Vehicle Architectures

This chapter describes the capabilities of self-navigating vehicle-processing architectures. It begins with a discussion of the types of sensors and data a self-driving vehicle receives as input while driving. We continue with a discussion of the requirements for ensuring the safety of the autonomous vehicle and its occupants, as well as for other vehicles, pedestrians, and stationary objects. Next is a description of the types of processing required for effective vehicle control. The chapter concludes with an overview of an example self-driving computer architecture.

After completing this chapter, you will have learned the basics of the computing architectures used by self-driving vehicles and will understand the types of sensors used by these vehicles. You will be able to describe the types of processing required by self-driving vehicles and will understand the safety issues associated with self-driving vehicles.

The following topics will be presented in this chapter:

- Overview of self-driving vehicles
- Safety concerns of self-driving vehicles
- Hardware and software requirements for self-driving vehicles
- Autonomous vehicle computing architecture

Technical requirements

The files for this chapter, including answers to the exercises, are available at https://github.com/PacktPublishing/Modern-Computer-Architecture-and-Organization-Second-Edition.

Overview of self-driving vehicles

Several major motor vehicle manufacturers and technology companies are actively pursuing the development and sale of fully self-driving, or autonomous, motor vehicles. The utopian vision of safe, entirely self-driving vehicles beckons us to a future in which commuters are free to relax, read, or even sleep while in transit and the likelihood of being involved in a serious traffic accident is drastically reduced from the hazardous situation of today.

While this is the dream, the current state of self-driving vehicles remains far from this goal. Experts in the field predict it will take decades to fully develop and deploy the technology to support the widespread use of fully autonomous transportation.

To understand the requirements of autonomous driving systems, we begin with the inputs a human driver provides to control the operation of a current-generation motor vehicle. These are:

- **Gear selection**: For simplicity, we assume the presence of an automatic transmission that allows the driver to select between *park*, *forward*, and *reverse*
- **Steering**: The input for this is the steering wheel
- **Accelerator**: Pressing a floor pedal accelerates the vehicle in the direction selected by the gearshift
- **Brake**: Whether the vehicle is moving forward or backward, pressing the brake pedal slows the vehicle and eventually brings it to a stop

To achieve the goals of fully autonomous driving, technology must advance to the point that control of all four of these inputs can be entrusted to the sensors and computing systems present in the automated vehicle.

To understand the current state of self-driving vehicle technology in relation to the goal of fully autonomous driving, it is helpful to refer to a scale that defines the transitional steps from entirely driver-controlled vehicles to a fully autonomous architecture. This is the subject of the next section.

Driving autonomy levels

The **Society of Automotive Engineers (SAE)** has defined six levels of driving automation (see `https://www.sae.org/standards/content/j3016_202104/`), covering the range from no automation whatsoever to fully automated vehicles that have no human driver. These levels are:

- **Level 0 – No driving automation**: This is the starting point, describing the way motor vehicles have operated since they were first invented. At Level 0, the driver is responsible for all aspects of vehicle operation, which includes reaching the intended destination while ensuring the safety of the vehicle, its occupants, and everything outside the vehicle.

Level 0 vehicles may contain safety features such as a forward collision warning and automated emergency braking. These features are considered Level 0 because they do not control the vehicle for sustained periods of time.

- **Level 1 – Driver assistance**: A Level 1 driving automation system can perform either steering control or acceleration/deceleration control for a sustained period, but not both functions simultaneously. When using Level 1 driver assistance, the driver must continuously perform all driving functions other than the single automated function. Level 1 steering control is called **Lane-Keeping Assistance (LKA)**. Level 1 acceleration/deceleration control is called **Adaptive Cruise Control (ACC)**. When using a Level 1 driver assistance feature, the driver is required to remain continuously alert and ready to take full control.

- **Level 2 – Partial driving automation**: Level 2 driving automation systems build upon the capabilities of Level 1 by performing simultaneous steering control and acceleration/deceleration control. As in Level 1, the driver is required to always remain alert and ready to take full control.

- **Level 3 – Conditional driving automation**: A Level 3 driving automation system can perform all driving tasks for sustained periods of time. A driver must always be present and must be ready to take control if the automated driving system requests human intervention. The primary difference between Level 2 and Level 3 is that in Level 3, the driver is not required to continuously monitor the performance of the automated driving system or maintain awareness of the situation outside the vehicle. Instead, the human driver must always be ready to respond to requests for intervention from the automated driving system.

- **Level 4 – High driving automation**: A Level 4 automated driving system can perform all driving tasks for sustained periods of time and is also capable of automatically reacting to unexpected conditions in a way that minimizes the risk to the vehicle, its occupants, and others outside the vehicle. This risk minimization process is referred to as *driving task fallback*, and it may involve actions that result in avoiding risk and resuming normal driving or performing other maneuvers such as bringing the vehicle to a stop at a safe location. A human driver may take over control of the vehicle in response to driving task fallback, or at other times if desired. However, unlike the lower driving automation levels, there is no requirement for anyone in the Level 4 vehicle to remain ready to take over control of the vehicle while it is in operation. It is also not required that vehicles with Level 4 automation include operating controls for human drivers. The primary use case for Level 4 driving automation is in applications such as taxis and public transportation systems where vehicle operation is constrained to a specific geographic region and a known set of roads.

- **Level 5 – Full driving automation:** In a Level 5 vehicle, all driving tasks are always performed by the automated driving system. There is no need for control inputs that would allow a human driver to operate the vehicle—all vehicle occupants are passengers. A Level 5 driving system must be capable of operating a vehicle on all road types, at all times of day, and in all weather conditions in which a responsible and typically skilled human driver would be able to drive safely. The only driving-related task the human occupants of the vehicle perform is selecting a destination.

In 2021, most vehicles on the road are at Level 0. Many newer vehicle models contain Level 1 and Level 2 automation features. There have been almost no Level 3 systems approved for operation in countries around the world. The Level 3 systems that have received regulatory approval are typically limited to specific operating conditions such as driving in heavy traffic on highways.

An important distinction in the performance requirements for autonomous vehicles compared to many traditional computing applications such as smartphones and web servers is the very real potential for autonomous vehicles to injure and kill vehicle occupants and others outside the vehicle. This is the topic of the next section.

Safety concerns of self-driving vehicles

At all times when some level of autonomous control is active in a moving motor vehicle, the algorithms behind the autonomous behavior must continuously apply a hierarchical set of requirements to meet the needs of passengers while making every effort to avoid negative outcomes, such as collisions with other objects.

The highest priority encoded in autonomous vehicle algorithms must always be to ensure the safety of the vehicle, its occupants, and others in the vicinity. Consider the alternative: if the vehicle's highest priority was to get its passengers to the requested destination, the vehicle would interpret this as a license to run through red lights and strike pedestrians if those actions result in the quickest path to the destination.

The vehicle not only needs to predict and manage its path through the numerous obstacles that present themselves; it must also ensure that all of its safety-critical components are operating properly and are receiving valid input data. This means that if vital sensors such as video cameras suffer degradation from a buildup of snow or dirt that prevents effective sensor operation, the vehicle algorithms must bring the system to a state of minimum risk. In this situation, the vehicle may take steps to ensure safety by notifying the driver to take control or by bringing the vehicle to a stop at a safe location.

Because it is critical that autonomous vehicles behave in a safe manner, approval for autonomous vehicles to operate on public roads is required by government regulatory agencies responsible for road safety. It is not generally possible for an individual developer or company to build an autonomous vehicle and allow it to operate on roads without this kind of approval. If someone does this and the vehicle ends up causing an accident or injury, the legal liability for the event may be assigned to the vehicle's occupants or to the party responsible for allowing the vehicle to operate on the road.

Various academic and commercial efforts have completed some of the steps of technological development required for autonomous driving, but a fully capable, fully autonomous vehicle design has yet to be fielded. The following four stages provide a rough guide to the types of autonomous driving capabilities that have been demonstrated and that will be required to reach Level 5 autonomy:

- **Stage 1 – Road following**: A road-following automated driving system can detect and follow road markings such as lane lines and can even detect texture changes between the road surface and the shoulder on unmarked roads. A driving system that merely maintains the position within a lane does not perform driving necessities like obeying traffic lights and avoiding other vehicles.
- **Stage 2 – Obeying traffic rules**: A driving automation system at this stage can perform lane keeping while also detecting and responding correctly to driving directions provided by road signs and signal lights. A system with this level of capability can reliably react to signs providing driving information such as a speed limit or a requirement to yield at an intersection.
- **Stage 3 – Obstacle avoidance**: A driving system capable of obstacle avoidance performs lane keeping and obeys traffic signals while also detecting all significant objects in the vehicle's vicinity, stationary or moving, and responds appropriately to minimize risk to all. Obstacles include other motor vehicles, cyclists, pedestrians, animals, road construction, debris on the roadway, and other unusual driving situations such as flooded or washed-out roads.
- **Stage 4 – Handling edge cases reliably**: While it may seem that an automated driving system that handles the first three stages of capability listed here would be of sufficiently high quality to put into operation, perhaps the most challenging aspect of automated driving may be ensuring the system can deal properly with rare but consequential situations where a human driver would be expected to respond in an appropriate manner. For example, suppose flooding has caused a portion of a bridge to collapse into a river.

A car driven by a human crossing the bridge would ideally come to a stop when the driver observes that a section of the bridge is missing. To be confident of the capabilities of an automated driving system, passengers in an autonomous vehicle must trust it will respond appropriately in all situations where a competent human driver would be able to effectively minimize risk to everyone inside and outside the vehicle.

- A more mundane example of these capabilities occurs when two vehicles pull up to a four-way stop from different directions at nearly the same time. One driver may motion to the other with a hand signal to proceed through the intersection. When the second driver is an autonomous system, will that system detect the gesture and respond appropriately?

While achieving the Stage 4 ability to respond to rare but dangerous events may seem to be a logical extension of the capabilities described in the first three stages, this scenario in fact presents a formidable challenge for an automated driving system. As we'll see later in this chapter, neural networks are the primary technology currently in use for implementing automated driving systems. Neural networks learn from a series of example situations presented to them along with the "correct" answer the network should produce in each situation. These networks have shown a tremendous ability to generalize from the situations presented to them and respond correctly to novel situations that lie within the scope of the learning situations the network has observed.

The process of dealing with situations that are in some sense "in between" the network learning scenarios is called **interpolation**. Problems arise when the neural networks attempt to generalize to scenarios that lie outside the scope of their learning scenarios. This is called **extrapolation**. A neural network's extrapolated response to a novel situation may or may not be something a competent human driver would consider to be correct. Dealing effectively with the very large number of rare but possible driving scenarios that human drivers experience daily may be the greatest challenge facing the development of autonomous driving systems.

The next section introduces the data inputs provided to autonomous driving systems by the various sensors installed in the vehicle.

Hardware and software requirements for self-driving vehicles

Human drivers must sense the state of their vehicles and constantly evaluate the surrounding environment, keeping track of stationary and moving obstacles. The primary means of gathering this information is through vision.

Using eyesight, a competent driver monitors vehicle instrumentation, principally the speedometer, and scans the surrounding environment to perform lane keeping, maintain appropriate spacing from other vehicles, obey traffic signs and signals, and avoid any obstacles on or near the road surface.

Human drivers rely on other senses to a lesser degree, including the use of hearing to detect signals such as car horns and railway crossings. The sense of touch comes into play as well, for example when bump strips are installed on a highway surface to warn of an upcoming intersection. The sense of touch can also assist when an inattentive driver drifts off the roadway and onto the shoulder, which typically has a significantly different texture from the road surface.

Sight, hearing, and touch are the only inputs a human driver uses while driving. Looked at in one way, this demonstrates that the information input provided by these senses enables human drivers to travel billions of miles each day, largely successfully. Viewed differently, significant gaps obviously remain in the processes of sensing and executing appropriate responses to hazardous situations as evidenced by the thousands of deaths that occur daily in traffic accidents.

To be accepted by the driving public, autonomous driving systems must not be merely as safe as human drivers—they must be demonstrably far superior to humans in terms of the rate of traffic accidents per mile traveled. This high level of performance will be necessary because many human drivers will be reluctant to turn over control to an automated system until they have been convinced the system is superior to their own (perhaps imagined) high level of driving skill.

The next section introduces the types of sensors used in autonomous driving systems and their contributions to meeting the goals of safety and reliability.

Sensing vehicle state and the surroundings

Autonomous driving sensors must accurately measure the state of the vehicle itself as well as the location and velocity (**velocity** is the combination of speed and direction of motion) of all significant objects in the vehicle's vicinity. The following sections describe the primary sensor types used in current generations of autonomous vehicle designs.

GPS, speedometer, and inertial sensors

An autonomous vehicle must continuously maintain awareness of its state, which includes its location, direction of movement, and speed. Vehicle location information is used for low-level driving tasks like lane keeping, and for higher-level functions such as developing a route to a destination.

Low-level location measurements enable information to be derived from lidar data or from video camera images that provide information including the vehicle's position relative to lane-marking lines. This information is high-resolution (meaning accuracy is measured in centimeters) and is updated at a high rate (perhaps dozens or even hundreds of times per second), enabling smooth and continuous responses by the driving system to changing conditions.

The information provided by a **Global Positioning System (GPS)** sensor is updated less frequently (perhaps a few times per second) and may have much lower accuracy, with position errors of multiple meters. The information provided by a GPS receiver is ideal for use in route planning, but it is likely to provide too coarse of a measurement and to be updated too infrequently to be of use for keeping a vehicle centered in a lane.

GPS receivers have a notable performance limitation because they rely on the continuous reception of satellite signals to enable their operation. In situations where visibility of the sky from the vehicle is impaired, perhaps on a road with dense tree cover, in the urban canyon of a city center, or in a tunnel, the GPS receiver may not work at all.

The vehicle's speedometer normally provides an accurate measurement of vehicle speed based on tire rotation rate. At times the speedometer measurement may be an inaccurate representation of the vehicle speed due to tire slippage, perhaps resulting from the presence of mud or ice on the roadway. As a backup to the speedometer, GPS provides an accurate measurement of vehicle speed whenever it is receiving the necessary satellite signals. In this situation, divergence between the vehicle speedometer reading and the speed measured by the GPS may be an indication of tire slippage, which is an unsafe condition.

Modern vehicles often contain inertial sensors in the form of accelerometers and, in some cases, a gyroscope. An accelerometer measures the acceleration, or rate of change of velocity, along a single axis of motion. Humans sense acceleration as the force pressing them back into the seat while a vehicle is speeding up and as the force pushing them sideways during an aggressive turn.

Motor vehicles typically contain two accelerometers: one to measure acceleration and deceleration along the forward-reverse axis of motion, and a second measuring acceleration along the side-to-side axis of motion. The accelerometer measuring side-to-side acceleration measures the inertial effects of turns. A gyroscope can be used to measure the vehicle's rate of turn directly.

Accelerometers can provide measurements at very high rates and are used for purposes such as deploying air bags within milliseconds after a collision has been detected. The accelerometers and gyroscope in a vehicle track its orientation and velocity, enabling it to maintain an accurate understanding of its relationship to its surroundings.

Using an algorithm called a **Kalman filter**, it is possible to combine measurements from multiple sources like the GPS receiver and inertial sensors with a mathematical model of the physical laws that constrain the motion of objects such as an automobile. The Kalman filtering process recognizes that each measurement contains some error and that the mathematical model is an imperfect representation of the dynamic behavior of the vehicle. By incorporating the statistical error characteristics of the sensors and the range of unpredictable behaviors a vehicle can experience (resulting from acceleration, braking, and steering inputs as well as from external effects such as driving up and down hills), the Kalman filter synthesizes the available information to produce an estimate of the vehicle's state that is significantly more accurate than the information received from any sensor or the mathematical model by itself.

The GPS, speedometer, and inertial sensors provide an estimate of the vehicle's state. To drive safely and reach a requested destination, an autonomous driving system must also sense the environment surrounding it. The sensors that perform this function are the subjects of the following sections.

Video cameras

Some autonomous driving systems employ an array of video cameras as the primary sensor of the external environment. The video cameras used in these vehicles have features that are familiar to users of consumer-grade digital video cameras, including handheld devices and the video cameras in smartphones. An autonomous vehicle video camera is a moderately high-resolution device, typically 1,920x1,080 pixels. A vehicle may have several cameras located around its periphery with overlapping fields of view.

The use of multiple cameras highlights one improvement of autonomous vehicles over human-driven vehicles: vision systems used in autonomous vehicles are capable of simultaneously and continuously monitoring activity around the vehicle in all directions. Human drivers, in comparison, can only look in one direction at a time, with some limited peripheral vision across a greater field of view. The presence of internal and external mirrors in human-driven vehicles helps to mitigate this inherent limitation of human vision, though there are typically substantial blind spots that human drivers must be cognizant of and incorporate into their situational awareness.

The greatest challenge associated with using video cameras in autonomous driving systems is that the video images output by these devices are not directly useful in performing driving tasks. Substantial processing must be performed to identify significant features in the images, combine those features to recognize individual objects, and then understand what the objects are doing and react appropriately. We will discuss the processing of video images in autonomous driving systems in the *Perceiving the environment* section later in this chapter.

Radar

One limitation of video cameras in autonomous driving systems is that they provide a flat, two-dimensional image of the scene viewed by the camera. Without substantial further processing, it is not possible to tell if the portion of the scene represented by any individual pixel is close to the vehicle (and therefore potentially an obstacle that requires an immediate reaction) or something that is far away and is of no relevance to the driving task.

The technology of **radio detection and ranging (radar)** provides a solution to part of this problem. A radar system repetitively sends out pulses of electromagnetic energy into the environment and listens for echoes of those signals bouncing off objects in its vicinity.

A radar system can detect objects and provide measurements of the direction to each object as well as the range to those objects. It can also measure the speed of objects relative to the vehicle carrying the radar system. A radar system can track a vehicle several hundred feet ahead and maintain a safe following distance when used as part of an ACC.

Radar systems have some significant limitations, however. Compared to a video camera, the scene perceived by a typical automotive radar system is much blurrier and has lower resolution. Radar systems are also prone to noisy measurements, which reduces confidence in the quality of the information they provide.

Despite those limitations, radar sensors perform without degradation in situations where video cameras often fail, such as in heavy rain, snow, and dense fog.

Lidar

Instead of using video cameras, some autonomous vehicle developers have elected to rely on lidar systems for determining vehicle location and orientation and detecting objects in the vehicle's vicinity. A **light detection and ranging (lidar)** sensor uses a laser to scan the area around the vehicle and collect data from the reflections received from surfaces and objects.

Each measurement by a lidar device represents the distance from the lidar laser illuminator to a surface that reflects some of the laser energy back to the lidar sensor. The direction of the laser beam relative to the vehicle at the time of each sample is known to the lidar processing software. The distance to the reflecting point and back is calculated by measuring the time between when the laser pulse is emitted and the time the echo arrives at the lidar sensor, which is referred to as the pulse's time of flight. The round-trip distance traveled by the pulse is equal to the pulse's time of flight multiplied by the speed of light through the atmosphere.

By collecting measurements in rapid succession in all directions around the vehicle, a three-dimensional set of points called a point cloud is developed. This **point cloud** represents the distance from the lidar sensor to surrounding surfaces, including the road surface, buildings, vehicles, trees, and other types of objects.

Sonar

Sound navigation and ranging (sonar) performs a function that is intuitively similar to radar and lidar systems, except sonar systems emit pulses of sound rather than electromagnetic waves.

Sonar systems generally operate at much shorter ranges than radar and lidar systems and are mainly used to detect obstacles in close-quarters situations, like when parking between other vehicles or for detecting when a vehicle in an adjacent lane on a highway is approaching uncomfortably close to the vehicle.

Perceiving the environment

As sensors collect raw information about the vehicle and its surroundings, the sensor output information is not immediately usable for performing effective vehicle control. Several stages of processing must be performed to convert raw sensor measurements into actionable information that can be used for autonomous driving. The following sections describe the processing steps needed to convert sensor data into driving decisions in autonomous vehicles that rely on video cameras and lidar systems as primary sensors. We begin with the processing of images from video cameras using convolutional neural networks.

Convolutional neural networks

Chapter 6, Specialized Computing Domains, briefly introduced the concepts of deep learning and artificial neural networks. To recap, artificial neurons are mathematical models intended to represent the behavior of biological neurons, which are the brain cells responsible for perception and decision making.

In modern autonomous vehicle systems that use video cameras to sense the environment, the leading technology used to extract decision-quality information from video images captured by cameras is the convolutional neural network.

A **Convolutional Neural Network (CNN)** is a specialized type of artificial neural network that performs a form of filtering called convolution on raw image data to extract information useful for detecting and identifying objects within the image.

An image captured by a video camera is a rectangular grid of pixels. The color of each pixel is represented by a set of red, green, and blue intensities. Each of these intensities is an 8-bit integer ranging from 0 to 255. An intensity of 0 means the corresponding color is not present in the pixel's color, while 255 represents the maximum intensity for that color. A pixel's color is represented by the three red, green, and blue values, abbreviated as **RGB**. Some examples of RGB colors are listed in *Table 16.1*:

Red Intensity	Green Intensity	Blue Intensity	Color
255	0	0	Red
0	255	0	Green
0	0	255	Blue
0	0	0	Black
255	255	255	White
128	128	128	Gray
255	255	0	Yellow
255	0	255	Magenta
0	255	255	Light Blue

Table 16.1: Examples of RGB colors

A CNN receives as its input the RGB image captured by a video camera as three separate two-dimensional arrays of 8-bit pixel colors, one each for red, blue, and green. We will refer to these as the three **color planes** of the image.

The CNN then performs convolution filtering on each of the color planes. A convolution filter consists of a rectangular grid of numbers that is typically small, perhaps 2 rows by 2 columns. The convolution operation begins at the top-left corner of each color plane. Mathematically, convolution is performed by multiplying the color intensity at each pixel location by the number at the corresponding location in the convolution filter and then summing the products of these operations over all elements in the filter. The result of this operation is one element of the convolution output. We'll go through an example shortly to help clarify this process.

The convolution continues by shifting the filter 1 or more pixels to the right, and then repeating the multiplication and summing operations with the image pixels covered at the filter's new location. This produces the convolution output at the second location in the top row. This process continues all the way across the top row of the image and then the filter shifts back to the left edge and moves downward by 1 or more pixels and repeats the process across the second row in the image. This sequence of shifting the filter, multiplying the filter elements by the pixel color intensities at the filter location, and summing the results continues across the entire image.

Example CNN implementation

We'll now present a simple example to demonstrate how CNNs work. Let's say we have an image that is 5x5 pixels and a convolution filter that is 2x2 pixels. We will only look at one of the three color planes, though in real filters, the same operations will be performed separately across all three color planes.

We will shift the filter by 1 pixel to the right and 1 pixel downward each time such movement is required. The distance (in terms of pixels) the convolution filter shifts each time it relocates is called the **stride**. For simplicity, all pixels and convolution filter elements in the example will be represented as single-digit integers. *Table 16.2* contains our example color plane data:

1	4	6	7	8
2	9	2	0	5
8	2	1	4	7
3	9	0	6	8
2	1	4	7	5

Table 16.2: Example color plane data

Table 16.3 contains the convolution filter. Each number in the table is one filter coefficient:

7	2
8	0

Table 16.3: Example convolution filter

We will be using a stride of 1, which means each sequential evaluation of the filter output will shift 1 pixel to the right and then 1 pixel downward and back to the left edge when the right edge of the image has been reached.

For the first filter evaluation, we multiply the pixels in the upper-left corner of the image by the corresponding elements in the convolution filter, as shown in *Table 16.4*:

1×7	4×2	6	7	8
2×8	9×0	2	0	5
8	2	1	4	7
3	9	0	6	8
2	1	4	7	5

Table 16.4: Multiplication step in convolution

After multiplying the color plane intensities by the convolution filter elements in the 2x2 grid, we add them together, as shown in *Equation 16.1*:

$$C_{1,1} = (1 \times 7) + (4 \times 2) + (2 \times 8) + (9 \times 0) = 31$$

Equation 16.1: Multiplication and summing steps in convolution

The result of the element-by-element multiplication of the color plane intensities by the filter coefficients, followed by summing each of those results, is one element of our output array, $C_{1,1}$, which we place in row 1, column 1 of our output array, as shown in *Table 16.5*:

31			

Table 16.5: First convolution output element

Next, we shift the filter grid one position to the right and perform the multiply-and-sum operation again, as shown in *Table 16.6*:

1	4×7	6×2	7	8
2	9×8	2×0	0	5
8	2	1	4	7
3	9	0	6	8
2	1	4	7	5

Table 16.6: Multiplication step in convolution

After multiplying the color plane intensities at the second filter location by the convolution filter elements, the result is shown in *Equation 16.2*:

$$C_{1,2} = (4 \times 7) + (6 \times 2) + (9 \times 8) + (2 \times 0) = 112$$

Equation 16.2: Summing step in convolution

Table 16.7 shows the convolution output after the second element has been inserted at row 1, column 2:

31	112		

Table 16.7: Second convolution output element

We continue shifting 1 pixel to the right for each evaluation across the top row. Because our filter is 2 pixels wide and the image is 5 pixels wide, we can only place the filter in 4 locations without running the filter off the image. This is the reason the filter output array has 4 rows and 4 columns.

After the top row of the image is complete, we bring the filter back to the left image edge and move it down 1 pixel to compute the element in row 2, column 1 ($C_{2,1}$). This process repeats across the entire image until it ends at the bottom-right corner. For our 5x5 image, the completed output of the convolution operation is shown in *Table 16.8*:

31	112	72	65
96	83	22	42
84	88	15	90
55	71	44	114

Table 16.8: Result of convolution operation

While this explanation of convolution may have seemed a bit tedious and unenlightening, we are now getting to the good part.

CNNs in autonomous driving applications

These are the most interesting points related to the use of convolution filtering in video image processing for autonomous driving systems:

- Mathematically, convolution consists of the simple operations of multiplication and addition.
- CNNs can be trained in the same way as the **Artificial Neural Networks (ANNs)** discussed in *Chapter 6, Specialized Computing Domains*. This training consists of repeated trials that present the neural network with a set of input data along with the "correct" output that should be produced when presented with that data. The training process adjusts the weights within the ANN (and in convolution filters of the CNN) to improve its ability to respond correctly to the given inputs on each iteration.
- As part of the training process, the CNN discovers the best data values to place at each location in its convolution filter tables.

To provide a concrete example, consider the problem of steering a vehicle to remain properly centered within its lane. To train a CNN to perform this task, the general procedure is to record a series of video segments from vehicle cameras together with appropriate steering commands that the autonomous system must attempt to replicate as its output.

The training dataset must provide sufficient variety to ensure the behavior learned by the CNN covers the full range of its intended operating conditions. This means the training data should include straight roads and roads with a variety of curves. The data should also include daytime and nighttime driving, and other conditions that may be encountered, such as different types of weather.

During training, each training example is presented to the network as a video image along with the correct steering response. As the CNN learns how to respond properly to the varying inputs, it adjusts the coefficients in its convolution filters to identify image features that are useful in performing the desired tasks.

One image feature that is particularly useful in autonomous driving is edge detection. **Edge detection** involves locating the dividing lines that separate distinct portions of the image from one another. For example, the lines painted on roadways are always in a very different color from the road surface to make them highly visible to drivers, both human and autonomous.

The most interesting, even fascinating, attribute of the CNN training process is that by simply presenting video frames of road surfaces to a CNN along with accurate examples of the desired outputs, the CNN will discover image attributes that are useful (such as edges) and build convolution filters automatically that are specifically tuned to identify useful features.

Because the CNN filter is applied repeatedly across the entire image, it is possible for the filter to detect objects wherever they may appear in the image. This is an important feature of autonomous driving because it is not possible to predict the location in a video image where an important object may appear.

Real-world applications of CNNs may contain several convolution filters, each containing different sets of coefficients. Each of these filters is tuned during the training process to seek out different types of relevant features that can be processed in later stages of the network to perform higher-level functions, like identifying stop signs, traffic lights, and pedestrians.

A complete CNN design contains one or more stages of convolution that each operates as described in this section. The initial convolution stage performs filtering to detect simple features like edges, while later stages combine these features to identify more complex objects like a stop sign.

To reliably detect a feature such as a stop sign under varying conditions, the training data must contain images of stop signs as seen from nearby and far away, as well as when looking at it straight on and from off to the side at an angle, all under a variety of lighting and weather conditions.

Within the CNN structure, the data elements computed in each convolution stage pass through an activation function to produce the final output of the stage. Although many types of functions can be used for this purpose, a common activation function is the **Rectified Linear Unit** function, abbreviated to **RELU**.

The RELU activation function is a very simple formula: given the table of output values from a convolution operation such as *Table 16.8*, examine each element. If the element is less than 0, replace it with 0. That's all there is to it. Our simple example in *Table 16.8* does not contain any negative elements, so applying RELU will not change it. In real applications, the RELU activation function provides important benefits, including improved training speed compared to other common activation functions.

Another type of stage used in CNN architectures is the pooling stage. High-resolution images contain a large number of pixels. To keep memory consumption and processing requirements within achievable limits, it is necessary to reduce the quantity of data flowing through the network while retaining the features that lead to the correct recognition of objects within the image. One way to do this is a technique called pooling. **Pooling** combines multiple elements in an input table into a single element in the output table. For example, each 2x2 pixel subset of *Table 16.8* could be combined to reduce the size of the output from that subset to a single numerical value.

There are several ways to combine the pixels within a subset into a single value. One obvious possibility is to compute the average. For the 2x2 pixel region in the upper-left corner of *Table 16.8*, the average is (31 + 112 + 96 + 83) / 4 = 80.5). While averaging provides a representation that includes effects from all the pixels in the region, this has not proven to be a pooling method providing the best performance in CNNs. Instead, in many cases, simply selecting the maximum value within the region as the pool result has been shown to give good performance. In this example, the maximum value of the selected region is 112.

This technique is referred to as **max pooling**. *Table 16.9* shows the results of applying max pooling with a region size of 2x2 to *Table 16.8*:

112	72
88	114

Table 16.9: Result of 2x2 max pooling on Table 16.8

To convert a layer from the two-dimensional structure of the video images and the convolutional layers that process the video data into a form suitable for input to a traditional ANN, it is necessary to rearrange the data into a one-dimensional format.

As with the other mathematical operations described earlier, this is a simple procedure. The conversion from a two-dimensional structure to a one-dimensional vector is called **flattening**. In a flattening layer, the coefficients in the two-dimensional structure are simply transferred sequentially into a vector, which serves as the input to a traditional ANN.

Table 16.10 presents the results of *Table 16.9* after the flattening operation has been performed:

Table 16.10: Result of flattening on Table 16.9

Table 16.10 is a one-dimensional vector of numeric values that is in a suitable format for use as input to a traditional ANN.

Following the convolutional, pooling, and flattening stages, a CNN implements one or more ANN layers, each of which forms a hidden layer of neurons. These hidden layers are usually fully connected sets of neurons, as shown in *Figure 6.5* in *Chapter 6, Specialized Computing Domains*. Following the hidden layer is an output layer, which presents the final outputs of the network for use in further processing.

When a sufficiently rich set of training data has been combined with an appropriately sized CNN structure capable of learning and retaining the necessary knowledge, it is possible to encode a tremendous amount of information that represents a wide variety of driving situations into the CNN and ANN coefficients.

In a roadworthy autonomous vehicle design, the result of the CNN design and training process is a system that can recognize and identify the object types encountered across the full range of driving situations encountered by drivers daily. This is a tremendously complex task, and it may take several years before the computing technology and software capability (both in the network structure and in the training process) are capable of consistently outperforming human drivers in the tasks of object detection and recognition.

Figure 16.1 presents a basic (Stage 1, as described above) CNN architecture that has shown to be capable of steering an autonomous driving system to remain somewhat centered in a lane on a curving road:

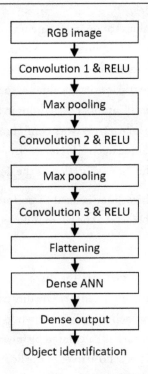

Figure 16.1: CNN layers used in object identification application

The exercises at the end of this chapter develop an example CNN based on the structure of *Figure 16.1* that can identify and categorize objects in low-resolution color images with a significant degree of accuracy.

You may wonder how the developer of a CNN for a particular application selects architectural features such as the types and number of layers and other parameters defining the network. In many cases, it is not at all clear what the best combination of layer types and the settings for other network parameters (such as convolution filter dimensions) is for achieving the overall goals of the CNN application.

The types and dimensions of layers and related parameters describing a neural network are referred to as hyperparameters. A **hyperparameter** is a high-level design feature of a neural network that defines some part of its structure. Hyperparameters are distinct from the weighting factors (see *Figure 6.4*) on the neural connections, which are defined automatically during the training process.

The selection of hyperparameter values to configure a neural network for a particular application can be considered a search process. The architect of the neural network design can use software tools to test a variety of network architectures composed of different sequences of neural network layer types along with different parameter values associated with each layer, such as the dimensions of convolution filters or the number of neurons in a fully connected layer. This search process comprises a training phase for each network configuration followed by a testing phase to evaluate the performance of the trained network. Network designs that perform the best during the testing phase are retained for further evaluation as the design process moves forward.

In the next section, we'll look at how lidar is used to perform the functions of determining a vehicle's location and identifying objects in the surroundings.

Lidar localization

As we discussed earlier, lidar sensors measure the distance to surfaces that reflect some of the transmitted laser energy back to the lidar sensor. The direction of the transmitted laser beam relative to the vehicle at the time of each sample is known to the lidar processing software. The distance to the reflecting point and back is calculated by measuring the time between when the laser pulse is emitted and the time the echo arrives at the lidar sensor, which is referred to as the pulse's time of flight. The round-trip distance traveled by the pulse is equal to the pulse's time of flight multiplied by the speed of light.

By collecting measurements in rapid succession in all directions around the vehicle, a point cloud can be formed, which contains the full set of measurements to points in the surroundings.

By comparing the point cloud produced by the lidar sensor to a stored map of surfaces and structures in the vicinity of the vehicle location, the lidar data processing software can adjust and align its measured point cloud to bring it into agreement with the stored map data. This process enables the lidar system to precisely determine its location and orientation relative to the surrounding environment. This process is referred to as **localization**.

In comparison to video image-based autonomous driving systems, lidar-based driving systems have the following advantages:

- Lidar sensors continuously produce precise three-dimensional maps of the surroundings. This information represents the distance from the sensor to each point in the point cloud. A single video camera produces a two-dimensional image with no information about the distance to any point in the scene.

- Video cameras can be blinded by sunlight or another bright light source. Lidar sensors are much less susceptible to such interference.
- The object detection function in video camera-based systems may fail to identify objects when the perceived color of the object lacks sufficient contrast from the surrounding environment. Lidar systems detect objects in their surroundings regardless of their color.

Lidar systems have the following limitations:

- Lidar systems have substantially less resolution than standard video cameras.
- As with video cameras, the performance of lidar sensors can be impaired by weather conditions such as fog and heavy rain.
- Lidar localization systems only work on roads and in other environments that have been precisely mapped. The database containing the map information must be available to the vehicle while operating. This necessitates an ongoing investment to collect mapping data on the roads used for autonomous driving. The mapping effort must repeat the data collection process to pick up changes such as the construction of new buildings near roadways. When driving on roads that have not been mapped, the lidar system is not available for use by the vehicle.
- Lidar systems tend to be significantly more expensive than video cameras.

There is ongoing competition between automakers and technology companies to develop and deploy autonomous driving systems using different sensing technologies. As of 2021, Tesla has focused its efforts on the use of multiple video cameras to provide the information necessary to perform autonomous driving functions without the need for a precision map of the surrounding environment.

Other companies, including Toyota and Waymo, rely on the use of lidar as the primary system for sensing the environment in concert with a database containing three-dimensional localization data.

Whether an autonomous driving system relies on video cameras or a lidar system, it must identify and track objects over time to perform the functions required for safe driving. This is the subject of the next section.

Object tracking

The output from the perception stage of an autonomous driving system includes a list of objects and their classification (such as a truck or cyclist) derived from video or lidar data. To support the task of driving, additional processing must be performed to maintain a history of objects identified in sequential sensor updates and track the behavior of those objects over time.

Object tracking measures object motion over time and enables the prediction of the future motion of those objects. The results of object motion prediction are used in path planning decisions to maintain safe separation distances from other objects and to minimize the risk of collisions.

Tracking objects over sequential updates permits errors such as spurious object detection within individual scene samples to be identified and corrected if the presence of the possible object cannot be confirmed in later updates.

After developing a dataset describing the state of the vehicle and all the objects relevant to driving the vehicle, the autonomous driving system must make a series of decisions, discussed next.

Decision processing

Using the information provided during the perception stage of sensor data processing, the autonomous driving system decides upon the actions it will execute at each moment to continue safe vehicle operation while making progress toward the requested destination. The basic functions it must support consist of lane keeping, obeying the rules of the road, avoiding collisions with other objects, and planning the vehicle path. These are the topics of the following sections.

Lane keeping

The lane-keeping task requires the vehicle driver (human or autonomous) to continuously monitor the vehicle's location within the width of the lane it occupies and maintain the vehicle's position within an acceptable deviation from the center of the lane.

Lane keeping is a straightforward task on a clearly marked road in good weather conditions. On the other hand, an autonomous driving system (or a human driver) may experience significant difficulty staying centered in a lane when lane markings are obscured by fresh snowfall or mud on the road's surface.

A lidar-based autonomous driving system should experience little or no impairment while navigating a road surface covered by a light layer of snow, as long as lidar measurements of the surroundings continue to provide valid information. A video camera-based system, however, may experience substantial difficulty in this context.

Complying with the rules of the road

Autonomous driving systems must obey all traffic laws and regulatory requirements while operating on public roads. This includes basic functions, like stopping for stop signs and responding appropriately to traffic lights, as well as other necessities, such as yielding the right of way to another vehicle when the situation requires it.

The driving task can become complex when autonomous vehicles must share the road with human-operated vehicles. Humans have developed a range of behaviors for use in driving situations that are intended to make driving easier for everyone. For example, a vehicle on a crowded highway may slow down to provide space for a merging vehicle to enter the roadway.

A successful autonomous driving system must comply with all legal requirements and must also align its behavior with the expectations of human drivers.

Avoiding objects

Autonomous driving systems are required to perceive and respond appropriately to the entire gamut of inanimate objects and living creatures that drivers encounter on roads every day. This includes not only other vehicles, pedestrians, and cyclists—random objects such as ladders, car hoods, tires, and tree limbs also wind up on roads every day. Animals such as squirrels, cats, dogs, deer, cattle, bears, and moose regularly attempt to cross the roads in various regions.

Human drivers generally attempt to avoid striking smaller animals on the road if it can be done without creating an unacceptably dangerous situation. When encountering a large animal, avoiding impact might be necessary to prevent killing the vehicle's occupants. Autonomous driving systems will be expected to outperform human drivers under all these conditions.

Planning the vehicle path

High-level path planning generates a sequence of connected road segments the vehicle intends to traverse while traveling from its starting point to its destination. GPS-based route planning systems are familiar to drivers of today's vehicles. Autonomous vehicles use the same approach to develop a planned path to their destination.

Low-level path planning encompasses all the driving actions along the path to the destination. An autonomous driving system must constantly evaluate its surroundings and make decisions such as when to change lanes, when it is safe to enter an intersection, and when to give up trying to turn left onto a very busy street and fall back to an alternate route.

As with all aspects of autonomous driving, the highest goals of path planning are to keep the vehicle occupants and others outside the vehicle safe, obey all traffic laws, and, while accomplishing all that, travel to the destination as quickly as possible.

Autonomous vehicle computing architecture

Figure 16.2 summarizes the hardware components and the processing stages in an autonomous driving system based on the current state of technology, as described in this chapter.

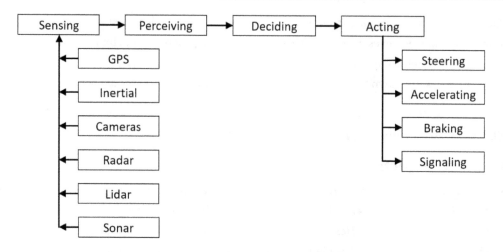

Figure 16.2: Components and processes of an autonomous driving system

We introduced sensor technologies that gather information about the state of the vehicle and its surroundings. This information flows into a *Sensing* process, which receives data from the sensors, validates that each sensor is performing properly, and prepares the data for perception. The process of *Perceiving* takes raw sensor data and extracts useful information from it, such as identifying objects in video images and determining their location and velocity. With an accurate understanding of the vehicle's state and all relevant surrounding objects, the *Deciding* process performs high-level navigation functions, like selecting which route to take to the destination, as well as low-level functions such as choosing what direction to go at an intersection. The process of *Acting* on these decisions sends commands to hardware units performing steering, controlling vehicle speed, and providing information to other drivers, most notably by operating the turn signals.

Tesla HW3 Autopilot

To maintain trade secrets, some of the automotive and technology companies working to develop autonomous driving systems have released very limited information about the designs of their systems. Tesla, on the other hand, has been more forthcoming with information about the autonomous driving computer hardware used in their vehicles. Tesla vehicles are currently operating on public roads with so-called full self-driving capabilities using a computer system called **Hardware 3.0 (HW3)**.

The Tesla HW3 processor board contains two fully redundant computing systems, each of which is capable of safely operating the vehicle on its own. The two systems operate in synchrony and continuously compare their outputs. If one of the systems experiences a failure, the second system takes over vehicle operation until a repair can be performed.

The HW3 computer is based on a custom SoC that Tesla developed to optimize system performance for the autonomous driving task. The SoC incorporates substantial resources for neural network computing in addition to traditional capabilities related to image processing and scalar computation.

These are some of the key features of the Tesla HW3 computer:

- The SoC integrated circuit is based on a 14 nm FinFET process with a total of 6 billion transistors. The **Fin Field-Effect Transistor (FinFET)** technology employs a significant vertical dimension (called the "fin"), which differs from the planar circuit structure of traditional CMOS technology. FinFETs can achieve faster switching times and carry a higher current than traditional CMOS transistors. HW3 was the first automotive application of the 14 nm FinFET technology.

- The HW3 computer uses **Low-Power DDR4 (LPDDR4)** DRAM. LPDDR4 is a variation of DDR4 DRAM optimized to reduce power consumption for applications such as smartphones and laptop computers. LPDDR4 has lower bandwidth than DDR4 and consumes substantially less power. While a Tesla vehicle has a very large battery compared to a smartphone, the power consumed by the vehicle computer must still be kept to a minimum, and the use of LPDDR4 DRAM helps with that.

- Each SoC contains two neural network accelerators, which together perform 72 **TOPS (tera-ops, or trillions of operations per second)**. These processors execute a CNN for object detection within video camera images.

- Each neural network accelerator has 32 KB of dedicated high-speed SRAM. As we saw in *Chapter 8, Performance-Enhancing Techniques*, SRAM is much faster than DRAM, but has a significantly higher cost per bit in terms of chip area. This substantial quantity of SRAM provides a performance boost compared to other general-purpose processors that Tesla might have employed instead of the custom SoC.

- For general-purpose computing, each HW3 processor contains three quad-core ARM Cortex A72 64-bit CPUs running at 2.2 GHz.

According to Tesla, the HW3 computer can process an amazing 2,300 frames per second of high-definition video through its CNN architecture. This level of performance is required to support progress toward Tesla's stated goal of using video camera-based sensing to enable Level 5 autonomous driving.

Autonomous vehicle technology is a rapidly developing field, but the current state of the art in vehicles available for purchase by the public remains significantly short of the goal of driverless operation. It may be several years before vehicles become available that are capable of carrying passengers on public roads without the need for someone to constantly monitor the surroundings through the windshield.

Summary

This chapter presented the capabilities required in self-navigating vehicle-processing architectures. It began by introducing driving autonomy levels and the requirements for ensuring the safety of the autonomous vehicle and its occupants, as well as the safety of other vehicles, pedestrians, and stationary objects. We continued with a discussion of the types of sensors and data a self-driving vehicle receives as input while driving. Next, we discussed the types of processing required for vehicle control. We ended with an overview of the Tesla HW3 computer architecture.

Having completed this chapter, you have learned the basics of the computing architectures used by self-driving vehicles and understand the types of sensors used by self-driving vehicles. You can describe the types of processing required by self-driving vehicles and understand the safety issues associated with self-driving vehicles.

In the next and final chapter, we will develop a view of the road ahead for computer architectures. The chapter will review the significant advances and ongoing trends that have led to the current state of computer architectures and extrapolate those trends to identify some possible future technological directions. Potentially disruptive technologies that could alter the path of future computer architectures will be considered as well. In closing, some approaches will be proposed for the professional development of a computer architect that are likely to result in a future-tolerant skill set.

Exercises

1. If you do not already have Python installed on your computer, visit `https://www.python.org/downloads/` and install the current version. Ensure Python is in your search path by typing `python --version` at a system command prompt. You should receive a response similar to `Python 3.10.3`.

Install TensorFlow (an open source platform for machine learning) with the command (also at the system command prompt) `pip install tensorflow`. You may need to use the *Run as administrator* option when opening the command prompt to get a successful installation. Install Matplotlib (a library for visualizing data) with the command `pip install matplotlib`.

2. Create a program using the TensorFlow library that loads the CIFAR-10 dataset and displays a subset of the images along with the label associated with each image. This dataset is a product of the **Canadian Institute for Advanced Research (CIFAR)** and contains 60,000 images, each consisting of 32x32 RGB pixels. The images have been randomly separated into a training set containing 50,000 images and a test set of 10,000 images. Each image has been labeled by humans as representing an item in one of 10 categories: airplane, automobile, bird, cat, deer, dog, frog, horse, ship, or truck. For more information on the CIFAR-10 dataset, see the technical report by Alex Krizhevsky at `https://www.cs.toronto.edu/~kriz/learning-features-2009-TR.pdf`.

3. Create a program using the TensorFlow library that builds a CNN using the structure shown in *Figure 16.1*. Use a 3x3 convolution filter in each convolutional layer. Use 32 filters in the first convolutional layer and 64 filters in the other two convolutional layers. Use 64 neurons in the hidden layer. Provide 10 output neurons representing an image's presence in one of the 10 CIFAR-10 categories.

4. Create a program using the TensorFlow library that trains the CNN developed in *Exercise 3* and test the resulting model using the CIFAR-10 test images. Determine the percentage of test images that the CNN classifies correctly.

Join our community Discord space

Join the book's Discord workspace for a monthly *Ask me Anything* session with the author: `https://discord.gg/7h8aNRhRuY`

17
Quantum Computing and Other Future Directions in Computer Architectures

This chapter anticipates the road ahead for computer architecture design. We will review the significant technological advances and ongoing trends that have led to the current state of computer architectures. We will then extrapolate current trends and identify some directions that computing system designs are likely to take in the future. We will also examine some potentially disruptive technologies that may substantially alter the evolution of future computer architectures.

This chapter provides some suggested approaches for the professional development of a computer architect. By following these recommendations, you should be able to maintain a skill set that remains relevant and tolerant of future advances, whatever they turn out to be.

After completing this chapter, you will understand the historical evolution of computer architecture that led to its current state and will be familiar with ongoing trends in computer design and their likely impact on future technological directions. You will be aware of some potentially disruptive technologies that might significantly alter future computer architectures. You will also have learned some useful techniques for maintaining an ongoing, current skill set in the field of computer architecture.

The following topics will be presented in this chapter:

- The ongoing evolution of computer architectures
- Extrapolating current trends into the future

- Potentially disruptive technologies
- Building a future-tolerant skill set

Technical requirements

The files for this chapter, including answers to the exercises, are available at `https://github.com/PacktPublishing/Modern-Computer-Architecture-and-Organization-Second-Edition`.

The ongoing evolution of computer architectures

Chapter 1, Introducing Computer Architecture, presented a brief history of automated computing devices from the mechanical design of Babbage's Analytical Engine to the advent of the x86 architecture that continues to serve as the basis for most modern personal computers. This progress relied on several groundbreaking technological achievements, most notably the invention of the transistor and the development of integrated circuit manufacturing processes.

Through the decades since the introduction of the Intel 4004 in 1971, processors have grown dramatically in terms of the sheer number of transistors and other circuit components integrated on a single-circuit die. In concert with the growth in the number of circuit elements per chip, the clock speed of modern devices has increased by several orders of magnitude.

The increase in processor capability and instruction execution speed has unleashed the growth of software development as an enormous, worldwide industry. In the early days of digital computers, software was developed by small teams of highly trained specialists in a research setting. Today, powerful personal computers are affordable and software development tools such as programming language compilers and interpreters are widely available, often for free. As processors have increased in capability, the availability of widespread computing power has created a strong demand for software to run on those devices.

Modern processors pack far more functionality into the processor's integrated circuit than early devices such as the 6502. The 6502 contains the basic component set required to perform useful processing: a control unit, a register set, an ALU, and with an external bus for accessing instructions, data, and peripherals.

The most sophisticated modern processors targeted at business and home users incorporate a base level of functionality similar to the capabilities of the 6502 with substantial added features and extensions such as the following:

- Up to 16 processor cores, each supporting simultaneous multithreading
- Multilevel instruction and data cache memories

- A µop cache to avoid the processing delay associated with repetitive instruction decode operations
- A memory management unit supporting paged virtual memory
- Integrated multichannel high-speed serial I/O capability
- An integrated graphics processor generating digital video output
- Support for running virtualized operating systems

To summarize the technological evolution from the 6502 processor to the modern x64 processor, modern processors provide multiple 64-bit cores operating in parallel compared to the 6502's single 8-bit core, and they implement numerous additional features specifically designed to accelerate instruction execution speed.

In addition to the raw computing capability of modern PC processors, the x86/x64 instruction set provides instructions to implement a wide variety of operations, ranging from simple to extremely complex. Modern RISC processors, such as ARM and RISC-V, on the other hand, implement intentionally slimmed-down instruction sets, with the goal of breaking complex operations into sequences of simpler steps, each of which executes at very high speed while working in the context of a larger register set.

The high-level configurations of computer architectures have, arguably, not undergone drastic changes since the days of the 6502. With each extension of the processor architecture's instruction set or the introduction of additional caching technology, these changes have incrementally expanded the functionality available to software developers or increased the speed at which algorithms execute. The expansion to multiple cores and to multithreading within a single core allows multiple independent execution threads to execute simultaneously rather than running in a time-sliced manner on a single core.

Much of the incrementalism during this evolution has been intentional, to avoid introducing changes in processor architectures that would inhibit backward compatibility with the universe of already-developed operating system and application software. The net result has been a series of processor generations that gradually become faster and more capable over time, but do not implement disruptive breaks from past technology.

In the next section, we will attempt to extrapolate from the current generation of high-performance computing systems discussed in *Chapter 13, Domain-Specific Computer Architectures*, to predict the advances in computer architectures likely to occur over the next one to two decades.

Extrapolating from current trends

The capabilities of current-generation processor technology are beginning to push against some significant physical limits that we can expect to constrain the rate of performance growth going forward. These limits certainly will not lead to an abrupt end of improvements in circuit density and clock speed; rather, capability improvements for future processor generations may take place in directions that differ from traditional semiconductor capability growth patterns. To look more closely at future processor performance growth expectations, we begin by returning to Moore's law and examining its applicability to the future of semiconductor technology.

Moore's law revisited

The revised version of Moore's law, published by Gordon Moore in 1975, predicted the number of integrated circuit components per device would double roughly every two years. This law has demonstrated remarkable predictive accuracy for several decades, but as of 2015, according to Intel, the growth rate had slowed to doubling approximately every two and a half years. This indicates the rate of growth in integrated circuit density has begun to slow, but it certainly has not ended, and is not expected to end in the foreseeable future.

Integrated circuit technology will continue to improve, resulting in denser and more highly capable devices for many years to come. We can, however, expect the rate of growth in circuit density to decrease over time because of the physical limits associated with the construction of single-digit nanometer-scale circuit components.

The slower rate of increase in circuit density does not mean the trend is near an end. As of 2022, current mass-produced integrated circuit technology is based on circuit features with dimensions as small as approximately 5 nm. Work is in progress to develop the next generation of circuit technology with 4 nm feature sizes. Future generations with feature sizes of 3 and even 2 nm are in the planning stages.

Although these increased circuit densities are likely to be realized at some point, each technological advance comes with increasing cost and technical challenges that result in delays in deployment to production lines. The most advanced integrated circuit production technologies are so costly to develop and difficult to implement that only a few massive semiconductor companies have the financial resources and technical expertise to bring such processes online.

Given the ongoing decline in the rate of improvement in circuit density, semiconductor manufacturers have begun to focus on alternative methods for packing smaller components together on a chip. Traditionally, integrated circuits have been viewed as primarily two-dimensional entities constructed in layers, as follows:

- Different types of material are laid out in a sequence of masking operations to create doped regions of transistors, as well as other circuit components like capacitors and diodes.
- Conductive metal traces serving as wires are deposited on the devices as additional layers.

Communication between circuit elements within a two-dimensional device layout involves electrical interactions between components placed some distance from each other on the chip's surface. The chip is small, so the time the electrical signal takes to propagate between components is often not significant in comparison to other constraints.

You may wonder if it is possible to organize the components of an integrated circuit in a manner other than spreading them around on a flat surface. It is indeed possible to stack components on top of one another on an integrated circuit die. We will look at this design approach in the next section.

The third dimension

By developing techniques for stacking components atop one another on a single integrated circuit die, semiconductor manufacturers have taken a step toward extending Moore's law. One of the early targets for stacked-component integrated circuit configurations is the ubiquitous n-channel and p-channel MOS transistor pair employed in CMOS circuit designs.

Intel publicly described advances achieved by its researchers in the area of stacked CMOS transistor pairs in early 2020. Not only has the company shown an ability to stack devices on a silicon die but it has also demonstrated how to use differing fabrication technologies in each layer to achieve maximum performance from the transistor pair. Silicon n-channel transistors exhibit good performance characteristics, but p-channel transistors constructed on silicon have a relatively slower switching speed. P-channel transistors implemented with a germanium transistor channel instead of silicon provide increased switching speed, improving the performance of the CMOS pair.

In a demonstration of Intel's mixed-technology device integration, silicon n-channel transistors were constructed on a base silicon die with germanium p-channel devices stacked on top of them. If this technique can be scaled to support integrated circuit production, it holds the promise of continued increases in device density and improved clock speeds.

Another density-increasing approach is to combine multiple separately constructed integrated circuit dies in a vertical stack with connections between the layers for power and communication. You can think of this technique as a method of soldering integrated circuit dies on top of each other.

Separately fabricated integrated circuits combined within a single package are referred to as **chiplets**. Chiplets can be laid out side by side on a silicon base or they can be stacked atop one another, depending on the needs of the device. This approach allows each of the chiplets in a complex device to be constructed using the most appropriate technology for that component. For example, one fabrication method may be most appropriate for a core processor, while a different process might be more suitable for a memory chiplet integrated with the processor. An integrated cellular radio interface in the same device package may be constructed using yet another process.

The use of the vertical dimension in the construction of individual integrated circuits and in the construction of complex devices composed of multiple chiplets within a single package enables a higher level of SoC integration and higher overall performance. As these techniques continue to be refined and rolled out on production lines, we can expect the increasing circuit complexity and functionality predicted by Moore's law to continue in future years, though likely at a reduced growth rate.

The next trend we will examine is the ongoing growth in the use of highly specialized processing devices in place of general-purpose processors.

Increased device specialization

In previous chapters, we explored a few specialized processing technologies targeted at application areas such as digital signal processing, three-dimensional graphical image generation, bitcoin mining, and neural network processing. It is certainly possible for all the computations performed by these applications to be carried out by ordinary, general-purpose processors. The important difference in the processing performed by more specialized devices is the increased execution speed, with throughput that is sometimes hundreds or even thousands of times faster than an ordinary processor could achieve.

The growing importance of machine learning and autonomous technologies will continue to drive innovation in the computer architectures that underpin future digital systems. As automobiles and other complex systems gain autonomous features that either augment or replace functionality traditionally performed by human operators, the underlying processing architectures will continue to evolve to provide higher levels of performance tailored to specific tasks while minimizing power consumption.

Specialized processors will take advantage of the advances discussed earlier in this chapter while optimizing individual device designs for particular application niches. The trend toward increased specialization of processing devices will continue and will most likely accelerate in the coming years.

This discussion has focused on the continuation of ongoing trends into future years. The next section will examine the possibility that a technological advancement may arise that substantially alters the path from continued incremental improvements in computer architecture to something that is entirely different.

Potentially disruptive technologies

So far, this chapter has focused on trends currently in progress and the potential effects of their extension into the future. As with the introduction of the transistor, we saw that it is always possible that some new technology will appear that creates a drastic break from past experience and leads the future of computing technology in a new direction.

In this section, we attempt to identify some potential sources of such technological advances in the coming years.

Quantum physics

Charles Babbage's Analytical Engine tried to take the capabilities of purely mechanical computing devices to an extreme that had not been achieved previously. His attempt, while ambitious, was ultimately unsuccessful. The development of practical automated computing devices had to wait until the introduction of vacuum tube technology provided a suitable basis for the implementation of complex digital logic.

Later, the invention of the transistor moved computing technology onto a trajectory of increasing capability and sophistication that ultimately brought us to the state of computing we enjoy today. Ever since the introduction of the Intel 4004, advances in computing capabilities have taken the form of incremental improvements to what is fundamentally the same underlying silicon transistor technology.

Transistor operation is based on the properties of semiconducting materials such as silicon and the application of those properties to implement digital switching circuits. Digital circuits constructed with semiconductors generally perform operations using binary data values. These devices are designed to generate reliably repeatable results when given the same data input on a subsequent execution using the same sequence of instructions.

As an alternative to this approach, numerous research efforts are underway around the world exploring the possibility of employing aspects of quantum physics in computing technology. **Quantum physics** describes the behavior of matter at the level of individual atoms and subatomic particles.

The behavior of particles at the subatomic level differs in significant and surprising ways from the familiar behaviors of the macro-scale objects we interact with every day under the laws of **classical physics**. The laws of quantum physics were discovered and described in theories beginning in the mid-1800s.

Quantum physics is rigorously defined by a set of theories that have demonstrated remarkable predictive powers. For example, Wolfgang Pauli postulated the existence of the neutrino particle within the framework of quantum physics in 1930. Neutrinos are comparatively tiny subatomic particles that have barely any interaction with other particles, making them extremely difficult to detect. Neutrinos were not proven to exist by scientific experiments until the 1950s.

Several other types of subatomic particles have been predicted by theory and ultimately shown to exist in experiments. Quantum physics, including the strange behaviors exhibited in the subatomic realm, offers a promising new direction for future computer architectures.

Physical parameters associated with macro-scale objects, for example the speed of a moving vehicle, seem to vary in a continuous manner as the car accelerates or slows. The electrons within an atom, on the other hand, can only exist at specific, discrete energy levels. The energy level of an electron in an atom corresponds roughly to the speed of a particle moving in an orbit around a central body in classical physics.

There is no possibility for an electron in an atom to exist at some point between two energy levels. It is always precisely at one energy level or another. These discrete energy levels lead to the use of the term **quantum** to describe such phenomena.

Spintronics

In addition to the energy level of an electron in an atom, electrons exhibit a property analogous to the spinning of an object in classical physics. The spin of an elementary atomic particle is a type of angular momentum conceptually similar to the momentum of a spinning basketball balanced on a fingertip. As with the energy level, this spin state is quantized.

Researchers have demonstrated the ability to control and measure the spin behavior of electrons in a manner that may prove suitable for use in practical digital switching circuits. The use of electron spin as a component of a digital switching circuit is referred to as **spintronics**, combining the terms *spin* and *electronics*. This technology uses the quantum spin state of electrons to hold information in a manner similar to the charge state of capacitors in traditional electronics.

There are some significant differences in the spin behavior of electrons compared to basketballs, however. Electrons do not actually rotate; rather, their spin behavior obeys the mathematical laws of angular momentum in a quantized form. A basketball can be made to spin at an arbitrarily selected rotational speed, while electrons can only exhibit spin at one discrete, quantized level. The spin of an elementary particle is determined by its particle type, and electrons always have a spin of $\frac{1}{2}$, representing a quantum number.

The spin of a basketball can be fully characterized by the combination of its rotational speed and the direction of the axis about which the rotation is taking place. A spinning ball balanced on a fingertip rotates about the vertical axis. The entirety of the ball's rotational motion can be described by a vector pointing along the axis of rotation (in this case, upward) with a magnitude equal to its rotational speed.

Electrons always have the same spin value of $\frac{1}{2}$, defining the angular momentum vector length, so the only way to differentiate the spin of one electron from another is the direction of the spin vector. Practical devices have been developed that can enable the alignment of electron spin vectors in two different orientations, referred to as *up* and *down*. The combination of the spin value of $\frac{1}{2}$ and the spin orientation forms the **spin quantum number**.

The spin of an electron generates a tiny magnetic field. Materials in which most electron spins are aligned directionally produce a magnetic field with the same orientation as the aligned electrons. The effect of these aligned electrons is apparent in common magnetic materials, such as refrigerator magnets.

The magnetic field produced by electron spin cannot be explained by classical physics. This type of magnetism is purely an effect of quantum physics.

A switching device called a **spin valve** can be constructed from a channel within an integrated circuit that has a magnetic layer at each end. The magnetic layers function as gates. If the gates are of the same spin polarity (up or down), a current consisting of spin-polarized electrons can flow through the device. If the gates have opposite polarities, the current is blocked. A spin valve can be switched on and off by reversing the polarity of one of the magnets by applying current to it with the opposite spin direction.

Switching electron spin directions has the potential to be much faster while consuming much less power than the process of charging and discharging capacitors that underlies the functioning of today's CMOS digital devices.

This key feature provides a glimpse into the potential for spintronics to eventually augment or replace CMOS circuitry in high-performance digital devices.

Spintronics is an area of ongoing, active research. The commercialization and production of digital devices that outperform today's CMOS processors are not likely to occur for several years, if the technology turns out to be viable at all.

Spintronics relies on the laws of quantum physics to perform digital switching. Quantum computing, the subject of the next section, directly exploits quantum-mechanical phenomena to perform analog and digital processing.

Quantum computing

Quantum computing holds the promise of dramatic execution speed improvements for certain classes of problems. **Quantum computing** uses quantum-mechanical phenomena to perform processing, and can employ analog or digital approaches to solve problems.

Digital quantum computing uses quantum logic gates to perform computing operations. Quantum logic gates are based on circuits called **quantum bits**, or **qubits**. Qubits are analogous in some ways to the bits in traditional digital computers, but there are significant differences. Traditional bits can take on only the states 0 and 1. A qubit can be in the 0 or 1 quantum state; however, it can also exist in a superposition of the 0 and 1 states, which means it has some probability of appearing as either 0 or 1 when it is observed. The principle of **quantum superposition** states that any two quantum states can be added together and the result is a valid quantum state.

Whenever the value of a qubit is read, the result returned is always either 0 or 1. This is due to the collapse of the superposition of quantum states to a single state. If, prior to the readout, the qubit held the quantum value corresponding to the binary value 0 or 1, the output of the read operation would equal the binary value. If, on the other hand, the qubit contained a superposition of states, the value returned by the read operation will be a probabilistic function of the superposition of states.

In other words, the likelihood of receiving a 0 or 1 as the result of reading the qubit depends on the characteristics of its quantum state. The value returned by the read operation will not be predictable. The reason for this unpredictability is not simply a lack of knowledge; in quantum physics, a particle simply does not have a defined state until a measurement has been taken. This is one of the counterintuitive and, frankly, mind-bending features of quantum physics.

A qubit state that is *close to* the binary value 1 will have a higher probability of returning a value of 1 when read than one that is closer to the binary value of 0. Performing a read operation on multiple qubits that all begin in identical quantum states will not always produce the same result because of the probabilistic nature of the read operation.

Qubit circuits can demonstrate and exploit the properties of quantum entanglement, a central principle of quantum physics. **Quantum entanglement** occurs when multiple particles are linked in a manner that causes the measurement of one of the particles to affect the measurement of linked particles. The most surprising aspect of this linkage is that it remains in effect even when the particles are separated by great distances. The entanglement effect appears to propagate instantaneously, unrestricted by the speed of light. While this behavior may seem like science fiction, it has been confirmed experimentally and has even been used in the communication technology of the NASA **Lunar Atmosphere Dust and Environment Explorer (LADEE)** that orbited the moon from 2013 to 2014.

Quantum computers are capable of exploiting entanglement in information processing. If you work through the examples at the end of this chapter, you will have an opportunity to develop a program for a quantum computer that exhibits the effects of quantum entanglement, and you will run this program on an actual quantum computer.

The somewhat unpredictable nature of the results returned by reading a qubit would seem to argue against using this technology as the basis for a digital computing system. This partial unpredictability is one reason why quantum computers are envisioned as useful for only certain classes of problems. Most customers would not appreciate a bank using a computer that calculates different account balances each time the computation is run due to quantum uncertainty.

Two key application categories currently envisioned for quantum computers are described in the following sections.

Quantum code-breaking

Quantum code-breaking uses digital quantum computing techniques to break modern cryptographic codes. Many cryptographic algorithms in use today are based on the assumption that it is computationally infeasible to determine the factors of a large number (containing perhaps hundreds of decimal digits) that is the product of two large prime numbers. Factoring such a number on modern computers, even with a supercomputer or relying on thousands of processors operating in parallel in a cloud environment, cannot be expected to produce a correct result in a reasonable period of time.

Shor's algorithm, developed by Peter Shor in 1994, describes the steps a quantum computer must perform to identify the prime factors of a given number. A quantum computer running Shor's algorithm can potentially factor a very large number in a much shorter time than ordinary computers, thereby rendering modern cryptographic systems based on public-key cryptography vulnerable to such attacks. To date, due to the limited number of qubits in quantum computers, quantum computing has only demonstrated the ability to factor relatively small numbers, such as 21, but the potential threat is recognized by organizations and governments that require high levels of communication security. The future may bring quantum computing systems capable of cracking the codes we use today for securing websites and online banking.

However, there is probably little reason to be concerned about the security of your bank account from the threat of quantum attacks. An assortment of quantum computing-resistant public-key encryption algorithms is under development. Collectively, these algorithms are referred to as **post-quantum cryptography**. We can expect a large-scale transition to quantum-resistant cryptographic algorithms in the event the quantum threat to current cryptography methods become real.

Adiabatic quantum computation

Adiabatic quantum computation is an analog quantum computing approach that holds the promise of efficiently solving a wide variety of practical optimization problems. Imagine that you are in a rectangular region of hilly terrain surrounded by a fence. You need to find the lowest point within the fenced boundary. In this scenario, it is very foggy, and you cannot see the surrounding terrain. The only clue you have is the slope of the surface under your feet. You can follow the slope downward, but when you reach a level area, you can't be sure if you're in a local basin or have truly found the lowest point in the entire bounded region.

This is an example of a simple two-dimensional optimization problem. The goal is to find the x and y coordinates of the lowest altitude in the entire region, called the **global minimum**, without being sidetracked and getting stuck in a basin at a higher altitude, referred to as a **local minimum**.

You don't need anything as fancy as quantum computing to find the lowest point in a hilly region, but many real-world optimization problems have a larger number of inputs, perhaps 20 to 30, that must all be adjusted in the search for the global minimum. The computational power required to solve such problems is beyond the capability of even today's fastest supercomputers.

The quantum computing approach to solving such problems begins by setting up a configuration of qubits containing the superposition of all possible solutions to the problem, then slowly reducing the superposition effect.

By constraining the state of the quantum circuit configuration during this process, it is possible to ensure that the solution that remains after superposition has been removed, and all of the quantum bits are resolved to discrete 0 or 1 values, is the global minimum.

The term **adiabatic** in the name of this method refers to an analogy between the process of removing the superposition and a thermodynamic system that neither loses nor gains heat as it operates.

Adiabatic quantum optimization is an area of active research. It remains to be seen what level of capability this technology can bring to the solution of complex optimization problems.

The future of quantum computing

The term **quantum supremacy** describes the transition point at which quantum computing exceeds the capability of traditional digital computing in a particular problem domain. There is spirited debate among researchers as to whether quantum supremacy has already been achieved by any of the major organizations developing quantum computing technologies; when this point may be reached at a future date; or whether such a transition is ever going to occur.

A number of substantial barriers stand in the way of the widespread deployment of quantum computing in a manner similar to the ubiquitous use of CMOS-based computing devices by users around the world today. Some of the most pressing issues to be addressed are:

- Increasing the number of qubits in a computer to support the solution of large, complex problems
- Providing the ability to initialize qubits to arbitrary values
- Providing mechanisms to reliably read the state of qubits
- The components required for quantum computers are hard to produce and are very expensive
- Eliminating the effects of quantum decoherence

Quantum decoherence refers to the loss of phase coherence in a quantum system. For a quantum computer to function properly, phase coherence must be maintained within the system. Quantum decoherence results from interference from the outside world in the internal operation of the quantum system, or from interference generated internally within the system. A quantum system that remains perfectly isolated can maintain phase coherence indefinitely. Disturbing the system, for example by reading its state, disrupts the coherence and may lead to decoherence. The management and correction of decoherence effects are referred to as **quantum error correction**.

The effective management of decoherence is one of the greatest challenges in quantum computing.

Current quantum computer designs rely on exotic materials such as Helium-3, which is produced by nuclear reactors, and they require superconducting cables, which have zero resistance. Quantum computing systems must be cooled to temperatures near absolute zero during operation. Current quantum computers are mostly laboratory-based systems and require a dedicated staff of experts for their construction and operation. This situation is somewhat analogous to the early days of vacuum tube-based computers. One major difference from the vacuum tube days is that today we have the internet, which provides ordinary users with a degree of access to quantum computing capabilities.

Current quantum computing systems contain at most between one and two hundred qubits and are mainly accessible to the commercial, academic, and government organizations that fund their development. There are, however, some unique opportunities for students and individuals to gain access to real quantum computers.

One example is IBM Quantum at https://www.ibm.com/quantum-computing/. With this free collection of resources, IBM provides a set of tools, including a quantum algorithm development environment called **Qisket**, available at https://www.qiskit.org/. Using the Qisket tools, developers can learn to code quantum algorithms and can even submit programs for execution in batch mode on a real quantum computer. The exercises at the end of this chapter lay out steps you can take to get started in this domain.

Quantum computing shows great promise for addressing particular categories of problems, though the widespread commercialization of the technology is most likely years in the future.

The next technology we will examine is the carbon nanotube, which has the potential to move digital processing at least partially away from the world of silicon.

Carbon nanotubes

The **carbon nanotube field-effect transistor** (**CNTFET**) is a transistor that uses either a single carbon nanotube or an array of carbon nanotubes as the gate channel rather than the silicon channel of the traditional MOSFET. A carbon nanotube is a tubular structure constructed from carbon atoms with a diameter of approximately 1 nanometer.

Carbon nanotubes are exceptionally good electrical conductors, exhibit high tensile strength, and conduct heat very well. A carbon nanotube can sustain current densities over 1,000 times greater than metals such as copper. Unlike in metals, electrical current can propagate only along the axis of the nanotube.

Compared to MOSFETs, CNTFETs have the following advantages:

- Higher drive current.
- Substantially reduced power dissipation.
- Resilience to high temperatures.
- Excellent heat dissipation, allowing for high-density packing of the devices.
- The performance characteristics of n-channel and p-channel CNTFET devices match closely. In CMOS devices, on the other hand, there can be substantial variation between the performance of n-channel and p-channel transistors. This limits overall circuit performance to the capabilities of the lower-performing device.

As with the other emerging technologies discussed in this chapter, CNTFET technology faces some substantial barriers to commercialization and widespread use:

- Production of CNTFETs is extremely challenging because of the need to place and manipulate the nanometer-scale tubes.
- Production of the nanotubes required for CNTFETs is also very difficult. The nanotubes can be thought of as starting out as flat sheets of carbon fabric that must be rolled into tubes along a specific axis in order to produce a material with the desired semiconducting properties.
- Carbon nanotubes degrade rapidly when exposed to oxygen. Fabrication technologies must take this limitation into account to ensure the resulting circuit is durable and reliable.

Given the challenges of mass-producing CNTFETs, it will likely be several years before commercial devices begin to make wide use of carbon nanotube-based transistors, if it happens at all.

The preceding sections have identified some advanced technologies (spintronics, quantum computing, and carbon-nanotube-based transistors) as promising areas that may someday contribute substantially to the future of computing. None of these technologies are in wide-scale use at the time of writing, but research has shown promising results, and many government, university, and commercial laboratories are hard at work developing these technologies and finding ways to put them to use in the computing devices of the future.

In addition to technologies such as these that are widely reported and appear to be advancing along at least a semi-predictable path, there is always the possibility that an organization or individual may announce an unanticipated technological breakthrough. This may occur at any time, and such an event may upend the conventional wisdom regarding the anticipated path for the future of computing technology. Only time will tell.

In the context of the uncertainty shrouding the road ahead for computer architectures, it is prudent for the architecture professional to devise a strategy that will ensure ongoing relevance regardless of the twists and turns future technology takes. The next section presents some suggestions for staying up to date with technological advances.

Building a future-tolerant skill set

Given the technological transitions that kicked off the era of transistor-based digital computing, and the possibility of similar future events, it is important for professionals in the field of computer architecture to keep up with ongoing advances and to develop some intuition regarding the likely directions the technology will take in the future. This section provides some recommended practices for keeping up with the state of the art in computing technology.

Continuous learning

Computer architecture professionals must embrace the idea that technology continues to evolve rapidly, and they must devote substantial ongoing effort to monitoring advances and factoring new developments into their day-to-day work and career-planning decisions.

The prudent professional relies on a wide variety of information sources to track technological developments and assess their impact on career goals. Some sources of information, such as traditional news reports, can be skimmed quickly and fully absorbed. Other sources, like scientific literature and websites curated by experts in particular technologies, require time to digest complex technical information. More advanced topics, such as quantum computing, may require extended study just to grasp the fundamentals and begin to appreciate potential applications of the technology.

Even with a clear understanding of a particular technology, it can be challenging, or even impossible, to accurately predict its impact on the industry and, ultimately, the ways it will be integrated into the architectures of computing systems used by businesses, governments, and the public.

A practical and easy-to-implement approach for information gathering is to develop a collection of trusted sources for both mainstream and technical news and keep up to date with the information they offer. Mainstream news organizations, including television news, newspapers, magazines, and websites, often publish articles about promising technological developments and the impacts digital devices are having on societies around the world. In addition to discussing the purely technical aspects of computing systems (to some degree), these sources provide information on the social impact of computing technologies, such as concerns about their use for government and corporate surveillance and their employment in the spread of disinformation.

Technical websites operated by research organizations, individual technology experts, and enthusiastic amateurs offer an immense quantity of information related to advances in computer architecture. As with all information accessible on the internet, it is advisable to consider the reliability of the source whenever you encounter surprising information. While there are many spirited debates underway regarding the efficacy of individual early-stage technologies, there are also some people who appear to disagree with published information just for the sake of arguing. It is ultimately up to you to determine how much credence you should grant to any opinions expressed on a web page.

Although individuals will have their own preferences, and the landscape of technology news sources is ever-changing, the following list provides a few fairly reliable sources for news on computing technology, in no particular order:

- `https://techcrunch.com/`: TechCrunch reports on the business of the tech industry.
- `https://www.wired.com/`: Wired is a monthly magazine and website that focuses on how emerging technologies affect culture, the economy, and politics.
- `https://arstechnica.com/`: Ars Technica, founded in 1998, publishes information targeted at technologists and information technology professionals.
- `https://www.tomshardware.com/`: Tom's Hardware provides news, articles, price comparisons, and reviews of computer hardware and high-technology devices.
- `https://www.engadget.com/`: Engadget, founded in 2004, covers the intersection of gaming, technology, and entertainment.
- `https://gizmodo.com/`: Gizmodo focuses on design, technology, and science fiction. The website tagline is "We come from the future."
- `https://thenextweb.com/`: TNW was started in 2006 to bring insight and meaning to the world of technology.

This list, while by no means complete, provides some starting points for gathering information on the current state and near future of computing technology and its applications.

Information retrieved online can, when approached from a reasonably skeptical viewpoint, provide current and accurate information on the state of advances in computer architecture. Information consumed in this manner does not, however, provide an education with the rigorousness associated with formal schooling, or provide any form of public declaration that you have absorbed this information and are capable of making use of it in a professional context.

A college degree, the subject of the next section, provides a thorough grounding in a discipline and is generally accepted by potential employers and clients as evidence of the attainment of professional skills.

College education

If may have been a few years since you last attended college, or if you began your career without a college degree, it may be time to consider enrolling in a degree program. If even the thought of undertaking such a journey seems out of the question because of work or family responsibilities, consider that many accredited institutions offering excellent programs in areas of study directly related to computer architecture provide fully online education experiences. Online classes, combined with proctored examinations, can lead to Bachelor's and Master's degrees in technical disciplines from some of the most respected universities in the world.

For workers with a degree who have been in the workforce for several years, the technology and analytical methods they learned in school may have become stale and obsolete to some degree. To restore relevance and remain fully informed about the forefront of technologies involved in the design and production of modern computer systems, the best approach may be a return to the classroom to gain a deeper understanding of the technical advances that have occurred in the intervening years.

If you are not prepared to commit to a degree program, many institutions offer online courses leading to a certificate in a subject area such as computer hardware engineering or computer engineering technology. While providing a lesser credential than a Bachelor's or Master's degree, completion of a technology certificate program nevertheless demonstrates a level of educational attainment and knowledge of the subject matter.

There will be some expenses for tuition and books when taking college courses, whether the learning venue is in-person or online. Some employers are willing to provide partial or complete funding for the participation of employees in accredited degree programs. This funding may be accompanied by a mandatory commitment by the student to remain with the employer for some period following completion of the coursework. Students should take care to fully understand any obligations they may incur if circumstances require them to withdraw from school or leave the employer.

Many websites are available to assist with a search for an online college degree or certificate program that meets your needs. Some US-specific examples are:

- `https://www.usnews.com/education/online-education`: U.S. News & World Report publishes annual rankings of accredited colleges, including online programs specifically, at this URL.
- `https://www.onlineu.com/`: The OnlineU website provides reviews from students taking online programs at hundreds of colleges.

Without trying to be too repetitive with my warnings, you should carefully scrutinize any information gleaned from the internet regarding online colleges. Ensure any institution under consideration is appropriately accredited and that the degrees it confers are accepted and valued by employers.

Those with the necessary resources, possibly with support provided by an employer, may even consider becoming a full-time student for the duration of a degree program. Employers who pay for degree programs will typically expect the student to agree to a binding commitment to the organization following completion of such a program. This approach can provide the quickest turnaround to gaining a college degree and, in many cases, presents opportunities for participation in cutting-edge research on some of the most advanced computing technologies under development.

While a college degree from a respected institution in a relevant field of study is the gold-standard credential sought by employers and recognized by peers, opportunities are also available to keep up with the latest research findings through participation in conferences and by reading scientific literature. These learning options are explored in the next section.

Conferences and literature

For professionals interested in keeping up with the cutting edge of research in technologies related to the computer architectures of the future, there may be no better forum than hearing about the latest developments from the researchers themselves. There are regular conferences at locations around the world on every advanced computing topic you can imagine. For example, a list of worldwide conferences on the subject of quantum behavior, including many focusing on aspects of quantum computing, is available at `http://quantum.info/conf/index.html`.

As with other information from the internet, it is helpful to view any unfamiliar conference with a degree of skepticism until you have vetted it thoroughly. There is, unfortunately, a phenomenon known as **junk conferences**, in which predatory individuals or organizations arrange conferences for the purpose of revenue generation rather than for sharing scientific knowledge. Be sure that any conference you sign up for and attend is overseen by a reputable organization and contains presentations by legitimate researchers in subject areas relevant to the conference.

There is a wide variety of scientific literature related to ongoing advances in technologies related to computer architecture. Professional organizations, such as IEEE, publish numerous scholarly journals devoted to the bleeding edge of current research. Journals like these are intended to communicate directly from researcher to researcher, so the level of technical knowledge expected of readers is quite high. If you have the necessary background and are willing to work to appreciate the details in the papers published in scientific journals, you can read these publications to establish and maintain a level of knowledge on par with that of the scientists and engineers developing the next generation of computing technology.

Summary

Let's briefly review the topics we've discussed and learned about in the chapters of this book:

- In *Chapter 1*, *Introducing Computer Architecture*, we began with the earliest design of an automated computing machine, Babbage's Analytical Engine, and traced the course of digital computer history from the earliest vacuum tube-based computers to the first generations of processors. We also learned about the architecture of an early, but still prevalent, microprocessor: the 6502.

- In *Chapter 2*, *Digital Logic*, we learned the basics of transistor technology, digital logic, registers, and sequential logic. We also discussed the use of hardware description languages in the development of complex digital devices.

- *Chapter 3*, *Processor Elements*, covered the fundamental components of processors, including the control unit, the ALU, and the register set. The chapter introduced concepts related to the processor instruction set, including details on 6502 addressing modes, instruction categories, interrupt processing, and I/O operations.

- *Chapter 4*, *Computer System Components*, introduced the MOSFET transistor and described its use in DRAM circuit technology. The chapter covered the processing and communication subsystems of modern computers, including the I/O subsystem, graphics displays, the network interface, and interfaces for the keyboard and mouse.

- In *Chapter 5*, *Hardware-Software Interface*, we learned about the inner workings of drivers and how the BIOS firmware of the original PC has transitioned to UEFI in modern computers. This chapter covered the boot process and the concepts associated with processes and threads in modern operating systems.

- *Chapter 6*, *Specialized Computing Domains*, introduced the unique features of real-time computing, digital signal processing, and GPU processing. Examples of specialized computing architectures relying on these processing capabilities were presented, including cloud computer servers, business desktop computers, and high-performance gaming computers.

- *Chapter 7, Processor and Memory Architectures*, addressed the unique features of the von Neumann, Harvard, and modified Harvard architectures. The chapter described the distinction between physical and virtual memory and introduced the architecture of paged virtual memory, including the functions of the MMU.
- In *Chapter 8, Performance-Enhancing Techniques*, we learned about a variety of techniques used in modern processors to accelerate instruction execution speed. Topics included cache memory, instruction pipelining, superscalar processing, simultaneous multithreading, and SIMD processing.
- *Chapter 9, Specialized Processor Extensions*, addressed several auxiliary processor capabilities, including privileged execution modes, floating-point mathematics, power management, and system security management.
- *Chapter 10, Modern Processor Architectures and Instruction Sets*, delved into the details of the architectures and instruction sets of the most prevalent 32-bit and 64-bit modern processors. For each of the x86, x64, 32-bit ARM, and 64-bit ARM processor architectures, the chapter introduced the register set, addressing modes, and instruction categories, and presented a short but functional assembly language program.
- *Chapter 11, The RISC-V Architecture and Instruction Set*, examined the features of the RISC-V architecture in detail. The chapter introduced the base 32-bit architecture, including the register set, instruction set, and standard extensions to the instruction set. Additional topics included the 64-bit version of the architecture and standard configurations available as commercially produced RISC-V processors. The chapter included some simple RISC-V assembly language programs and provided guidance for implementing a RISC-V processor in a low-cost FPGA device.
- *Chapter 12, Processor Virtualization*, introduced concepts associated with processor virtualization, including challenges that virtualization tools must overcome. The techniques used to implement virtualization in modern processor families, including x86, ARM, and RISC-V, were discussed. Several popular virtualization tools were described and virtualization approaches used in cloud computing environments were presented.
- *Chapter 13, Domain-Specific Computer Architectures*, examined some specific computer architectures, including smartphones, personal computers, warehouse-scale cloud computing environments, and neural networks. The unique processing requirements associated with each of these domains were examined and the tailoring of processor hardware to optimize the trade-off between cost, performance, and power consumption in each case was discussed.

- *Chapter 14, Cybersecurity and Confidential Computing Architectures,* introduced computing architectures suitable for applications that require an exceptional assurance of security. Critical application areas like national security systems and financial transaction processing demand this high level of protection. These systems must be resilient against a broad range of cybersecurity threats including malicious code, covert channel attacks, and attacks enabled by physical access to the computing hardware.

- *Chapter 15, Blockchain and Bitcoin Mining Architectures,* began with a brief introduction to the concepts associated with blockchain, a public, cryptographically secured ledger recording a sequence of transactions. We continued with an overview of the process of bitcoin mining, which appends transactions to the bitcoin blockchain and rewards those who complete this task with payment in the form of bitcoin. Bitcoin mining requires high-performance computing hardware, which was presented in terms of a current-generation bitcoin mining computer architecture.

- *Chapter 16, Self-Driving Vehicle Architectures,* presented the capabilities required in self-navigating vehicle processing architectures. It began with a discussion of the requirements for ensuring the safety of an autonomous vehicle and its occupants, as well as for other vehicles, pedestrians, and stationary objects. We continued with a discussion of the types of sensors and data a self-driving vehicle receives as input while driving. Next was a description of the types of processing required for effective vehicle control. The chapter concluded with an overview of an example self-driving computer architecture.

In this chapter, we attempted to gain some perspective on the road ahead for computer architectures. We reviewed the major advances and ongoing trends that have led to the current state of computer design and attempted to extrapolate to identify the directions the development of computing system architectures is likely to take in the future. We also examined some potentially disruptive technologies that could alter the path of future computer architectures. To get a glimpse into this future, if you work through the exercises at the end of this chapter, you will develop a quantum computing algorithm and run it on an actual quantum computer, for free!

This chapter also reviewed some suggested approaches for professional development for a computer architect that should lead to a skill set that remains relevant and tolerant of future advances, whatever they may be.

Having completed this chapter, and this book, you now have a good understanding of the evolution of computer architecture design from the earliest days to its current state, and will be familiar with ongoing trends in computer architecture that are likely to indicate future technological directions.

You are also aware of some potentially disruptive technologies that may substantially alter computer architectures in the near future. Finally, you will have learned some useful techniques for maintaining a current skill set in the field of computer architecture.

This brings us to the end of the book. I hope you have enjoyed reading it and working through the exercises as much as I have enjoyed writing it and working through the exercises myself.

Exercises

1. Install the Qiskit quantum processor software development framework by following the instructions at https://qiskit.org/documentation/getting_started.html. The instructions suggest the installation of the Anaconda (https://www.anaconda.com/) data science and machine learning toolset. After installing Anaconda, create a Conda virtual environment named qisketenv to contain your work on quantum code and install Qisket in this environment with the command pip install qiskit. Make sure that you install the optional visualization dependencies with the pip install qiskit-terra[visualization] command.

2. Create a free IBM Quantum account at https://quantum-computing.ibm.com/. Locate your IBM Quantum Services API token at https://quantum-computing.ibm.com/account and install it into your local environment using the instructions at https://qiskit.org/documentation/stable/0.24/install.html.

3. Work through the example quantum program at https://qiskit.org/documentation/tutorials/circuits/1_getting_started_with_qiskit.html. This example creates a quantum circuit containing three qubits that implements a **Greenberger–Horne–Zeilinger (GHZ)** state. The GHZ state exhibits key properties of quantum entanglement. Execute the code in a simulation environment on your computer.

4. Execute the code from *Exercise 3* on an IBM quantum computer.

Join our community Discord space

Join the book's Discord workspace for a monthly *Ask me Anything* session with the author: https://discord.gg/7h8aNRhRuY

Appendix

Answers to Exercises

Chapter 1: Introducing Computer Architecture

Exercise 1

Using your favorite programming language, develop a simulation of a single-digit decimal adder that operates in the same manner as in Babbage's Analytical Engine. First, prompt the user for two digits in the range 0-9: the addend and the accumulator. Display the addend, the accumulator, and the carry, which is initially zero. Perform a series of cycles as follows:

1. If the addend is zero, display the values of the addend, accumulator, and carry and terminate the program
2. Decrement the addend by one and increment the accumulator by one
3. If the accumulator incremented from nine to zero, increment the carry
4. Go back to *step 1*

Test your code with these sums: 0+0, 0+1, 1+0, 1+2, 5+5, 9+1, and 9+9.

Answer

The Ex__1_single_digit_adder.py Python file contains the adder code:

```
#!/usr/bin/env python


"""Ex__1_single_digit_adder.py: Answer to Ch 1 Ex 1."""


import sys


# Perform one step of the Analytical Engine addition
# operation. a and b are the digits being added, c is the
# carry
def increment_adder(a, b, c):
    a = a - 1          # Decrement addend
    b = (b + 1) % 10 # Increment accum, wrap to 0 if necessary
```

```python
        if b == 0:         # If accumulator is 0, increment carry
            c = c + 1

    return a, b, c


# Add two decimal digits passed on the command line.
# The sum is returned as digit2 and the carry is 0 or 1.
def add_digits(digit1, digit2):
    carry = 0

    while digit1 > 0:
        [digit1, digit2, carry] = increment_adder(
            digit1, digit2, carry)

    return digit2, carry
```

The Ex__1_test_single_digit_adder.py file contains the test code:

```python
#!/usr/bin/env python

"""Ex__1_test_single_digit_adder.py: Tests for answer to
chapter 1 exercise 1."""

import unittest
import Ex__1_single_digit_adder

class TestSingleDigitAdder(unittest.TestCase):
    def test_1(self):
        self.assertEqual(Ex__1_single_digit_adder.add_digits(
        0, 0), (0, 0))

    def test_2(self):
        self.assertEqual(Ex__1_single_digit_adder.add_digits(
        0, 1), (1, 0))

    def test_3(self):
        self.assertEqual(Ex__1_single_digit_adder.add_digits(
        1, 0), (1, 0))
```

```
        def test_4(self):
            self.assertEqual(Ex__1_single_digit_adder.add_digits(
            1, 2), (3, 0))

        def test_5(self):
            self.assertEqual(Ex__1_single_digit_adder.add_digits(
            5, 5), (0, 1))

        def test_6(self):
            self.assertEqual(Ex__1_single_digit_adder.add_digits(
            9, 1), (0, 1))

        def test_7(self):
            self.assertEqual(Ex__1_single_digit_adder.add_digits(
            9, 9), (8, 1))

    if __name__ == '__main__':
        unittest.main()
```

To execute the tests, assuming Python is installed and is in your path, execute the following command:

```
python Ex__1_test_single_digit_adder.py
This is the output of a test run:
C:\>python Ex__1_test_single_digit_adder.py
.......
----------------------------------------------------------------------
Ran 7 tests in 0.001s
OK
```

Exercise 2

Create arrays of 40 decimal digits each for the addend, accumulator, and carry. Prompt the user for two decimal integers of up to 40 digits each. Perform the addition digit by digit using the cycles described in *Exercise 1*, and collect the carry output from each digit position in the carry array. After the cycles are complete, insert carries and, where necessary, ripple them across digits to complete the addition operation. Display the results after each cycle and at the end. Test with the same sums as in *Exercise 1* and test 99+1, 999999+1, 49+50, and 50+50.

Answer

The `Ex__2_40_digit_adder.py` Python file contains the adder code:

```python
#!/usr/bin/env python


"""Ex__2_40_digit_adder.py: Answer to Ch 1 Ex 2."""

import sys
import Ex__1_single_digit_adder

# Add two decimal numbers of up to 40 digits and return the
# sum. Input and output numeric values are represented as
# strings.
def add_40_digits(str1, str2):
    max_digits = 40

    # Convert str1 into a 40 decimal digit value
    num1 = [0]*max_digits
    for i, c in enumerate(reversed(str1)):
        num1[i] = int(c) - int('0')

    # Convert str2 into a 40 decimal digit value
    num2 = [0]*max_digits
    for i, c in enumerate(reversed(str2)):
        num2[i] = int(c) - int('0')

    # Sum the digits at each position and record the
    # carry for each position
    sum = [0]*max_digits
    carry = [0]*max_digits
    for i in range(max_digits):
        (sum[i], carry[i]) = Ex__1_single_digit_adder. \
        add_digits(num1[i], num2[i])

    # Ripple the carry values across the digits
    for i in range(max_digits-1):
        if (carry[i] == 1):
            sum[i+1] = (sum[i+1] + 1) % 10
```

```python
            if (sum[i+1] == 0):
                carry[i+1] = 1

    # Convert the result into a string with leading zeros
    # removed
    sum.reverse()
    sum_str = "".join(map(str, sum))
    sum_str = sum_str.lstrip('0') or '0'
    return sum_str
```

The Ex__2_test_40_digit_adder.py file contains the test code:

```python
#!/usr/bin/env python

"""Ex__2_test_40_digit_adder.py: Tests for answer to
 chapter 1 exercise 2."""

import unittest
import Ex__2_40_digit_adder

class Test40DigitAdder(unittest.TestCase):
    def test_1(self):
        self.assertEqual(Ex__2_40_digit_adder.add_40_digits(
        "0", "0"), "0")

    def test_2(self):
        self.assertEqual(Ex__2_40_digit_adder.add_40_digits(
        "0", "1"), "1")

    def test_3(self):
        self.assertEqual(Ex__2_40_digit_adder.add_40_digits(
        "1", "0"), "1")

    def test_4(self):
        self.assertEqual(Ex__2_40_digit_adder.add_40_digits(
        "1", "2"), "3")

    def test_5(self):
```

```python
            self.assertEqual(Ex__2_40_digit_adder.add_40_digits(
                "5", "5"), "10")

    def test_6(self):
            self.assertEqual(Ex__2_40_digit_adder.add_40_digits(
                "9", "1"), "10")

    def test_7(self):
            self.assertEqual(Ex__2_40_digit_adder.add_40_digits(
                "9", "9"), "18")

    def test_8(self):
            self.assertEqual(Ex__2_40_digit_adder.add_40_digits(
                "99", "1"), "100")

    def test_9(self):
            self.assertEqual(Ex__2_40_digit_adder.add_40_digits(
                "999999", "1"), "1000000")

    def test_10(self):
            self.assertEqual(Ex__2_40_digit_adder.add_40_digits(
                "49", "50"), "99")

    def test_11(self):
            self.assertEqual(Ex__2_40_digit_adder.add_40_digits(
                "50", "50"), "100")

if __name__ == '__main__':
    unittest.main()
```

To execute the tests, assuming Python is installed and is in your path, execute the following command:

```
python Ex__2_test_40_digit_adder.py
```

This is the output of a test run:

```
C:\>python Ex__2_test_40_digit_adder.py
...........
```

Appendix 479

```
-----------------------------------------------------------------
Ran 11 tests in 0.002s
OK
```

Exercise 3

Modify the programs of *Exercise 1* and *Exercise 2* to implement the subtraction of 40-digit decimal values. Perform borrowing as required. Test with 0-0, 1-0, 1000000-1, and 0-1. What is the result for 0-1?

Answer

The Ex__3_single_digit_subtractor.py Python file contains the single-digit subtractor code:

```python
#!/usr/bin/env python

"""Ex__3_single_digit_subtractor.py: Answer to Ch 1 Ex 3
(single digit subtractor)."""

import sys

# Perform one step of the Analytical Engine subtraction
# operation. a and b are the digits being subtracted (a - b),
# c is the carry: 0 = borrow, 1 = not borrow
def decrement_subtractor(a, b, c):
    a = (a - 1) % 10  # Decrement left operand, to 9 if wrapped
    b = b - 1         # Decrement accumulator

    if a == 9:        # If accum reached 9, decrement carry
        c = c - 1

    return a, b, c

# Subtract two decimal digits. The difference is returned as
# digit1 and the carry output is 0 (borrow) or 1 (not borrow).
def subtract_digits(digit1, digit2):
    carry = 1

    while digit2 > 0:
```

```python
        [digit1, digit2, carry] = decrement_subtractor(
        digit1, digit2, carry)

    return digit1, carry
```

The `Ex__3_test_single_digit_subtractor.py` file contains the test code for the single-digit subtractor:

```python
#!/usr/bin/env python

"""Ex__3_test_single_digit_subtractor.py: Tests for answer
to chapter 1 exercise 3 (tests for single digit
subtractor)."""

import unittest
import Ex__3_single_digit_subtractor

class TestSingleDigitSubtractor(unittest.TestCase):
    def test_1(self):
        self.assertEqual(Ex__3_single_digit_subtractor.
        subtract_digits(0, 0), (0, 1))

    def test_2(self):
        self.assertEqual(Ex__3_single_digit_subtractor.
        subtract_digits(0, 1), (9, 0))

    def test_3(self):
        self.assertEqual(Ex__3_single_digit_subtractor.
        subtract_digits(1, 0), (1, 1))

    def test_4(self):
        self.assertEqual(Ex__3_single_digit_subtractor.
        subtract_digits(1, 2), (9, 0))

    def test_5(self):
        self.assertEqual(Ex__3_single_digit_subtractor.
        subtract_digits(5, 5), (0, 1))
```

```python
    def test_6(self):
        self.assertEqual(Ex__3_single_digit_subtractor.
            subtract_digits(9, 1), (8, 1))

    def test_7(self):
        self.assertEqual(Ex__3_single_digit_subtractor.
            subtract_digits(9, 9), (0, 1))

if __name__ == '__main__':
    unittest.main()
```

The `Ex__3_40_digit_subtractor.py` Python file contains the 40-digit subtractor code:

```python
#!/usr/bin/env python

"""Ex__3_40_digit_subtractor.py: Answer to Ch 1 Ex 3."""

import sys
import Ex__3_single_digit_subtractor

# Subtract two decimal numbers of up to 40 digits and
# return the result. Input and output numeric values are
# represented as strings.
def subtract_40_digits(str1, str2):
    max_digits = 40

    # Convert str1 into a 40 decimal digit value
    num1 = [0]*max_digits
    for i, c in enumerate(reversed(str1)):
        num1[i] = int(c) - int('0')

    # Convert str2 into a 40 decimal digit value
    num2 = [0]*max_digits
    for i, c in enumerate(reversed(str2)):
        num2[i] = int(c) - int('0')

    # Subtract the digits at each position and record the
    # carry for each position
```

```python
        diff = [0]*max_digits
        carry = [0]*max_digits
        for i in range(max_digits):
            (diff[i], carry[i]) = Ex__3_single_digit_subtractor. \
            subtract_digits(num1[i], num2[i])

        # Ripple the carry values across the digits
        for i in range(max_digits-1):
            if (carry[i] == 0):
                diff[i+1] = (diff[i+1] - 1) % 10
                if (diff[i+1] == 9):
                    carry[i+1] = 0

        # Convert the result into a string with leading zeros
        # removed
        diff.reverse()
        diff_str = "".join(map(str, diff))
        diff_str = diff_str.lstrip('0') or '0'
        return diff_str
```

The Ex__3_test_40_digit_subtractor.py file contains the test code for the 40-digit subtractor:

```python
#!/usr/bin/env python

"""Ex__3_test_40_digit_subtractor.py: Tests for answer to
chapter 1 exercise 3."""

import unittest
import Ex__3_40_digit_subtractor

class Test40DigitSubtractor(unittest.TestCase):
    def test_1(self):
        self.assertEqual(Ex__3_40_digit_subtractor.
        subtract_40_digits("0", "0"), "0")

    def test_2(self):
        self.assertEqual(Ex__3_40_digit_subtractor.
        subtract_40_digits("1", "0"), "1")
```

```
    def test_3(self):
        self.assertEqual(Ex__3_40_digit_subtractor.
        subtract_40_digits("1000000", "1"), "999999")

    def test_4(self):
        self.assertEqual(Ex__3_40_digit_subtractor.
        subtract_40_digits("0", "1"),
        "9999999999999999999999999999999999999999")

if __name__ == '__main__':
    unittest.main()
```

To execute the tests, assuming Python is installed and is in your path, execute the following commands:

```
python Ex__3_test_single_digit_subtractor.py
python Ex__3_test_40_digit_subtractor.py
```

This is the output of a test run of `Ex__3_test_single_digit_subtractor.py`:

```
C:\>python Ex__3_test_single_digit_subtractor.py
.......
----------------------------------------------------------------------
Ran 7 tests in 0.001s
OK
```

This is the output of a test run of `Ex__3_test_40_digit_subtractor.py`:

```
C:\>python Ex__3_test_40_digit_subtractor.py
....
----------------------------------------------------------------------
Ran 4 tests in 0.001s
OK
```

The result for 0-1 is 9 with a carry of 0.

Exercise 4

6502 assembly language references data in memory locations using an operand value containing the address (without the # character that indicates an immediate value).

For example, the LDA $00 instruction loads the byte at memory address $00 into A. STA $01 stores the byte in A in address $01. Addresses can be any value in the range 0 to $FFFF, assuming memory exists at the address and the address is not already in use for some other purpose. Using your preferred 6502 emulator, write 6502 assembly code to store a 16-bit value in addresses $00-$01, store a second value in addresses $02-$03, and then add the two values and store the result in $04-$05. Be sure to propagate any carry between the 2 bytes. Ignore any carry from the 16-bit result. Test with $0000+$0001, $00FF+$0001, and $1234+$5678.

Answer

The 6502 assembly file Ex__4_16_bit_addition.asm contains the 16-bit addition code:

```
; Ex__4_16_bit_addition.asm
; Try running this code at
; https://skilldrick.github.io/easy6502/

; Set up the values to be added
; Remove the appropriate semicolons to select the bytes to add:
; ($0000 + $0001) or ($00FF + $0001) or ($1234 + $5678)

LDA #$00
;LDA #$FF
;LDA #$34
STA $00

LDA #$00
;LDA #$00
;LDA #$12
STA $01

LDA #$01
;LDA #$01
;LDA #$78
STA $02

LDA #$00
;LDA #$00
;LDA #$56
```

```
    STA $03

    ; Add the two 16-bit values
    CLC
    LDA $00
    ADC $02
    STA $04

    LDA $01
    ADC $03
    STA $05
```

Try running this code at https://skilldrick.github.io/easy6502/.

Exercise 5

Write 6502 assembly code to subtract two 16-bit values in a manner similar to *Exercise 4*. Test with $0001-$0000, $0001-$0001, $0100-$00FF, and $0000-$0001. What is the result for $0000-$0001?

Answer

The 6502 assembly file Ex__5_16_bit_subtraction.asm contains the 16-bit subtraction code:

```
; Ex__5_16_bit_subtraction.asm
; Try running this code at
; https://skilldrick.github.io/easy6502/

; Set up the values to be subtracted
; Remove the appropriate semicolons to select the bytes to
; subtract:
; ($0001 - $0000) or ($0001 - $0001) or ($0001 - $00FF) or
; ($0000 - $0001)

LDA #$01
;LDA #$01
;LDA #$01
;LDA #$00
STA $00

LDA #$00
```

```
;LDA #$00
;LDA #$00
;LDA #$00
STA $01

LDA #$00
;LDA #$01
;LDA #$FF
;LDA #$01
STA $02

LDA #$00
;LDA #$00
;LDA #$00
;LDA #$00
STA $03

; Subtract the two 16-bit values
SEC
LDA $00
SBC $02
STA $04

LDA $01
SBC $03
STA $05
```

Try running this code at https://skilldrick.github.io/easy6502/.

The result for $0000-$0001 is $FFFF.

Exercise 6

Write 6502 assembly code to store two 32-bit integers in addresses $00-03 and $04-$07, and then add them, storing the results in $08-$0B. Use a looping construct, including a label and a branch instruction, to iterate over the bytes of the two values to be added. Search the internet for the details of the 6502 decrement and branch instructions and the use of labels in assembly language. Hint: the 6502 zero-page indexed addressing mode works well in this application.

Answer

The 6502 assembly file `Ex__6_32_bit_addition.asm` contains the 32-bit addition code:

```
; Ex__6_32_bit_addition.asm
; Try running this code at
; https://skilldrick.github.io/easy6502/

; Set up the values to be added
; Remove the appropriate semicolons to select the bytes to
; add:
; ($00000001 + $00000001) or ($0000FFFF + $00000001) or
; ($FFFFFFFE + $00000001) or ($FFFFFFFF + $00000001)

LDA #$01
;LDA #$FF
;LDA #$FE
;LDA #$FF
STA $00

LDA #$00
;LDA #$FF
;LDA #$FF
;LDA #$FF
STA $01

LDA #$00
;LDA #$00
;LDA #$FF
;LDA #$FF
STA $02

LDA #$00
;LDA #$00
;LDA #$FF
;LDA #$FF
STA $03
```

```
    LDA #$01
    STA $04

    LDA #$00
    STA $05
    STA $06
    STA $07

    ; Add the two 32-bit values using absolute indexed
    ; addressing mode
    LDX #$00
    LDY #$04
    CLC

ADD_LOOP:
    LDA $00, X
    ADC $04, X
    STA $08, X
    INX
    DEY
    BNE ADD_LOOP
```

Try running this code at https://skilldrick.github.io/easy6502/.

Chapter 2: Digital Logic

Exercise 1

Rearrange the circuit in *Figure 2.5* to convert the AND gate to a NAND gate. Hint: there is no need to add or remove components.

Answer

Relocate the *R2* resistor and the output signal connection point as follows:

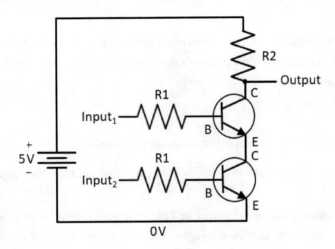

Figure 1: NAND gate circuit

Exercise 2

Create a circuit implementation of an OR gate by modifying the circuit in *Figure 2.5*. Wires, transistors, and resistors can be added as needed.

Answer

The OR gate circuit is as follows:

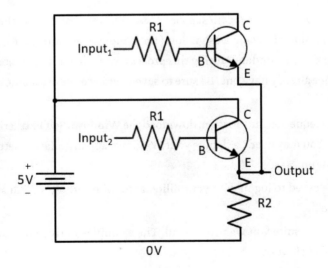

Figure 2: OR gate circuit

Exercise 3

Search the internet for free VHDL development software suites that include a simulator. Get one of these suites, set it up, and build any simple demo projects that come with the suite to ensure it is working properly.

Answer

Some freely available VHDL development suites are as follows:

1. Xilinx Vivado Design Suite is available at https://www.xilinx.com/support/download.html.

2. Intel® Quartus® Prime Software Lite Edition is available at https://www.intel.com/content/www/us/en/software/programmable/quartus-prime/download.html.

3. The open source GHDL simulator for VHDL is available at https://github.com/ghdl/ghdl.

4. Mentor ModelSim PE Student Edition is available at https://www.mentor.com/company/higher_ed/modelsim-student-edition.

5. Electronic Design Automation (EDA) Playground is available at https://www.edaplayground.com/.

Vivado Design Suite will be used for the examples in *Chapter 2, Digital Logic,* and the following chapters, including installing circuit designs in a low-cost FPGA development board. These steps describe the installation and setup process for Windows 10:

1. Visit https://www.xilinx.com/support/download.html and select the web installer for the latest version of Vivado Design Suite for Windows. Be sure to select the full Vivado installer and not an update. During this process, you will need to create a Xilinx account if you do not already have one. Be sure to save your account username and password for later use.

2. Provide the requested information, download the **Windows Self Extracting Web Installer**, and run it. You may need to change your Windows app installation settings to allow the installer to run.

3. You will be asked to log in with your Xilinx account information and accept the license agreements.

4. Select the tool suite you want to install. The examples in this book use Vivado. Select **Vivado** and click **Next**.

5. Select **Vivado HL WebPack** (this is the free version). Click **Next**.
6. Accept the default design tools, devices, and installation options for Vivado HL Webpack. Click **Next**.
7. Accept the default installation directory and other options. Click **Next**.
8. On the **Installation Summary** page, click **Install**. Downloading and installation will take some time. The time required depends on your internet connection speed. Plan for a few hours.

After the installation completes, follow these steps to build an example project:

1. You should find an icon on your desktop with a name similar to **Vivado 2021.2**. Double-click this icon (and not the icon that says **Vivado HLS**) to start the application.
2. In the Vivado main window, click **Open Example Project**.
3. Click through to the **Select Project Template** screen and select **CPU (HDL)**.
4. Click through and accept the defaults on the following screens and click **Finish** to create the project.
5. On the **Project Manager** page, you'll find the **Sources** panel. Expand the tree listing and double-click some of the files to open them in the editor. Most of the files in this design are in the Verilog hardware design language.
6. Click **Run Synthesis** in the **Project Manager** panel. The **Design Runs** panel will update the status as synthesis proceeds. This may take several minutes.
7. After synthesis completes, a dialog will appear offering to run the implementation. Click **Cancel**.
8. Click **Run Simulation** in the Vivado main dialog **Project Manager** section, and then select **Run behavioral simulation**. Again, this may take several minutes.
9. After the simulation completes, you will see a timing diagram in the **Simulation** window showing the simulated CPU signals using the input data provided by the simulation source files.
10. This completes the exercise. You may close Vivado.

Exercise 4

Using your VHDL toolset, implement the 4-bit adder using the code listings presented in *Chapter 2, Digital Logic*.

Answer

Follow these steps to implement the 4-bit adder:

1. Double-click the **Vivado 2021.2** (or similar) icon to start Vivado.
2. Click **Create Project** in the Vivado main dialog.
3. Click through and accept the default project name and location.
4. Select **RTL Project**, the default project type.
5. On the **Default Part** page, select the **Boards** tab. Type Arty in the search field, select **Arty A7-35**, and then click **Next**. If **Arty** does not appear after searching, click **Update Board Repositories** and then search again.
6. Click **Finish** to create the project.
7. Click **Add Sources** in the **Project Manager** panel, select **Add or create design sources**, add Ex__4_adder4.vhdl and Ex__4_fulladder.vhdl, and then click **Finish**.
8. Expand the tree in the **Design Sources** window in the **Project Manager** dialog and locate the two files you added. Double-click each of them and expand the source code window to view the code.
9. Click **Run Synthesis** in the **Project Manager** panel. Leave the options in the **Launch Runs** dialog at their defaults and click **OK**. The **Design Runs** panel will update the status as synthesis proceeds.
10. Wait for the synthesis to complete and then select **View Reports** in the **Synthesis Completed** dialog. Double-click some of the reports produced during the synthesis process. Only the reports that have an icon with a green dot are present.
11. This completes the exercise. You may close Vivado.

Exercise 5

Add test driver code (search the internet for *VHDL testbench* to find examples) to your 4-bit adder to drive it through a limited set of input sets and verify that the outputs are correct.

Answer

Follow these steps to test the 4-bit adder project created in *Exercise 4*:

1. Double-click the **Vivado 2021.2** (or similar) icon to start Vivado.
2. Click **Open Project** in the Vivado main dialog and open the project you created in *Exercise 4*. You will need to select the project filename ending in **.xpr**.

3. Click **Add Sources** in the **Project Manager** panel, select **Add or create simulation sources**, add Ex__5_adder4_testbench.vhdl, and then click **Finish**.

4. Expand the tree in the **Simulation Sources** window in the **Project Manager** dialog and locate the file you added. Double-click the file and expand the source code window to view the code. Observe the six test cases present in the code.

5. Click **Run Simulation** in the Vivado main dialog **Project Manager** section, and then select **Run behavioral simulation**.

6. Wait for the simulation to complete, then expand the windows with the timing diagram (probably labeled **Untitled 1**).

7. Use the magnifying glass icons and the window's horizontal scroll bar to view the six test cases in the first 60 **nanoseconds (ns)** of execution. Determine whether the sum and carry for each addition operation are correct. You can drag the yellow marker to update the information in the **Value** column.

8. This completes the exercise. You may close Vivado.

The VHDL file Ex__5_adder4_testbench.vhdl contains the testbench code:

```
library IEEE;
  use IEEE.STD_LOGIC_1164.ALL;

entity ADDER4_TESTBENCH is
end entity ADDER4_TESTBENCH;

architecture BEHAVIORAL of ADDER4_TESTBENCH is

  component ADDER4 is
    port (
      A4        : in   std_logic_vector(3 downto 0);
      B4        : in   std_logic_vector(3 downto 0);
      SUM4      : out  std_logic_vector(3 downto 0);
      C_OUT4    : out  std_logic
    );
  end component;

  signal a     : std_logic_vector(3 downto 0);
  signal b     : std_logic_vector(3 downto 0);
  signal s     : std_logic_vector(3 downto 0);
```

```vhdl
    signal c_out : std_logic;

begin

  TESTED_DEVICE : ADDER4
    port map (
      A4     => a,
      B4     => b,
      SUM4   => s,
      C_OUT4 => c_out
    );

  TEST : process
  begin
    a <= "0000";
    b <= "0000";

    wait for 10 ns;
    a <= "0110";
    b <= "1100";

    wait for 10 ns;
    a <= "1111";
    b <= "1100";

    wait for 10 ns;
    a <= "0110";
    b <= "0111";

    wait for 10 ns;
    a <= "0110";
    b <= "1110";

    wait for 10 ns;
    a <= "1111";
    b <= "1111";
```

```
      wait;

   end process TEST;

end architecture BEHAVIORAL;
```

Exercise 6

Expand the test driver code and verify that the 4-bit adder produces correct results for all possible combinations of inputs.

Answer

Follow these steps to test the 4-bit adder project created in *Exercise 4*:

1. Double-click the Vivado **2021.2** (or similar) icon to start Vivado.
2. Click **Open Project** in the Vivado main dialog and open the project you created in *Exercise 4* and modified in *Exercise 5*. You will need to select the project filename ending in **.xpr**.
3. We're going to replace the test driver code from *Exercise 5* with a different test driver. Expand the tree in the **Simulation Sources** window in the **Project Manager** dialog and locate the module you added in *Exercise 5* (ADDER4_TESTBENCH). Right-click the module name, select **Remove File from Project**, and then click **OK** to confirm the removal.
4. Click **Add Sources** in the **Project Manager** panel, select **Add or create simulation sources**, add Ex__6_adder4_fulltestbench.vhdl, and then click **Finish**.
5. Expand the tree in the **Simulation Sources** window in the **Project Manager** dialog and locate the file you added. Double-click the file and expand the source code window to view the code. Observe the loop with 256 test cases in the code.
6. Click **Run Simulation** in the Vivado main dialog **Project Manager** section, and then select **Run behavioral simulation**.
7. Wait for the simulation to complete, and then expand the windows with the timing diagram (probably labeled **Untitled 1**).
8. Use the magnifying glass icons and the window horizontal scroll bar to view the test cases. *Uh-oh!* The run stops after 1,000 ns, which isn't enough time for all of the tests to execute.
9. Right-click **Simulation** in the **Project Manager** panel, and then select **Simulation Settings....**
10. Click the **Simulation** tab and change the value for xsim.simulate.runtime to 3000ns. Click **OK**.

11. Click the **X** on the **Simulation** window to close the simulation.
12. Re-run the simulation.
13. After expanding and scaling the timing diagram, you will be able to see all 256 test cases. See if the error signal has a value of 1 anywhere along the trace. This would indicate that the adder's output did not match the expected output.
14. This completes the exercise. You may close Vivado.

The VHDL file Ex__6_adder4_fulltestbench.vhdl contains the testbench code:

```vhdl
library IEEE;
  use IEEE.STD_LOGIC_1164.ALL;
  use IEEE.NUMERIC_STD.ALL;

entity ADDER4_TESTBENCH is
end entity ADDER4_TESTBENCH;

architecture BEHAVIORAL of ADDER4_TESTBENCH is

  component ADDER4 is
    port (
      A4        : in     std_logic_vector(3 downto 0);
      B4        : in     std_logic_vector(3 downto 0);
      SUM4      : out    std_logic_vector(3 downto 0);
      C_OUT4    : out    std_logic
    );
  end component;

  signal a                : std_logic_vector(3 downto 0);
  signal b                : std_logic_vector(3 downto 0);
  signal s                : std_logic_vector(3 downto 0);
  signal c_out            : std_logic;

  signal expected_sum5    : unsigned(4 downto 0);
  signal expected_sum4    : unsigned(3 downto 0);
  signal expected_c       : std_logic;
  signal error            : std_logic;

begin
```

```vhdl
TESTED_DEVICE : ADDER4
  port map (
    A4     => a,
    B4     => b,
    SUM4   => s,
    C_OUT4 => c_out
  );

TEST : process
begin

  -- Test all combinations of two 4-bit addends (256 total tests)
  for a_val in 0 to 15 loop
    for b_val in 0 to 15 loop
      -- Set the inputs to the ADDER4 component
      a <= std_logic_vector(to_unsigned(a_val, a'length));
      b <= std_logic_vector(to_unsigned(b_val, b'length));
      wait for 1 ns;

      -- Compute the 5-bit sum of the two 4-bit values
      expected_sum5 <= unsigned('0' & a) + unsigned('0' & b);
      wait for 1 ns;

      -- Break the sum into a 4-bit output and a carry bit
      expected_sum4 <= expected_sum5(3 downto 0);
      expected_c    <= expected_sum5(4);
      wait for 1 ns;

      -- The 'error' signal will only go to 1 if an error occurs
      if ((unsigned(s) = unsigned(expected_sum4)) and
          (c_out = expected_c)) then
        error <= '0';
      else
        error <= '1';
      end if;
```

```
            -- Each pass through the inner loop takes 10 ns
            wait for 7 ns;

        end loop;
    end loop;

    wait;

  end process TEST;

end architecture BEHAVIORAL;
```

Chapter 3: Processor Elements

Exercise 1

Consider the addition of two signed 8-bit numbers (that is, numbers in the range -128 to +127) where one operand is positive and the other is negative. Is there any pair of 8-bit numbers of different signs that, when added together, will exceed the range -128 to +127? This would constitute a signed overflow. Note: we're only looking at addition here because, as we've seen, subtraction in the 6502 architecture is the same as addition with the right operand's bits inverted.

Answer

The range of the positive (or non-negative) numbers is 0 to 127. The range of negative numbers is -128 to -1. It is only necessary to consider the extremes of each of these ranges to cover all possibilities:

Sum	Result
0 + -128	-128
127 + -128	-1
0 + -1	-1
127 + -1	126

In the preceding table, we can see that there is no pair of 8-bit numbers of different signs that, when added together, exceeds the range -128 to +127.

Exercise 2

If the answer to *Exercise 1* is *no*, this implies the only way to create a signed overflow is to add two numbers of the same sign. If an overflow occurs, what can you say about the result of performing XOR between the most significant bit of each operand with the most significant bit of the result? In other words, what will be the result of the expressions left(7) XOR result(7) and right(7) XOR result(7)? In these expressions, (7) indicates bit 7, the most significant bit.

Answer

Bit 7 is the sign bit. Since overflow can only occur when both operands are of the same sign, left(7) must equal right(7) when an overflow occurs.

When overflow occurs, the sign of the result differs from the sign of the two operands. This means result(7) differs from bit 7 of both of the operands.

Therefore, left(7) XOR result(7) = 1 and right(7) XOR result(7) = 1 whenever overflow occurs.

Exercise 3

Review the VHDL listing in the *Arithmetic Logic Unit* section in *Chapter 3, Processor Elements*, and determine whether the logic for setting or clearing the V flag is correct for addition and subtraction operations. Check the results of adding 126+1, 127+1, -127+(-1), and -128+(-1).

Answer

The listing of the VHDL implementation of a portion of a 6502-like **Arithmetic Logic Unit (ALU)** in *Chapter 3, Processor Elements*, implements the computation of the overflow flag with the following code:

```
if (((LEFT(7) XOR result8(7)) = '1') AND
    ((right_op(7) XOR result8(7)) = '1')) then  -- V flag
  V_OUT <= '1';
else
  V_OUT <= '0';
end if;
```

The following table shows the results of this code for the four test cases in the question:

left	right	left(7)	right(7)	result8(7)	V_OUT	Correct?
126	1	0	0	0	0	Yes
127	1	0	0	1	1	Yes
-127	-1	1	1	1	0	Yes
-128	-1	1	1	0	1	Yes

The logic for setting or clearing the V flag is correct for these test cases.

Exercise 4

When transferring blocks of data over an error-prone transmission medium, it is common to use a **checksum** to determine whether any data bits were lost or corrupted during transmission. The checksum is typically appended to the transferred data record. One checksum algorithm uses these steps:

1. Add all of the bytes in the data record together, retaining only the lowest 8 bits of the sum
2. The checksum is the two's complement of the 8-bit sum
3. Append the checksum byte to the data record

After receiving a data block with the appended checksum, the processor can determine whether the checksum is valid by simply adding all of the bytes in the record, including the checksum, together. The checksum is valid if the lowest 8 bits of the sum are zero. Implement this checksum algorithm using 6502 assembly language. The data bytes begin at the memory location store in addresses $10-$11 and the number of bytes (including the checksum byte) is provided as an input in the X register. Set the A register to 1 if the checksum is valid, and to 0 if it is invalid.

Answer

The Ex__4_checksum_alg.asm file contains the following checksum code:

```
; Ex__4_checksum_alg.asm
; Try running this code at https://skilldrick.github.io/easy6502/

; Set up the array of bytes to be checksummed
LDA #$01
STA $00


LDA #$72
STA $01
```

Appendix

```
LDA #$93
STA $02

LDA #$F4
STA $03

LDA #$06 ; This is the checksum byte
STA $04

; Store the address of the data array in $10-$11
LDA #$00
STA $10
STA $11

; Store the number of bytes in X
LDX #5

; Entering the checksum algorithm
; Move X to Y
TXA
TAY

; Compute the checksum
LDA #$00
DEY

LOOP:
CLC
ADC ($10), Y
DEY
BPL LOOP

CMP #$00
BNE ERROR

; The sum is zero: Checksum is correct
```

```
  LDA #1
  JMP DONE


; The sum is nonzero: Checksum is incorrect
ERROR:
  LDA #0


; A contains 1 if checksum is correct, 0 if it is incorrect
DONE:
```

Exercise 5

Make the checksum validation code from *Exercise 4* into a labeled subroutine that can be called with a JSR instruction and that ends with an RTS instruction.

Answer

The Ex__5_checksum_subroutine.asm file implements the checksum algorithm as a subroutine:

```
; Ex__5_checksum_subroutine.asm
; Try running this code at https://skilldrick.github.io/easy6502/

; Set up the array of bytes to be checksummed
LDA #$01
STA $00
LDA #$72
STA $01
LDA #$93
STA $02
LDA #$F4
STA $03
LDA #$06 ; This is the checksum byte
STA $04

; Store the address of the data array in $10-$11
LDA #$00
STA $10
STA $11
```

```
; Store the number of bytes in X
LDX #5


; Call the checksum calculation subroutine
JSR CALC_CKSUM


; Halt execution
BRK


; ================================================
; Compute the checksum
CALC_CKSUM:
; Move X to Y
TXA
TAY

LDA #$00
DEY

LOOP:
CLC
ADC ($10), Y
DEY
BPL LOOP


CMP #$00
BNE CKSUM_ERROR

; The sum is zero: Checksum is correct
LDA #1
JMP DONE


; The sum is nonzero: Checksum is incorrect
CKSUM_ERROR:
LDA #0

; A contains 1 if checksum is correct, 0 if it is incorrect
```

```
DONE:
RTS
```

Exercise 6

Write and execute a set of tests to verify the correct operation of the checksum testing subroutine you implemented in *Exercise 4* and *Exercise 5*. What is the shortest block of data your code can perform checksum validation upon? What is the longest block?

Answer

The Ex__6_checksum_tests.asm file implements the following checksum test code:

```
; Ex__6_checksum_tests.asm
; Try running this code at https://skilldrick.github.io/easy6502/

; After tests complete, A=$AA if success, A=$EE if error detected

; Store the address of the data array in $10-$11
LDA #$00
STA $10
STA $11


; =================================================
; Test 1: 1 byte; Checksum: 00 Checksum should pass? Yes
LDA #$00
STA $00

; Store the number of bytes in X
LDX #1

; Call the checksum calculation subroutine
JSR CALC_CKSUM

CMP #$01
BEQ TEST2
JMP ERROR

TEST2:
```

```
; ==============================================
; Test 2: 1 byte; Checksum: 01 Checksum should pass? No
LDA #$01
STA $00

; Store the number of bytes in X
LDX #1

; Call the checksum calculation subroutine
JSR CALC_CKSUM

CMP #$00
BEQ TEST3
JMP ERROR

TEST3:
; ==============================================
; Test 3: 2 bytes: 00 Checksum: 00 Checksum should pass? Yes
LDA #$00
STA $00
STA $01

; Store the number of bytes in X
LDX #2

; Call the checksum calculation subroutine
JSR CALC_CKSUM

CMP #$01
BEQ TEST4
JMP ERROR

TEST4:
; ==============================================
; Test 4: 2 bytes: 00 Checksum: 01 Checksum should pass? No
LDA #$00
STA $00
```

```
LDA #$01
STA $01


; Store the number of bytes in X
LDX #2


; Call the checksum calculation subroutine
JSR CALC_CKSUM


CMP #$00
BEQ TEST5
JMP ERROR


TEST5:
; ===============================================
; Test 5: 2 bytes: 01 Checksum: 00 Checksum should pass? No
LDA #$01
STA $00
LDA #$00
STA $01


; Store the number of bytes in X
LDX #1


; Call the checksum calculation subroutine
JSR CALC_CKSUM


CMP #$00
BEQ TEST6
JMP ERROR


TEST6:
; ===============================================
; Test 6: 3 bytes: 00 00 Checksum: 00 Checksum should pass? Yes
LDA #$00
STA $00
STA $01
```

```
STA $02

; Store the number of bytes in X
LDX #3

; Call the checksum calculation subroutine
JSR CALC_CKSUM

CMP #$01
BEQ TEST7
JMP ERROR

TEST7:
; ==============================================
; Test 7: 3 bytes: 00 00 Checksum: 00 Checksum should pass? Yes
LDA #$00
STA $00
STA $01
STA $02

; Store the number of bytes in X
LDX #3

; Call the checksum calculation subroutine
JSR CALC_CKSUM

CMP #$01
BEQ TEST8
JMP ERROR

TEST8:
; ==============================================
; Test 8: 3 bytes: 00 00 Checksum: 01 Checksum should pass? No
LDA #$00
STA $00
LDA #$00
STA $01
```

```
        LDA #$01
        STA $02


; Store the number of bytes in X
LDX #3

; Call the checksum calculation subroutine
JSR CALC_CKSUM

CMP #$00
BEQ TEST9
JMP ERROR

TEST9:
; ===============================================
; Test 9: 3 bytes: 00 01 Checksum: FF Checksum should pass? Yes
LDA #$00
STA $00
LDA #$01
STA $01
LDA #$FF
STA $02

; Store the number of bytes in X
LDX #3

; Call the checksum calculation subroutine
JSR CALC_CKSUM

CMP #$01
BEQ TEST10
JMP ERROR

TEST10:
; ===============================================
; Test 10: 5 bytes: 01 72 93 F4 Checksum: 06 Checksum should pass? Yes
LDA #$01
```

```
        STA $00
        LDA #$72
        STA $01
        LDA #$93
        STA $02
        LDA #$F4
        STA $03
        LDA #$06 ; This is the checksum byte
        STA $04

        ; Store the number of bytes in X
        LDX #5

        ; Call the checksum calculation subroutine
        JSR CALC_CKSUM

        CMP #$01
        BEQ PASSED

ERROR:
; ===============================================
; Error occurred; Halt execution with $EE in A
        LDA #$EE
        BRK


PASSED:
; ===============================================
; All tests passed; Halt execution with $AA in A
        LDA #$AA
        BRK


; ===============================================
; Compute the checksum
CALC_CKSUM:
        ; Move X to Y
        TXA
        TAY
```

```
        LDA #$00
        DEY

LOOP:
        CLC
        ADC ($10), Y
        DEY
        BPL LOOP

        CMP #$00
        BNE CKSUM_ERROR

        ; The sum is zero: Checksum is correct
        LDA #1
        JMP DONE

        ; The sum is nonzero: Checksum is incorrect
CKSUM_ERROR:
        LDA #0

        ; A contains 1 if checksum is correct, 0 if it is incorrect
DONE:
        RTS
```

The checksum routine works for byte sequences with lengths from 1 to 255 bytes.

Chapter 4: Computer System Components

Exercise 1

Create a circuit implementation of a NAND gate using two CMOS transistor pairs. Unlike NPN transistor gate circuits, no resistors are required for this circuit.

Answer

The diagram for this circuit is as follows:

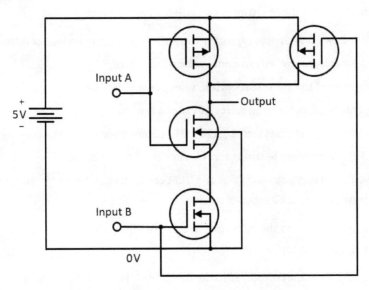

Figure 3: NAND gate circuit

Exercise 2

A 16-gigabit DRAM integrated circuit has two bank group selection inputs, two bank selection inputs, and 17 row address inputs. How many bits are in each row of a bank in this device?

Answer

The DRAM circuit contains 16 gigabits = 16×2^{30} bits.

The number of address bits is 2 bank group bits + 2 bank bits + 17 row address bits = 21 bits.

The row dimension of each bank is therefore $(16 \times 2^{30}) \div 2^{21}$ = **8,192 bits**.

Chapter 5: Hardware-Software Interface

Exercise 1

Restart your computer and enter the BIOS or UEFI settings. Examine each of the menus available in this environment. Does your computer have a BIOS or does it use UEFI? Does your motherboard support overclocking? When you are finished, be sure to select the option to quit without saving changes unless you are absolutely certain you want to make changes.

Answer

In Windows, you can enter the BIOS/UEFI settings by changing the startup options while Windows is running. To access these settings, perform the following steps:

1. In the Windows search box, type `startup` and select **Change advanced startup options**.
2. Select the **Restart now** button under **Advanced startup**.
3. When prompted to **Choose an option**, select **Troubleshoot**.
4. On the **Troubleshoot** screen, select **Advanced options**.
5. On the **Advanced options** screen, select **UEFI Firmware Settings**.
6. On the **UEFI Firmware Settings** screen, click the **Restart** button.
7. The system will restart and display the UEFI configuration main screen. Use the left and right arrow keys on the keyboard to move between the screens.

The following is in response to the questions in this exercise for a specific computer system (an Asus ZenBook UX303LA laptop, in this example):

1. Although the messages displayed in the menus use the term "BIOS" frequently, mentions of "EFI applications" and its age indicate it is actually **UEFI**.
2. **No overclocking options** are available.

After you've finished examining the UEFI information, exit without saving any changes by following these steps:

1. Move to the **Save & Exit** page.
2. Use the up and down arrow keys to select **Discard Changes and Exit**.
3. Press **Enter**.
4. Select **Yes** and press **Enter** on the **Exit Without Saving** dialog.
5. The system will reboot.

Exercise 2

Run the appropriate command on your computer to display the currently running processes. What is the **process ID (PID)** of the process you are using to run this command?

Answer

In Windows, open a Command Prompt window (type `command` in the Windows search box to locate the application) and then type the `tasklist` command as follows:

```
C:\>tasklist

Image Name PID Session Name Session# Mem Usage
=================== ===== ============ ======== ============
System Idle Process 0 Services 0 8 K
System 4 Services 0 9,840 K
Registry 120 Services 0 85,324 K
smss.exe 544 Services 0 640 K
csrss.exe 768 Services 0 4,348 K
wininit.exe 852 Services 0 4,912 K
services.exe 932 Services 0 8,768 K
lsass.exe 324 Services 0 18,160 K
svchost.exe 1044 Services 0 2,308 K
svchost.exe 1068 Services 0 27,364 K
.
.
.
svchost.exe 12184 Services 0 8,544 K
cmd.exe 16008 Console 3 3,996 K
conhost.exe 21712 Console 3 18,448 K
tasklist.exe 15488 Console 3 10,096 K
```

The current process is the one running the tasklist.exe application. The PID of this process is 15488.

Chapter 6: Specialized Computing Domains

Exercise 1

Rate monotonic scheduling (RMS) is an algorithm for assigning thread priorities in preemptive, hard, real-time applications in which threads execute periodically. RMS assigns the highest priority to the thread with the shortest execution period, the next-highest priority to the thread with the next-shortest execution period, and so on. An RMS system is schedulable, meaning all tasks are guaranteed to meet their deadlines (assuming no inter-thread interactions or other activities such as interrupts, resulting in processing delays) if the following condition is met:

$$\sum_{k=1}^{n} \frac{C_i}{T_i} \leq n(2^{1/n} - 1)$$

This formula represents the maximum fraction of available processing time that can be consumed by n threads. In this formula, C_i is the maximum execution time required for thread i, and T_i is the execution period of thread i.

Is the following system composed of three threads schedulable?

Thread	Execution Time(C_i), ms	Execution period(T_i), ms
Thread 1	50	100
Thread 2	100	500
Thread 3	120	1,000

Answer

First, evaluate the left side of the RMS formula using the data from the table:

$$\frac{50}{100} + \frac{100}{500} + \frac{120}{1000} = 0.82$$

Then, evaluate the right side of the RMS formula:

$$3(2^{\frac{1}{3}} - 1) = 0.7798$$

Because *0.82* is not less than or equal to *0.7798*, this set of tasks is not schedulable in RMS.

Exercise 2

A commonly used form of the one-dimensional **discrete cosine transform (DCT)** is as follows:

$$X_k = \sum_{n=0}^{N-1} x_n \cos\left[\frac{\pi}{N}\left(n + \frac{1}{2}\right)k\right]$$

In this formula, k, the index of the DCT coefficient, runs from 0 to N-1.

Write a program to compute the DCT of the sequence:

$$x = \{0.5, 0.2, 0.7, -0.6, 0.4, -0.2, 1.0, -0.3\}$$

The cosine terms in the formula depend only on the indexes n and k and do not depend on the input data sequence x. This means the cosine terms can be computed one time and stored as constants for later use. Using this as a preparatory step, the computation of each DCT coefficient reduces to a sequence of MAC operations.

Appendix

This formula represents the unoptimized form of the DCT computation, requiring N^2 iterations of the MAC operation to compute all N DCT coefficients.

Answer

The Ex__2_dct_formula.py Python file contains the DCT code:

```python
#!/usr/bin/env python

"""Ex__2_dct_formula.py: Answer to chapter 6 exercise 2."""

# Output produced by this program:
# Index       0       1       2       3       4       5       6       7
# x         0.5000  0.2000  0.7000 -0.6000  0.4000 -0.2000  1.0000 -0.3000
# DCT(x)    1.7000  0.4244  0.6374  0.4941 -1.2021  0.5732 -0.4936  2.3296

import math

# Input vector
x = [0.5, 0.2, 0.7, -0.6, 0.4, -0.2, 1.0, -0.3]

# Compute the DCT coefficients
dct_coef = [[i for i in range(len(x))] for j in range(len(x))]
for n in range(len(x)):
    for k in range(len(x)):
        dct_coef[n][k] = math.cos((math.pi/len(x))*(n + 1/2)*k)

# Compute the DCT
x_dct = [i for i in range(len(x))]
for k in range(len(x)):
    x_dct[k] = 0;
    for n in range(len(x)):
        x_dct[k] += x[n]*dct_coef[n][k]

# Print the results
print('Index', end=' ')
for i in range(len(x)):
    print("%8d" % i, end=' ')
```

```
    print('\nx         ', end='')
    for i in range(len(x)):
        print("%8.4f" % x[i], end=' ')

    print('\nDCT(x) ', end='')
    for i in range(len(x)):
        print("%8.4f" % x_dct[i], end=' ')
```

To run the code, assuming Python is installed and is in your path, execute the following command:

```
python Ex__2_dct_formula.py
```

This is the output produced by the program:

```
C:\>Ex__2_dct_formula.py
Index      0       1       2        3        4        5       6        7
x       0.5000  0.2000  0.7000  -0.6000   0.4000  -0.2000  1.0000  -0.3000
DCT(x)  1.7000  0.4244  0.6374   0.4941  -1.2021   0.5732 -0.4936   2.3296
```

Exercise 3

The hyperbolic tangent is often used as an activation function in **Artificial Neural Networks** (**ANNs**). The hyperbolic tangent function is:

$$\tanh(x) = \frac{e^x - e^{-x}}{e^x + e^{-x}}$$

Given a neuron with inputs from three preceding neurons as depicted in *Figure 6.4*, compute the neuron's output with the hyperbolic tangent as the activation function $F(x)$ using the following neuron outputs and path weights:

Neuron	Neuron output	Weight
N1	0.6	0.4
N2	-0.3	0.8
N3	0.5	-0.2

Answer

The Ex__3_activation_func.py Python file contains the following code:

```
#!/usr/bin/env python
```

```python
"""Ex__3_activation_func.py: Answer to Ch 6 Ex 3."""

# Output produced by this program:
# Neuron output = -0.099668

import math

# Neuron signal and weight vectors
neuron = [0.6, -0.3,  0.5]
weight = [0.4,  0.8, -0.2]

sum = 0
for i in range(len(neuron)):
    sum = sum + neuron[i] * weight[i]

output = math.tanh(sum)

# Print the results
print('Neuron output = %8.6f' % output)
```

To run the code, assuming Python is installed and is in your path, execute the following command:

```
python Ex__3_activation_func.py
```

This is the output produced by the program:

```
C:\>Ex__3_activation_func.py
Neuron output = -0.099668
```

Chapter 7: Processor and Memory Architectures

Exercise 1

A 16-bit embedded processor has separate memory regions for code and data. Code is stored in flash memory and modifiable data is stored in RAM. Some data values, such as constants and initial values for RAM data items, are stored in the same flash memory region as the program instructions. RAM and ROM reside in the same address space. Which of the processor architectures discussed in *Chapter 7, Processor and Memory Architectures*, best describes this processor?

Answer

Because the code and data are located in the same address space, this is a **von Neumann architecture**.

The fact that the code and some data items are stored in ROM and other data items reside in RAM is not relevant to determining the architecture category.

Exercise 2

The processor described in *Exercise 1* has memory security features that prevent executed code from modifying program instruction memory. The processor uses physical addresses to access instructions and data. Does this processor contain an MMU?

Answer

While the protection of memory regions is a feature of MMUs, the presence of memory protection alone does not mean an MMU is in use. *This processor does not contain an MMU.*

MMUs generally perform virtual-to-physical address translation, which does not occur in the processor described here.

Exercise 3

The order of accessing sequential elements in a large data structure can have a measurable impact on processing speed due to factors such as the reuse of TLB entries. Accessing distant array elements in sequence (that is, elements that are not in the same page frame as previously accessed elements) requires frequent soft faults as new TLB entries are loaded and old TLB entries are discarded.

Write a program that creates a two-dimensional array of numbers with a large size, such as 10,000 rows by 10,000 columns. Iterate through the array in column-major order, assigning each element the sum of the row and column indices. Column-major means the column index increments fastest. In other words, the column index increments in the inner loop. Measure precisely how long this procedure takes. Note, you may need to take steps to ensure your programming language does not optimize away the entire calculation if the results from the array are not used later. It may suffice to print one of the array values after the timing is complete, or you may need to do something like sum all the array elements and print that result.

Repeat the process, including the timing, exactly as explained before, except change the inner loop to iterate over the row index and the outer loop to iterate over the column index, making the access sequence row-major.

Appendix

Since general-purpose computers perform many other tasks while running your code, you may need to perform both procedures a number of times to get a statistically valid result. You might start by running the experiment 10 times and averaging the times for column-major and row-major array access.

Are you able to determine a consistently superior array access method? Which order is fastest on your system using the language you selected? Note that the difference between the column-major and row-major access order may not be dramatic — it might be just a few percent.

Answer

The Ex__3_row_column_major_order.py file contains the following Python implementation of a solution to this exercise:

```
#!/usr/bin/env python

"""Ex__3_row_column_major_order.py: Answer to chapter 7 exercise 3."""

# Typical output from a run of this script:
# Average row-major time    : 16.68 sec
# Average column-major time: 15.94 sec
# Average time difference   : 0.74 sec
# Winner is column-major indexing; It is faster by 4.42%

import time

dim = 10000
matrix = [[0] * dim] * dim

num_passes = 10
row_major_time = 0
col_major_time = 0

for k in range(num_passes):
    print('Pass %d of %d:' % (k+1, num_passes))

    t0 = time.time()
    for i in range(dim):
        for j in range(dim):
```

```python
            matrix[i][j] = i + j

    t1 = time.time()

    total_time = t1 - t0
    col_major_time = col_major_time + total_time
    print('  Column-major time to fill array: %.2f sec' %
            total_time)

    t0 = time.time()
    for i in range(dim):
        for j in range(dim):
            matrix[j][i] = i + j

    t1 = time.time()

    total_time = t1 - t0
    row_major_time = row_major_time + total_time
    print('  Row-major time to fill array: %.2f sec' %
            total_time)
    print('')

row_major_average = row_major_time / num_passes
col_major_average = col_major_time / num_passes

if (row_major_average < col_major_average):
    winner = 'row'
    pct_better = 100 * (col_major_average -
        row_major_average) / col_major_average
else:
    winner = 'column'
    pct_better = 100 * (row_major_average -
        col_major_average) / row_major_average

print('Average row-major time   : %.2f sec' % row_major_average)
print('Average column-major time: %.2f sec' % col_major_average)
print('Average time difference  : %.2f sec' % (
```

Appendix

```
        (row_major_time-col_major_time) / num_passes))
    print(('Winner is ' + winner +
        '-major indexing; It is faster by %.2f%%') % pct_better)
```

This program takes a few minutes to run on a Windows PC.

This is the typical output from running this program:

```
Average row-major time    : 16.68 sec
Average column-major time: 15.94 sec
Average time difference   : 0.74 sec
Winner is column-major indexing; It is faster by 4.42%
```

Chapter 8: Performance-Enhancing Techniques

Exercise 1

Consider a direct-mapped L1-I cache of 32 KB. Each cache line consists of 64 bytes and the system address space is 4 GB. How many bits are in the cache tag? Which bit numbers (bit 0 is the least significant bit) are they within the address word?

Answer

The cache contains 32,768 bytes with 64 bytes in each line. There are 32,768 ÷ 64 = 512 sets in the cache. 512 = 2^9. The set number is thus 9 bits in length.

Each cache line contains 64 (2^6) bytes, which means the lower 6 bits of each address represent the byte offset within the cache line.

A 4 GB address space requires 32-bit addresses. Subtracting the 9 bits in the set number and the 6 bits in the byte offset from the 32-bit address results in 32 − (9 + 6) = 17 bits in the cache tag.

The cache tag lies in the 17 most significant bits of the address, so **the range of these bits within a 32-bit address runs from bit 15 to bit 31.**

Exercise 2

Consider an 8-way set-associative L2 instruction and data cache of 256 KB, with 64 bytes in each cache line. How many sets are in this cache?

Answer

The number of lines in the cache is 262,144 ÷ 64 = 4,096.

Each set contains 8 lines.

The number of sets = 4,096 lines ÷ 8 lines per set = **512 sets**.

Exercise 3

A processor has a 4-stage pipeline with maximum delays of 0.8, 0.4, 0.6, and 0.3 nanoseconds in stages 1-4, respectively. If the first stage is replaced with two stages that have maximum delays of 0.5 and 0.3 nanoseconds, respectively, how much will the processor clock speed increase in percentage terms?

Answer

The maximum clock speed is determined by the slowest pipeline stage. The slowest stage of the 4-stage pipeline takes 0.8 ns. The maximum clock frequency is:

$$1 \div (0.8 \times 10^{-9}) = 1.25 \text{ GHz}$$

The 5-stage pipeline has a slowest stage of 0.6 ns. The maximum clock frequency is:

$$1 \div (0.6 \times 10^{-9}) = 1.667 \text{ GHz}$$

The clock frequency increase resulting from the addition of the pipeline stage is:

$$100 \times (1.667 \times 10^9 - 1.25 \times 10^9) \div (1.25 \times 10^9) = \mathbf{33.3\%}$$

Chapter 9: Specialized Processor Extensions

Exercise 1

Using a programming language that allows access to the byte representation of floating-point data types (such as C or C++), write a function that accepts a 32-bit single-precision variable as input. Extract the sign, exponent, and mantissa from the bytes of the floating-point variable and display them. Remove the bias term from the exponent before displaying its value and display the mantissa as a decimal number. Test the program with the values 0, -0, 1, -1, 6.674e-11, 1.0e38, 1.0e39, 1.0e-38, and 1.0e-39. The numeric values listed here containing *e* are using the C/C++ text representation of floating-point numbers. For example, 6.674e-11 means 6.674×10^{-11}.

Answer

The Ex__1_float_format.cpp C++ file contains the code for this exercise:

```
// Ex__1_float_format.cpp
```

```cpp
#include <iostream>
#include <cstdint>

void print_float(float f)
{
    const auto bytes = static_cast<uint8_t*>(static_cast<void*>(&f));

    printf(" Float | %9g | ", f);

    for (int i = sizeof(float) - 1; i >= 0; i--)
        printf("%02X", bytes[i]);

    printf(" | ");

    const auto sign = bytes[3] >> 7;
    const auto exponent = ((static_cast<uint16_t>(bytes[3] & 0x7F)
                                        << 8) | bytes[2]) >> 7;
    auto exp_unbiased = exponent - 127;

    uint32_t mantissa = 0;
    for (auto i = 0; i < 3; i++)
        mantissa = (mantissa << 8) | bytes[2 - i];

        mantissa &= 0x7FFFFF; // Clear upper bit

    double mantissa_dec;
    if (exponent == 0) // This is zero or a subnormal number
    {
        mantissa_dec = mantissa / static_cast<double>(0x800000);
        exp_unbiased++;
    }
    else
        mantissa_dec = 1.0 + mantissa /static_cast<double>(0x800000);

    printf(" %d | %4d | %lf\n", sign, exp_unbiased, mantissa_dec);
}
```

```c
int main(void)
{
    printf(" Type | Number | Bytes | Sign | Exponent | Mantissa\n");
    printf(" -------|----------|-------------------|------|----------|---------\n");
    print_float(0);
    print_float(-0); // Minus sign is ignored
    print_float(1);
    print_float(-1);
    print_float(6.674e-11f);
    print_float(1.0e38f);
    //print_float(1.0e39f); // Compile-time error
    print_float(1.0e-38f);
    print_float(1.0e-39f);

    return 0;
}
```

This is the output of the program:

Type	Number	Bytes	Sign	Exponent	Mantissa
Float	0	00000000	0	-126	0.000000
Float	0	00000000	0	-126	0.000000
Float	1	3F800000	0	0	1.000000
Float	-1	BF800000	1	0	1.000000
Float	6.674e-11	2E92C348	0	-34	1.146585
Float	1e+38	7E967699	0	126	1.175494
Float	1e-38	006CE3EE	0	-126	0.850706
Float	1e-39	000AE398	0	-126	0.085071

These are some notes about the results:

1. Zero in IEEE 754 can have a positive or negative sign. The zero passed to the `print_float` function in the second row of the table is preceded by a minus sign, but the sign is ignored during the conversion to a floating point.

2. The value `1.0e39f` is not shown because using it causes a compile-time error: the floating constant is out of range.

3. Zero is represented as a mantissa of zero and a biased exponent of zero.

4. The last two rows contain numbers that cannot be represented with an implicit leading 1 bit because the exponent would underflow. These numbers are called **subnormals** and contain the special biased exponent of 0. Subnormals have reduced precision because not all bits of the mantissa contain meaningful digits.
5. Numerically, subnormal floats actually use a biased exponent of 1, which translates to an unbiased exponent of -126.

Exercise 2

Modify the program from *Exercise 1* to also accept a double-precision, floating-point variable and print the sign, exponent (with the bias removed), and mantissa from the variable. Test with the same input values as in *Exercise 1*, and also with the values 1.0e308, 1.0e309, 1.0e-308, and 1.0e-309.

Answer

The `Ex__2_double_format.cpp` C++ file contains the code for this exercise:

```
// Ex__2_double_format.cpp
#include <iostream>
#include <cstdint>

void print_float(float f)
{
    const auto bytes = static_cast<uint8_t*>(static_cast<void*>(&f));

    printf(" Float | %9g | ", f);

    for (int i = sizeof(float) - 1; i >= 0; i--)
        printf("%02X", bytes[i]);

    printf(" | ");

    const auto sign = bytes[3] >> 7;
    const auto exponent = ((static_cast<uint16_t>(bytes[3] & 0x7F) << 8) | bytes[2]) >> 7;
    auto exp_unbiased = exponent - 127;

    uint32_t mantissa = 0;
    for (auto i = 0; i < 3; i++)
```

```cpp
        mantissa = (mantissa << 8) | bytes[2 - i];

    mantissa &= 0x7FFFFF; // Clear upper bit

    double mantissa_dec;
    if (exponent == 0) // This is zero or a subnormal number
    {
        mantissa_dec = mantissa / static_cast<double>(0x800000);
        exp_unbiased++;
    }
    else
        mantissa_dec = 1.0 + mantissa /static_cast<double>(0x800000);

    printf(" %d | %4d | %lf\n", sign, exp_unbiased, mantissa_dec);
}

void print_double(double d)
{
    const auto bytes = static_cast<uint8_t*>(static_cast<void*>(&d));

    printf(" Double | %9g | ", d);

    for (int i = sizeof(double) - 1; i >= 0; i--)
        printf("%02X", bytes[i]);

    printf(" | ");

    const auto sign = bytes[7] >> 7;
    const auto exponent = ((static_cast<uint16_t>(bytes[7] & 0x7F) << 8) | bytes[6]) >> 4;
    auto exp_unbiased = exponent - 1023;

    uint64_t mantissa = 0;
    for (auto i = 0; i < 7; i++)
        mantissa = (mantissa << 8) | bytes[6 - i];

    mantissa &= 0xFFFFFFFFFFFFF; // Save the low 52 bits
```

```cpp
    double mantissa_dec;
    if (exponent == 0) // This is zero or a subnormal number
    {
        mantissa_dec = mantissa /static_cast<double>(0x10000000000000);
        exp_unbiased++;
    }
    else
        mantissa_dec = 1.0 + mantissa /static_cast<double>(0x10000000000000);

    printf(" %d | %5d | %lf\n", sign, exp_unbiased, mantissa_dec);
}

int main(void)
{
    printf(" Type | Number | Bytes | Sign | Exponent | Mantissa\n");
    printf(" -------|-----------|--------------------|------|----------|---------\n");

    print_float(0);
    print_float(-0); // The minus sign is ignored
    print_float(1);
    print_float(-1);
    print_float(6.674e-11f);
    print_float(1.0e38f);
    //print_float(1.0e39f); // Compile-time error
    print_float(1.0e-38f);
    print_float(1.0e-39f);

    print_double(0);
    print_double(-0); // The minus sign is ignored
    print_double(1);
    print_double(-1);
    print_double(6.674e-11);
    print_double(1.0e38);
    print_double(1.0e39);
    print_double(1.0e-38);
    print_double(1.0e-39);
```

```
    print_double(1.0e308);
    //print_double(1.0e309); // Compile-time error
    print_double(1.0e-308);
    print_double(1.0e-309);

    return 0;
}
```

This is the output of the program:

Type	Number	Bytes	Sign	Exponent	Mantissa
Float	0	00000000	0	-126	0.000000
Float	0	00000000	0	-126	0.000000
Float	1	3F800000	0	0	1.000000
Float	-1	BF800000	1	0	1.000000
Float	6.674e-11	2E92C348	0	-34	1.146585
Float	1e+38	7E967699	0	126	1.175494
Float	1e-38	006CE3EE	0	-126	0.850706
Float	1e-39	000AE398	0	-126	0.085071
Double	0	0000000000000000	0	-1022	0.000000
Double	0	0000000000000000	0	-1022	0.000000
Double	1	3FF0000000000000	0	0	1.000000
Double	-1	BFF0000000000000	1	0	1.000000
Double	6.674e-11	3DD25868F4DEAE16	0	-34	1.146584
Double	1e+38	47D2CED32A16A1B1	0	126	1.175494
Double	1e+39	48078287F49C4A1D	0	129	1.469368
Double	1e-38	380B38FB9DAA78E4	0	-127	1.701412
Double	1e-39	37D5C72FB1552D83	0	-130	1.361129
Double	1e+308	7FE1CCF385EBC8A0	0	1023	1.112537
Double	1e-308	000730D67819E8D2	0	-1022	0.449423
Double	1e-309	0000B8157268FDAF	0	-1022	0.044942

These are some notes about the results:

1. Zero in IEEE 754 can have a positive or negative sign. The zero passed to the print_double function in the second row of the table containing the Double type is preceded by a minus sign, but the sign is ignored during the conversion to a floating-point value.

2. The value `1.0e309` is not shown because using it causes a compile-time error: **the floating constant is out of range**.
3. Zero is represented as a mantissa of zero and a biased exponent of zero.
4. The last two rows contain numbers that cannot be represented by an implicit leading 1 bit because the exponent would underflow. These numbers are called subnormals and contain the special biased exponent of 0. Subnormals have reduced precision because not all bits of the mantissa contain meaningful digits.
5. Numerically, subnormal doubles actually use a biased exponent of 1, which translates to an unbiased exponent of -1,022.

Exercise 3

Search the internet for information about the NXP Semiconductors i.MX RT1060 processor family. Download the product family datasheet and answer the following questions about these processors.

Answer

Introductory information about the i.MX RT1060 processor family is available at `https://www.nxp.com/docs/en/nxp/data-sheets/IMXRT1060CEC.pdf`.

The complete i.MX RT1060 reference manual is available only after you create an account at `https://www.nxp.com`.

While logged in to your account, search for `i.MX RT1060 Processor Reference Manual` to locate the reference manual and download it. The filename is `IMXRT1060RM.pdf`.

Exercise 4

Do the i.MX RT1060 processors support the concept of supervisor-mode instruction execution? Explain your answer.

Answer

Performing a search for `supervisor` in the i.MX RT1060 processor reference manual produces a few hits. However, all of these uses refer to access restrictions related to a particular subsystem, such as the FlexCAN module.

Supervisor mode in the i.MX RT1060 processor does not operate at the instruction execution level, so *these processors do not implement supervisor mode instruction execution* as described in *Chapter 9, Specialized Processor Extensions*.

Exercise 5

Do the i.MX RT1060 processors support the concept of paged virtual memory? Explain your answer.

Answer

The i.MX RT1060 processors use physical memory addressing with up to 16 memory protection regions. *These processors do not support the concept of paged virtual memory.*

Exercise 6

Do the i.MX RT1060 processors support floating-point operations in hardware? Explain your answer.

Answer

Section 1.3, **Features**, in the reference manual lists the following capability: **Single-precision and double-precision FPU (Floating Point Unit)**.

The **ARM Cortex-M7 Processor Technical Reference Manual**, available at http://infocenter.arm.com/help/topic/com.arm.doc.ddi0489b/DDI0489B_cortex_m7_trm.pdf, states that the FPU provides *"floating-point computation functionality that is compliant with the ANSI/IEEE Std 754-2008, IEEE Standard for Binary Floating-Point Arithmetic, referred to as the IEEE 754 standard."*

The i.MX RT1060 processors support floating-point operations in hardware.

Exercise 7

What power management features do the i.MX RT1060 processors support?

Answer

Section 12.4 of the reference manual describes the processor power management subsystem. Some of the key features are as follows:

1. Separate power domains for the processor, memory, and the remainder of the system
2. Integrated secondary power supplies that independently support powering a variety of subsystems
3. Voltage and clock frequency control enabling **dynamic voltage and frequency scaling (DVFS)**
4. Temperature sensors
5. Voltage sensors

Exercise 8

What security features do the i.MX RT1060 processors support?

Answer

Chapter 6, Specialized Computing Domains, in the reference manual describes the system security components. Some of the key features are as follows:

1. Secure boot, enforcing digital signature verification of an encrypted code image
2. On-chip, one-time programmable ROM for storing security-related information
3. Hardware cryptographic coprocessor supporting the AES-128, SHA-1, and SHA-256 encryption algorithms
4. True random number generator for creating secure cryptographic keys
5. JTAG debug controller with password-enabled secure debug capability
6. Memory interface supporting on-the-fly decryption of encrypted ROM instruction data

Chapter 10: Modern Processor Architectures and Instruction Sets

Exercise 1

Install the free Visual Studio Community edition, available at https://visualstudio.microsoft.com/vs/community/, on a Windows PC. After installation is complete, open the Visual Studio IDE and select **Get Tools and Features...** under the **Tools** menu. Install the **Desktop development with C++** workload:

1. In the Windows search box in the taskbar, begin typing `Developer Command Prompt for VS 2022`. When the app appears in the search menu, select it to open Command Prompt.
2. Create a file named `hello_x86.asm` with the content shown in the source listing in the *x86 assembly language* section of *Chapter 10, Modern Processor Architectures and Instruction Sets*.
3. Build the program using the command shown in the *x86 assembly language* section of *Chapter 10, Modern Processor Architectures and Instruction Sets*, and run it. Verify that the output **Hello, Computer Architect!** appears on the screen.

Answer

Install Visual Studio Community as described in the question and then install the **Desktop development with C++** workload within Visual Studio Community:

1. Create your assembly language source file. The Ex__1_hello_x86.asm file contains the following example solution to this exercise:

   ```
   .386
   .model FLAT,C
   .stack 400h

   .code
   includelib libcmt.lib
   includelib legacy_stdio_definitions.lib

   extern printf:near
   extern exit:near

   public main
   main proc
       ; Print the message
       push    offset message
       call    printf

       ; Exit the program with status 0
       push    0
       call    exit
   main endp

   .data
   message db "Hello, Computer Architect!",0

   end
   ```

2. Open **Developer Command Prompt for VS 2022** and change to the directory containing your source file.

3. Build the executable with this command:

```
ml /Fl /Zi /Zd Ex__1_hello_x86.asm
```

4. This is the output produced by the program:

```
C:\>Ex__1_hello_x86.exe
Hello, Computer Architect!
```

This is the listing file created by the build procedure:

```
Microsoft (R) Macro Assembler Version 14.31.31104.0        02/21/22 07:39:20
Ex__1_hello_x86.asm                    Page 1 - 1



                .386
                .model FLAT,C
                .stack 400h

 00000000       .code
                includelib libcmt.lib
                includelib legacy_stdio_definitions.lib

                extern printf:near
                extern exit:near

                public main
 00000000       main proc
                ; Print the message
 00000000  68 00000000 R        push    offset message
 00000005  E8 00000000 E        call    printf

                ; Exit the program with status 0
 0000000A  6A 00                push    0
 0000000C  E8 00000000 E        call    exit
 00000011       main endp


 00000000       .data
 00000000  48 65 6C 6C 6F       message db "Hello, Computer Architect!",0
           2C 20 43 6F 6D
           70 75 74 65 72
```

```
                    20 41 72 63 68
                    69 74 65 63 74
                    21 00


            end
Microsoft (R) Macro Assembler Version 14.31.31104.0        02/21/22 07:39:20
Ex__1_hello_x86.asm                Symbols 2 - 1
```

Segments and Groups:

```
                N a m e              Size     Length   Align   Combine Class

FLAT . . . . . . . . . . . . . .  GROUP
STACK . . . . . . . . . . . . .   32 Bit   00000400 DWord   Stack   'STACK'
_DATA . . . . . . . . . . . . .   32 Bit   0000001B DWord   Public  'DATA'
_TEXT . . . . . . . . . . . . .   32 Bit   00000011 DWord   Public  'CODE'
```

Procedures, parameters, and locals:

```
                N a m e              Type      Value    Attr

main . . . . . . . . . . . . . .  P Near   00000000 _TEXT  Length= 00000011
```

Symbols:

```
                N a m e              Type      Value    Attr

@CodeSize . . . . . . . . . . .   Number   00000000h
@DataSize . . . . . . . . . . .   Number   00000000h
@Interface  . . . . . . . . . .   Number   00000001h
@Model  . . . . . . . . . . . .   Number   00000007h
@code . . . . . . . . . . . . .   Text     _TEXT
@data . . . . . . . . . . . . .   Text     FLAT
@fardata? . . . . . . . . . . .   Text     FLAT
```

@fardata	Text	FLAT		
@stack	Text	FLAT		
exit	L Near	00000000 FLAT	External	C
message	Byte	00000000 _DATA		
printf	L Near	00000000 FLAT	External	C

```
    0 Warnings
    0 Errors
```

Exercise 2

Write an x86 assembly language program that computes the following expression and prints the result as a hexadecimal number: $[(129 - 66) \times (445 + 136)] \div 3$. As part of this program, create a callable function to print 1 byte as 2 hex digits.

Answer

1. Create your assembly language source file. The Ex__2_expr_x86.asm file contains the following example solution to this exercise:

    ```
    .386
    .model FLAT,C
    .stack 400h

    .code
    includelib libcmt.lib
    includelib legacy_stdio_definitions.lib

    extern printf:near
    extern exit:near

    public main
    main proc
        ; Print the leading output string
        push    offset msg1
        call    printf

        ; Compute [(129 - 66) * (445 + 136)] / 3
        mov     eax, 129
        sub     eax, 66
        mov     ebx, 445
    ```

```
        add     ebx, 136
        mul     bx
        mov     bx, 3
        div     bx

        ; Print the most significant byte
        push    eax
        mov     bl, ah
        call    print_byte

        ; Print the least significant byte
        pop     ebx
        call    print_byte

        ; Print the trailing output string
        push    offset msg2
        call    printf

        push    0
        call    exit
main endp

; Pass the byte to be printed in ebx
print_byte proc
        ; x86 function prologue
        push    ebp
        mov     ebp, esp

        ; Use the C library printf function
        and     ebx, 0ffh
        push    ebx
        push    offset fmt_str
        call    printf

        ; x86 function epilogue
        mov     esp, ebp
        pop     ebp
        ret
print_byte endp
```

```
        .data
        fmt_str db "%02X", 0
        msg1    db "[(129 - 66) * (445 + 136)] / 3 = ", 0
        msg2    db "h", 9

        end
```

2. Open **Developer Command Prompt for VS 2022** and change to the directory containing your source file.

3. Build the executable with this command:

```
ml /Fl /Zi /Zd Ex__1_hello_x86.asm
```

4. This is the output produced by the program:

```
C:\>Ex__2_expr_x86.exe
[(129 - 66) * (445 + 136)] / 3 = 2FA9h
```

This is the listing file created by the build procedure:

```
Microsoft (R) Macro Assembler Version 14.23.28107.0        01/26/20 20:45:09
Ex__2_expr_x86.asm                     Page 1 - 1



                    .386
                    .model FLAT,C
                    .stack 400h

 00000000           .code
                    includelib libcmt.lib
                    includelib legacy_stdio_definitions.lib

                    extern printf:near
                    extern exit:near

                    public main
 00000000           main proc
                    ; Print the leading output string
 00000000  68 00000005 R      push    offset msg1
 00000005  E8 00000000 E      call    printf
```

```
                    ; Compute [(129 - 66) * (445 + 136)] / 3
0000000A  B8 00000081        mov     eax, 129
0000000F  83 E8 42           sub     eax, 66
00000012  BB 000001BD        mov     ebx, 445
00000017  81 C3 00000088     add     ebx, 136
0000001D  66| F7 E3          mul     bx
00000020  66| BB 0003        mov     bx, 3
00000024  66| F7 F3          div     bx

                    ; Print the most significant byte
00000027  50                 push    eax
00000028  8A DC              mov     bl, ah
0000002A  E8 00000017        call    print_byte

                    ; Print the least significant byte
0000002F  5B                 pop     ebx
00000030  E8 00000011        call    print_byte

                    ; Print the trailing output string
00000035  68 00000027 R      push    offset msg2
0000003A  E8 00000000 E      call    printf

0000003F  6A 00              push    0
00000041  E8 00000000 E      call    exit
00000046          main endp

          ; Pass the byte to be printed in ebx
00000046          print_byte proc
                  ; x86 function prologue
00000046  55                 push    ebp
00000047  8B EC              mov     ebp, esp

                  ; Use the C library printf function
00000049  81 E3 000000FF     and     ebx, 0ffh
0000004F  53                 push    ebx
00000050  68 00000000 R      push    offset fmt_str
00000055  E8 00000000 E      call    printf
                  ; x86 function epilogue
0000005A  8B E5              mov     esp, ebp
```

```
 0000005C  5D              pop     ebp
 0000005D  C3              ret
 0000005E          print_byte endp

 00000000          .data
 00000000 25 30 32 58 00   fmt_str db "%02X", 0
 00000005 5B 28 31 32 39   msg1    db "[(129 - 66) * (445 + 136)] / 3 = ",
0
          20 2D 20 36 36
          29 20 2A 20 28
          34 34 35 20 2B
          20 31 33 36 29
          5D 20 2F 20 33
          20 3D 20 00
 00000027 68 09            msg2    db "h", 9

         end
```

Microsoft (R) Macro Assembler Version 14.23.28107.0   01/26/20 20:45:09
Ex__2_expr_x86.asm   Symbols 2 - 1

Segments and Groups:

                N a m e   Size   Length   Align   Combine Class

FLAT . . . . GROUP
STACK . . . . 32 Bit   00000400 DWord   Stack   'STACK'
_DATA . . . . 32 Bit   00000029 DWord   Public   'DATA'
_TEXT . . . . 32 Bit   0000005E DWord   Public   'CODE'

Procedures, parameters, and locals:

                    N a m e   Type   Value   Attr

main . . . .   P Near   00000000 _TEXT Length= 00000046
print_byte . . . .   P Near   00000046 _TEXT Length= 00000018

Symbols:

                    N a m e   Type   Value   Attr

@CodeSize . . . .   Number   00000000h
@DataSize . . . .   Number   00000000h
@Interface . . . .   Number   00000001h
@Model . . . .   Number   00000007h
@code . . . .   Text   _TEXT
@data . . . .   Text   FLAT
@fardata? . . . .   Text   FLAT
@fardata . . . .   Text   FLAT
@stack . . . .   Text   FLAT
exit . . . .   L Near   00000000 FLAT   External C
fmt_str . . . .   Byte   00000000 _DATA
msg1 . . . .   Byte   00000005 _DATA
msg2 . . . .   Byte   00000027 _DATA
printf . . . .   L Near   00000000 FLAT   External C

       0 Warnings
       0 Errors

Exercise 3

In the Windows search box in the taskbar, begin typing x64 Native Tools Command Prompt for VS 2022. When the app appears in the search menu, select it to open Command Prompt:

Appendix

1. Create a file named hello_x64.asm with the content shown in the source listing in the *x64 assembly language* section of *Chapter 10, Modern Processor Architectures and Instruction Sets*.
2. Build the program using the command shown in the *x64 assembly language* section of *Chapter 10, Modern Processor Architectures and Instruction Sets* and run it. Verify that the output **Hello, Computer Architect!** appears on the screen.

Answer

1. Create your assembly language source file. The Ex__3_hello_x64.asm file contains the following example solution to this exercise:

```
.code
includelib libcmt.lib
includelib legacy_stdio_definitions.lib

extern printf:near
extern exit:near

public main
main proc
    ; Reserve stack space
    sub     rsp, 40

    ; Print the message
    lea     rcx, message
    call    printf

    ; Exit the program with status 0
    xor     rcx, rcx
    call    exit
main endp

.data
message db "Hello, Computer Architect!",0

end
```

2. Open the **x64 Native Tools Command Prompt for VS 2019** and change to the directory containing your source file.

3. Build the executable with this command:

```
ml64 /Fl /Zi /Zd Ex__3_hello_x64.asm
```

4. This is the output produced by the program:

```
C:\>Ex__3_hello_x64.exe
Hello, Computer Architect!
```

This is the listing file created by the build procedure:

```
Microsoft (R) Macro Assembler (x64) Version 14.31.31104.0    02/21/22 07:47:41
Ex__3_hello_x64.asm                 Page 1 - 1


 00000000              .code
                       includelib libcmt.lib
                       includelib legacy_stdio_definitions.lib

                       extern printf:near
                       extern exit:near

                       public main
 00000000              main proc
                       ; Reserve stack space
 00000000  48/ 83 EC 28        sub      rsp, 40

                       ; Print the message
 00000004  48/ 8D 0D           lea      rcx, message
           00000000 R
 0000000B  E8 00000000 E       call     printf

                       ; Exit the program with status 0
 00000010  48/ 33 C9           xor      rcx, rcx
 00000013  E8 00000000 E       call     exit
 00000018              main endp
```

```
 00000000                    .data
 00000000 48 65 6C 6C 6F     message db "Hello, Computer Ar-chitect!",0
          2C 20 43 6F 6D
          70 75 74 65 72
          20 41 72 63 68
          69 74 65 63 74
          21 00


                end
```

Microsoft (R) Macro Assembler (x64) Version 14.31.31104.0   02/21/22 07:47:41
Ex__3_hello_x64.asm   Symbols 2 - 1

Procedures, parameters, and locals:

                N a m e   Type   Value   Attr

main . . . . P   00000000 _TEXT Length= 00000018 Public

Symbols:

                N a m e   Type   Value   Attr

exit L   00000000 _TEXT External
message . . . . Byte   00000000 _DATA
printf L   00000000 _TEXT External

      0 Warnings
      0 Errors

Exercise 4

Write an x64 assembly language program that computes the following expression and prints the result as a hexadecimal number: *[(129 – 66) × (445 + 136)] ÷ 3*. As part of this program, create a callable function to print 1 byte as 2 hex digits.

Answer

1. Create your assembly language source file. The Ex__4_expr_x64.asm file contains the following example solution to this exercise:

```
.code
includelib libcmt.lib
includelib legacy_stdio_definitions.lib

extern printf:near
extern exit:near

public main
main proc
    ; Reserve stack space
    sub     rsp, 40

    ; Print the leading output string
    lea     rcx, msg1
    call    printf

    ; Compute [(129 - 66) * (445 + 136)] / 3
    mov     eax, 129
    sub     eax, 66
    mov     ebx, 445
    add     ebx, 136
    mul     bx
    mov     bx, 3
    div     bx

    ; Print the most significant byte
    push    rax
    mov     bl, ah
```

```
        and     ebx, 0ffh
        call    print_byte

        ; Print the least significant byte
        pop     rbx
        and     ebx, 0ffh
        call    print_byte

        ; Print the trailing output string
        lea     rcx, msg2
        call    printf

        ; Exit the program with status 0
        xor     rcx, rcx
        call    exit
main endp

; Pass the byte to be printed in ebx
print_byte proc
        ; x64 function prologue
        sub     rsp, 40

        ; Use the C library printf function
        mov     rdx, rbx
        lea     rcx, fmt_str
        call    printf

        ; x64 function epilogue
        add     rsp, 40

        ret
print_byte endp

.data
fmt_str db "%02X", 0
msg1    db "[(129 - 66) * (445 + 136)] / 3 = ", 0
msg2    db "h", 9
```

end

2. Open the **x64 Native Tools Command Prompt for VS 2019** and change to the directory containing your source file.

3. Build the executable with this command:

```
ml64 /Fl /Zi /Zd Ex__3_hello_x64.asm
```

4. This is the output produced by the program:

```
C:\>Ex__4_expr_x64.exe
[(129 - 66) * (445 + 136)] / 3 = 2FA9h
```

This is the listing file created by the build procedure:

```
Microsoft (R) Macro Assembler (x64) Version 14.31.31104.0   02/21/22 07:49:37
Ex__4_expr_x64.asm                      Page 1 - 1



 00000000           .code
           includelib libcmt.lib
           includelib legacy_stdio_definitions.lib

           extern printf:near
           extern exit:near

           public main
 00000000       main proc
           ; Reserve stack space
 00000000  48/ 83 EC 28        sub     rsp, 40


           ; Print the leading output string
 00000004  48/ 8D 0D           lea     rcx, msg1
    00000005 R
 0000000B  E8 00000000 E       call    printf


           ; Compute [(129 - 66) * (445 + 136)] / 3
 00000010  B8 00000081         mov     eax, 129
```

```
00000015  83 E8 42              sub    eax, 66
00000018  BB 000001BD           mov    ebx, 445
0000001D  81 C3 00000088        add    ebx, 136
00000023  66| F7 E3             mul    bx
00000026  66| BB 0003           mov    bx, 3
0000002A  66| F7 F3             div    bx

                ; Print the most significant byte
0000002D  50                    push   rax
0000002E  8A DC                 mov    bl, ah
00000030  81 E3 000000FF        and    ebx, 0ffh
00000036  E8 00000020           call   print_byte

                ; Print the least significant byte
0000003B  5B                    pop    rbx
0000003C  81 E3 000000FF        and    ebx, 0ffh
00000042  E8 00000014           call   print_byte

                ; Print the trailing output string
00000047  48/ 8D 0D             lea    rcx, msg2
   00000027 R
0000004E  E8 00000000 E         call   printf

                ; Exit the program with status 0
00000053  48/ 33 C9             xor    rcx, rcx
00000056  E8 00000000 E         call   exit
0000005B          main endp

        ; Pass the byte to be printed in ebx
0000005B      print_byte proc
            ; x64 function prologue
0000005B  48/ 83 EC 28          sub    rsp, 40

            ; Use the C library printf function
0000005F  48/ 8B D3             mov    rdx, rbx
00000062  48/ 8D 0D             lea    rcx, fmt_str
   00000000 R
```

```
                 00000069  E8 00000000 E           call    printf

                            ; x64 function epilogue
                 0000006E  48/ 83 C4 28             add     rsp, 40


                 00000072  C3                       ret
                 00000073                           print_byte endp


 00000000                                        .data
 00000000 25 30 32 58 00     fmt_str db "%02X", 0
 00000005 5B 28 31 32 39     msg1    db "[(129 - 66) * (445 + 136)] / 3 = ",
0
          20 2D 20 36 36
          29 20 2A 20 28
          34 34 35 20 2B
          20 31 33 36 29
          5D 20 2F 20 33
          20 3D 20 00
 00000027 68 09              msg2    db "h", 9


                  end
Microsoft (R) Macro Assembler (x64) Version 14.31.31104.0    02/21/22 07:49:37
Ex__4_expr_x64.asm                      Symbols 2 - 1




Procedures, parameters, and locals:


                N a m e                 Type     Value    Attr

main  . . . . . . . . . . . . . . .  P     00000000 _TEXT Length= 0000005B
print_byte  . . . . . . . . . . . .  P     0000005B _TEXT Length= 00000018

Symbols:
```

```
                N a m e            Type     Value       Attr

    exit . . . . . . . . . . . . . .  L        00000000  _TEXT External
    fmt_str . . . . . . . . . . . .   Byte     00000000  _DATA
    msg1 . . . . . . . . . . . . . .  Byte     00000005  _DATA
    msg2 . . . . . . . . . . . . . .  Byte     00000027  _DATA
    printf . . . . . . . . . . . . .  L        00000000  _TEXT External

        0 Warnings
        0 Errors
```

Exercise 5

Install the free Android Studio IDE, available at https://developer.android.com/studio/. After installation is complete, open the Android Studio IDE and select **SDK Manager** under the **Tools** menu. Select the **SDK Tools** tab and check the **NDK** option, which may be called **NDK (Side by side)**. Complete the installation of the NDK (**NDK** stands for **native development kit**):

1. Locate the following files under the SDK installation directory (the default location is %LOCALAPPDATA%\Android) and add their directories to your PATH environment variable: arm-linux-androideabi-as.exe and adb.exe. Hint: the following command works for one specific version of Android Studio (your path may vary):

    ```
    set PATH=%PATH%;%LOCALAPPDATA%\Android\ Sdk\ndk\23.0.7599858\
    toolchains\llvm\prebuilt\windows-x86_64\bin;%LOCALAPPDATA%\Android\
    Sdk\platform-tools
    ```

2. Create a file named hello_arm.s with the content shown in the source listing in the *32-bit ARM assembly language* section of *Chapter 10, Modern Processor Architectures and Instruction Sets*.

3. Build the program using the commands shown in the *32-bit ARM assembly language* section of *Chapter 10, Modern Processor Architectures and Instruction Sets*.

4. Enable **Developer Options** on an Android phone or tablet. Search the internet for instructions on how to do this.

5. Connect your Android device to the computer with a USB cable.

6. Copy the program executable image to the phone using the commands shown in the *32-bit ARM assembly language* section of *Chapter 10, Modern Processor Architectures and Instruction Sets* and run the program. Verify that the output **Hello, Computer Architect!** appears on the host computer screen.

Answer

1. Create your assembly language source file. The `Ex__5_hello_arm.s` file contains the following example solution to this exercise:

    ```
    .text
    .global _start
    _start:
    // Print the message to file 1 (stdout) with syscall 4
    mov r0, #1
    ldr r1, =msg
    mov r2, #msg_len
    mov r7, #4
    svc 0
    // Exit the program with syscall 1, returning status 0
    mov r0, #0
    mov r7, #1
    svc 0
    .data
    msg:
    .ascii "Hello, Computer Architect!"
    msg_len = . - msg
    ```

2. Build the executable with these commands:

    ```
    arm-linux-androideabi-as -al=Ex__5_hello_arm.lst -o Ex__5_hello_arm.o Ex__5_hello_arm.s
    arm-linux-androideabi-ld -o Ex__5_hello_arm Ex__5_hello_arm.o
    ```

3. This is the output produced by copying the program to an Android device and running it:

    ```
    C:\>adb devices
    * daemon not running; starting now at tcp:5037
    * daemon started successfully
    List of devices attached
    9826f541374f4b4a68   device
    ```

```
C:\>adb push Ex__5_hello_arm /data/local/tmp/Ex__5_hello_arm
Ex__5_hello_arm: 1 file pushed. 0.0 MB/s (868 bytes in 0.059s)
C:\>adb shell chmod +x /data/local/tmp/Ex__5_hello_arm
C:\>adb shell /data/local/tmp/Ex__5_hello_arm
Hello, Computer Architect!
```

This is the listing file created by the build procedure:

```
ARM GAS  Ex__5_hello_arm.s          page 1


   1                      .text
   2                      .global _start
   3
   4                  _start:
   5                      // Print the message to file 1 (stdout) with syscall 4
   6 0000 0100A0E3        mov     r0, #1
   7 0004 14109FE5        ldr     r1, =msg
   8 0008 1A20A0E3        mov     r2, #msg_len
   9 000c 0470A0E3        mov     r7, #4
  10 0010 000000EF        svc     0
  11
  12                      // Exit the program with syscall 1, returning status 0
  13 0014 0000A0E3        mov     r0, #0
  14 0018 0170A0E3        mov     r7, #1
  15 001c 000000EF        svc     0
  16
  17                      .data
  18                  msg:
  19 0000 48656C6C        .ascii      "Hello, Computer Architect!"
  19      6F2C2043
  19      6F6D7075
  19      74657220
  19      41726368
  20                      msg_len = . - msg
```

Exercise 6

Write a 32-bit ARM assembly language program that computes the following expression and prints the result as a hexadecimal number: $[(129 - 66) \times (445 + 136)] \div 3$. As part of this program, create a callable function to print 1 byte as 2 hex digits.

Answer

1. Create your assembly language source file. The Ex__6_expr_arm.s file contains the following example solution to this exercise:

```
    .text
    .global _start

_start:
    // Print the leading output string
    ldr     r1, =msg1
    mov     r2, #msg1_len
    bl      print_string


    // Compute [(129 - 66) * (445 + 136)] / 3
    mov     r0, #129
    sub     r0, r0, #66
    ldr     r1, =#445
    add     r1, r1, #136
    mul     r0, r1, r0
    mov     r1, #3
    udiv    r0, r0, r1

    // Print the upper byte of the result
    push    {r0}
    lsr     r0, r0, #8
    bl      print_byte

    // Print the lower byte of the result
    pop     {r0}
    bl      print_byte

    // Print the trailng output string
```

```
        ldr     r1, =msg2
        mov     r2, #msg2_len
        bl      print_string

        // Exit the program with syscall 1, returning status 0
        mov     r0, #0
        mov     r7, #1
        svc     0

// Print a string; r1=string address, r2=string length
print_string:
        mov     r0, #1
        mov     r7, #4
        svc     0
        mov     pc, lr

// Convert the low 4 bits of r0 to an ascii character in r0
nibble2ascii:
        and     r0, #0xF
        cmp     r0, #10
        addpl   r0, r0, #('A' - 10)
        addmi   r0, r0, #'0'
        mov     pc, lr

// Print a byte in hex
print_byte:
        push    {lr}
        push    {r0}
        lsr     r0, r0, #4
        bl      nibble2ascii
        ldr     r1, =bytes
        strb    r0, [r1], #1

        pop     {r0}
        bl      nibble2ascii
        strb    r0, [r1]
```

```
        ldr     r1, =bytes
        mov     r2, #2
        bl      print_string

        pop     {lr}
        mov     pc, lr

.data
msg1:
    .ascii  "[(129 - 66) * (445 + 136)] / 3 = "
msg1_len = . - msg1

bytes:
    .ascii  "??"

msg2:
    .ascii  "h"
msg2_len = . - msg2
```

2. Build the executable with these commands:

```
arm-linux-androideabi-as -al=Ex__6_expr_arm.lst -o Ex__6_expr_arm.o
Ex__6_expr_arm.s
arm-linux-androideabi-ld -o Ex__6_expr_arm Ex__6_expr_arm.o
```

3. This is the output produced by copying the program to an Android device and running it:

```
C:\>adb devices
* daemon not running; starting now at tcp:5037
* daemon started successfully
List of devices attached
9826f541374f4b4a68       device


C:\>adb push Ex__6_expr_arm /data/local/tmp/Ex__6_expr_arm
Ex__6_expr_arm: 1 file pushed. 0.2 MB/s (1188 bytes in 0.007s)


C:\>adb shell chmod +x /data/local/tmp/Ex__6_expr_arm
C:\>adb shell /data/local/tmp/Ex__6_expr_arm
[(129 - 66) * (445 + 136)] / 3 = 2FA9h
```

This is the listing file created by the build procedure:

```
ARM GAS   Ex__6_expr_arm.s          page 1


   1                        .text
   2                        .global _start
   3
   4                    _start:
   5                        // Print the leading output string
   6 0000 A8109FE5         ldr     r1, =msg1
   7 0004 2120A0E3         mov     r2, #msg1_len
   8 0008 110000EB         bl      print_string
   9
  10                        // Compute [(129 - 66) * (445 + 136)] / 3
  11 000c 8100A0E3         mov     r0, #129
  12 0010 420040E2         sub     r0, r0, #66
  13 0014 98109FE5         ldr     r1, =#445
  14 0018 881081E2         add     r1, r1, #136
  15 001c 910000E0         mul     r0, r1, r0
  16 0020 0310A0E3         mov     r1, #3
  17 0024 10F130E7         udiv    r0, r0, r1
  18
  19                        // Print the upper byte of the result
  20 0028 04002DE5         push    {r0}
  21 002c 2004A0E1         lsr     r0, r0, #8
  22 0030 100000EB         bl      print_byte
  23
  24                        // Print the lower byte of the result
  25 0034 04009DE4         pop     {r0}
  26 0038 0E0000EB         bl      print_byte
  27
  28                        // Print the trailng output string
  29 003c 74109FE5         ldr     r1, =msg2
  30 0040 0120A0E3         mov     r2, #msg2_len
  31 0044 020000EB         bl      print_string
  32
  33                        // Exit the program with syscall 1, returning
status 0
```

```
34 0048 0000A0E3        mov     r0, #0
35 004c 0170A0E3        mov     r7, #1
36 0050 000000EF        svc     0
37
38                      // Print a string; r1=string address, r2=string length
39                      print_string:
40 0054 0100A0E3        mov     r0, #1
41 0058 0470A0E3        mov     r7, #4
42 005c 000000EF        svc     0
43 0060 0EF0A0E1        mov     pc, lr
44
45                      // Convert the low 4 bits of r0 to an ascii character in r0
46                      nibble2ascii:
47 0064 0F0000E2        and     r0, #0xF
48 0068 0A0050E3        cmp     r0, #10
49 006c 37008052        addpl   r0, r0, #('A' - 10)
50 0070 30008042        addmi   r0, r0, #'0'
51 0074 0EF0A0E1        mov     pc, lr
52
53                      // Print a byte in hex
54                      print_byte:
55 0078 04E02DE5        push    {lr}
56 007c 04002DE5        push    {r0}
57 0080 2002A0E1        lsr     r0, r0, #4
```

ARM GAS   Ex__6_expr_arm.s   page 2

```
58 0084 F6FFFFEB        bl      nibble2ascii
59 0088 2C109FE5        ldr     r1, =bytes
60 008c 0100C1E4        strb    r0, [r1], #1
61
62 0090 04009DE4        pop     {r0}
63 0094 F2FFFFEB        bl      nibble2ascii
64 0098 0000C1E5        strb    r0, [r1]
```

```
65
66 009c 18109FE5        ldr     r1, =bytes
67 00a0 0220A0E3        mov     r2, #2
68 00a4 EAFFFFEB        bl      print_string
69
70 00a8 04E09DE4        pop     {lr}
71 00ac 0EF0A0E1        mov     pc, lr
72
73                      .data
74                      msg1:
75 0000 5B283132            .ascii  "[(129 - 66) * (445 + 136)] / 3 = "
75      39202D20
75      36362920
75      2A202834
75      3435202B
76                      msg1_len = . - msg1
77
78                      bytes:
79 0021 3F3F                .ascii  "??"
80
81                      msg2:
82 0023 68                  .ascii  "h"
83                      msg2_len = . - msg2
```

Exercise 7

Locate the following files under the Android SDK installation directory (the default location is %LOCALAPPDATA%\Android) and add their directories to your PATH environment variable: aarch64-linux-android-as.exe and adb.exe. Hint: the following command works for one version of Android Studio (your path may vary):

```
set PATH=%PATH%;%LOCALAPPDATA%\Android\Sdk\ndk\23.0.7599858\toolchains\
llvm\prebuilt\windows-x86_64\bin;%LOCALAPPDATA%\Android\Sdk\platform-tools
```

1. Create a file named hello_arm64.s with the content shown in the source listing in the *64-bit ARM assembly language* section of *Chapter 10, Modern Processor Architectures and Instruction Sets*.
2. Build the program using the commands shown in the *64-bit ARM assembly language* section of *Chapter 10, Modern Processor Architectures and Instruction Sets*.

3. Enable **Developer Options** on an Android phone or tablet.
4. Connect your Android device to the computer with a USB cable.
5. Copy the program executable image to the phone using the commands shown in the *64-bit ARM assembly language* section of *Chapter 10, Modern Processor Architectures and Instruction Sets* and run the program. Verify that the output **Hello, Computer Architect!** appears on the host computer screen.

Answer

1. Create your assembly language source file. The Ex__7_hello_arm64.s file contains the following example solution to this exercise:

   ```
   .text
   .global _start

   _start:
       // Print the message to file 1 (stdout) with syscall 64
       mov     x0, #1
       ldr     x1, =msg
       mov     x2, #msg_len
       mov     x8, #64
       svc     0

       // Exit the program with syscall 93, returning status 0
       mov     x0, #0
       mov     x8, #93
       svc     0

   .data
   msg:
       .ascii      "Hello, Computer Architect!"
   msg_len = . - msg
   ```

2. Build the executable with these commands:

   ```
   arm-linux-androideabi-as -al=Ex__6_expr_arm.lst -o Ex__6_expr_arm.o
   Ex__6_expr_arm.s
   arm-linux-androideabi-ld -o Ex__6_expr_arm Ex__6_expr_arm.o
   ```

3. This is the output produced by copying the program to an Android device and running it:

```
C:\>adb devices
* daemon not running; starting now at tcp:5037
* daemon started successfully
List of devices attached
9826f541374f4b4a68       device


C:\>adb push Ex__7_hello_arm64 /data/local/tmp/Ex__7_hello_arm64
Ex__7_hello_arm64: 1 file pushed. 0.0 MB/s (1152 bytes in 0.029s)

C:\>adb shell chmod +x /data/local/tmp/Ex__7_hello_arm64
C:\>adb shell /data/local/tmp/Ex__7_hello_arm64
Hello, Computer Architect!
```

This is the listing file created by the build procedure:

```
AARCH64 GAS   Ex__7_hello_arm64.s           page 1


     1                      .text
     2                      .global _start
     3
     4              _start:
     5                      // Print the message to file 1 (stdout) with syscall 64
     6 0000 200080D2        mov     x0, #1
     7 0004 E1000058        ldr     x1, =msg
     8 0008 420380D2        mov     x2, #msg_len
     9 000c 080880D2        mov     x8, #64
    10 0010 010000D4        svc     0
    11
    12                      // Exit the program with syscall 93, returning status 0
    13 0014 000080D2        mov     x0, #0
    14 0018 A80B80D2        mov     x8, #93
    15 001c 010000D4        svc     0
    16
```

```
17                  .data
18              msg:
19 0000 48656C6C    .ascii      "Hello, Computer Architect!"
19      6F2C2043
19      6F6D7075
19      74657220
19      41726368
20              msg_len = . - msg
```

Exercise 8

Write a 64-bit ARM assembly language program that computes the following expression and prints the result as a hexadecimal number: [(129 − 66) × (445 + 136)] ÷ 3. As part of this program, create a callable function to print 1 byte as 2 hex digits.

Answer

1. Create your assembly language source file. The Ex__8_expr_arm64.s file contains the following example solution to this exercise:

```
    .text
    .global _start

_start:
    // Print the leading output string
    ldr     x1, =msg1
    mov     x2, #msg1_len
    bl      print_string

    // Compute [(129 - 66) * (445 + 136)] / 3
    mov     x0, #129
    sub     x0, x0, #66
    mov     x1, #445
    add     x1, x1, #136
    mul     x0, x1, x0
    mov     x1, #3
    udiv    x0, x0, x1

    // Print the upper byte of the result
```

```
        mov     x19, x0
        lsr     x0, x0, #8
        bl      print_byte

        // Print the lower byte of the result
        mov     x0, x19
        bl      print_byte

        // Print the trailng output string
        ldr     x1, =msg2
        mov     x2, #msg2_len
        bl      print_string

        // Exit the program with syscall 93, returning status 0
        mov     x0, #0
        mov     x8, #93
        svc     0

// Print a string; x1=string address, x2=string length
print_string:
        mov     x0, #1
        mov     x8, #64
        svc     0
        ret     x30

// Convert the low 4 bits of x0 to an ascii character in x0
nibble2ascii:
        and     x0, x0, #0xF
        cmp     x0, #10
        bmi     lt10

        add     x0, x0, #('A' - 10)
        b       done

lt10:
        add     x0, x0, #'0'
```

```
done:
    ret     x30


// Print a byte in hex
print_byte:
    mov     x21, x30
    mov     x20, x0
    lsr     x0, x0, #4
    bl      nibble2ascii
    ldr     x1, =bytes
    strb    w0, [x1], #1

    mov     x0, x20
    bl      nibble2ascii
    strb    w0, [x1]

    ldr     x1, =bytes
    mov     x2, #2
    bl      print_string

    mov     x30, x21
    ret     x30

.data
msg1:
    .ascii  "[(129 - 66) * (445 + 136)] / 3 = "
msg1_len = . - msg1

bytes:
    .ascii  "??"

msg2:
    .ascii  "h"
msg2_len = . - msg2
```

2. Build the executable with these commands:

```
aarch64-linux-android-as -al=Ex__8_expr_arm64.lst -o Ex__8_expr_
arm64.o Ex__8_expr_arm64.s
aarch64-linux-android-ld -o Ex__8_expr_arm64 Ex__8_expr_arm64.o
```

3. This is the output produced by copying the program to an Android device and running it:

```
C:\>adb devices
* daemon not running; starting now at tcp:5037
* daemon started successfully
List of devices attached
9826f541374f4b4a68      device


C:\>adb push Ex__8_expr_arm64 /data/local/tmp/Ex__8_expr_arm64
Ex__8_expr_arm64: 1 file pushed. 0.1 MB/s (1592 bytes in 0.015s)

C:\>adb shell chmod +x /data/local/tmp/Ex__8_expr_arm64
C:\>adb shell /data/local/tmp/Ex__8_expr_arm64
[(129 - 66) * (445 + 136)] / 3 = 2FA9h
```

This is the listing file created by the build procedure:

```
AARCH64 GAS   Ex__8_expr_arm64.s          page 1


   1                    .text
   2                    .global _start
   3
   4                    _start:
   5                        // Print the leading output string
   6 0000 C1050058      ldr    x1, =msg1
   7 0004 220480D2      mov    x2, #msg1_len
   8 0008 13000094      bl     print_string
   9
  10                        // Compute [(129 - 66) * (445 + 136)] / 3
  11 000c 201080D2      mov    x0, #129
```

```
12 0010 000801D1          sub      x0, x0, #66
13 0014 A13780D2          mov      x1, #445
14 0018 21200291          add      x1, x1, #136
15 001c 207C009B          mul      x0, x1, x0
16 0020 610080D2          mov      x1, #3
17 0024 0008C19A          udiv     x0, x0, x1
18
19                        // Print the upper byte of the result
20 0028 F30300AA          mov      x19, x0
21 002c 00FC48D3          lsr      x0, x0, #8
22 0030 14000094          bl       print_byte
23
24                        // Print the lower byte of the result
25 0034 E00313AA          mov      x0, x19
26 0038 12000094          bl       print_byte
27
28                        // Print the trailng output string
29 003c 21040058          ldr      x1, =msg2
30 0040 220080D2          mov      x2, #msg2_len
31 0044 04000094          bl       print_string
32
33                        // Exit the program with syscall 93, returning status 0
34 0048 000080D2          mov      x0, #0
35 004c A80B80D2          mov      x8, #93
36 0050 010000D4          svc      0
37
38                        // Print a string; x1=string address, x2=string length
39                        print_string:
40 0054 200080D2          mov      x0, #1
41 0058 080880D2          mov      x8, #64
42 005c 010000D4          svc      0
43 0060 C0035FD6          ret      x30
44
45                        // Convert the low 4 bits of x0 to an ascii character in x0
46                        nibble2ascii:
47 0064 000C4092          and      x0, x0, #0xF
```

```
48 0068 1F2800F1        cmp     x0, #10
49 006c 64000054        bmi     lt10
50
51 0070 00DC0091        add     x0, x0, #('A' - 10)
52 0074 02000014        b       done
53
54              lt10:
55 0078 00C00091        add     x0, x0, #'0'
56
57              done:
```

AARCH64 GAS   Ex__8_expr_arm64.s   page 2

```
58 007c C0035FD6        ret     x30
59
60              // Print a byte in hex
61              print_byte:
62 0080 F5031EAA        mov     x21, x30
63 0084 F40300AA        mov     x20, x0
64 0088 00FC44D3        lsr     x0, x0, #4
65 008c F6FFFF97        bl      nibble2ascii
66 0090 C1010058        ldr     x1, =bytes
67 0094 20140038        strb    w0, [x1], #1
68
69 0098 E00314AA        mov     x0, x20
70 009c F2FFFF97        bl      nibble2ascii
71 00a0 20000039        strb    w0, [x1]
72
73 00a4 21010058        ldr     x1, =bytes
74 00a8 420080D2        mov     x2, #2
75 00ac EAFFFF97        bl      print_string
76
77 00b0 FE0315AA        mov     x30, x21
78 00b4 C0035FD6        ret     x30
79
```

```
    80                      .data
    81                  msg1:
    82 0000 5B283132        .ascii    "[(129 - 66) * (445 + 136)] / 3 = "
    82      39202D20
    82      36362920
    82      2A202834
    82      3435202B
    83                  msg1_len = . - msg1
    84
    85                  bytes:
    86 0021 3F3F            .ascii    "??"
    87
    88                  msg2:
    89 0023 68             .ascii    "h"
    90                  msg2_len = . - msg2
```

Chapter 11: The RISC-V Architecture and Instruction Set

Exercise 1

Visit https://www.sifive.com/software/ and download **Freedom Studio**. Freedom Studio is an Eclipse IDE-based development suite with a complete set of tools for building an RISC-V application and running it on a hardware RISC-V processor or in the emulation environment included with Freedom Studio. Follow the instructions in the *Freedom Studio User Manual* to complete the installation. Start Freedom Studio and create a new Freedom E SDK project. In the project creation dialog, select qemu-sifive-u54 as the target (this is a single-core 64-bit RISC-V processor in the RV64GC configuration). Select the hello example program and click on the **Finish** button. This will start a build of the example program and the RISC-V emulator. After the build completes, the **Edit Configuration** dialog box will appear. Click on **Debug** to start the program in the emulator debug environment. Single-step through the program and verify that the text **Hello, World!** appears in the console window.

Answer

Install **Freedom Studio** as described. Note that the directory path for your workspace cannot include spaces. Start Freedom Studio:

1. In the **Welcome to SiFive FreedomStudio! Let's Get Started...** dialog, select **I want to create a new Freedom E SDK Project**.
2. In the **Create a Freedom E SDK Project** dialog, select **qemu-sifive-u54** as the target.
3. Select the **hello** example program.
4. Click the **Finish** button.
5. After the build completes, the **Edit Configuration** dialog box will appear.
6. Click **Debug** to start the program in the emulator debug environment.
7. Single-step through the program and verify that the text **Hello, World!** appears in the console window.

Exercise 2

With the project from *Exercise 1* still open, locate the hello.c file in the src folder in the **Project** window. Right-click on the file and rename it to hello.s. Open hello.s in the editor and delete the entire contents. Insert the assembly language program shown in the *RISC-V assembly language* section in *Chapter 11, The RISC-V Architecture and Instruction Set*. Perform a cleaning operation and then rebuild the project (press *Ctrl + 9* to initiate the cleaning operation). Select **Debug** under the **Run** menu. Once the debugger starts, open Windows to display the hello.s source file, the **Disassembly** window, and the **Registers** window. Expand the **Registers** tree to display the RISC-V processor registers. Single-step through the program and verify that the text **Hello, Computer Architect!** appears in the console window.

Answer

With the project from *Exercise 1* still open, locate the hello.c file in the src folder in the **Project** window and then do the following:

1. Right-click on the file and rename it to hello.s.
2. Open hello.s in the editor and delete the entire contents.
3. Insert the assembly language program shown in the *RISC-V assembly language* section in *Chapter 11, The RISC-V Architecture and Instruction Set*. This is the assembly code, also available in the Ex__2_riscv_assembly.s file:

   ```
   .section .text
   .global main

   main:
       # Reserve stack space and save the return address
   ```

```
        addi    sp, sp, -16
        sd      ra, 0(sp)

        # Print the message using the C library puts function
1:      auipc   a0, %pcrel_hi(msg)
        addi    a0, a0, %pcrel_lo(1b)
        jal     ra, puts

        # Restore the return address and sp, and return to caller
        ld      ra, 0(sp)
        addi    sp, sp, 16
        jalr    zero, ra, 0

        .section .rodata
msg:
        .asciz "Hello, Computer Architect!\n"
```

4. Perform a clean and then rebuild the project (press *Ctrl + 9* to initiate the clean operation).
5. Select **Debug** under the **Run** menu.
6. Once the debugger starts, open Windows to display the hello.s source file, the **Disassembly** window, and the **Registers** window.
7. Expand the Registers tree to display the RISC-V processor registers.
8. Single-step through the program and verify that the text **Hello, Computer Architect!** appears in the console window.

Exercise 3

Write an RISC-V assembly language program that computes the following expression and prints the result as a hexadecimal number:

$$[(129 - 66) \times (445 + 136)] \div 3$$

As part of this program, create a callable function to print 1 byte as 2 hex digits.

Answer

Create a new Freedom Studio project using the same steps as in *Exercise 1* in *Chapter 11, The RISC-V Architecture and Instruction Set*. Locate the hello.c file in the src folder in the **Project** window:

1. Right-click on the file and rename it to hello.s.

2. Create your assembly language source code within the hello.s file. The Ex__3_riscv_expr.s file contains the following example solution to this exercise:

```
.section .text
.global main

main:
    # Reserve stack space and save the return address
    addi    sp, sp, -16
    sd      ra, 0(sp)

    # Print the leading output string
    la      a0, msg1
    jal     ra, puts

    # Compute [(129 - 66) * (445 + 136)] / 3
    addi    a0, zero, 129
    addi    a0, a0, -66
    addi    a1, zero, 445
    add     a1, a1, 136
    mul     a0, a1, a0
    addi    a1, zero, 3
    divu    a0, a0, a1

    # Print the upper byte of the result
    sw      a0, 8(sp)
    srl     a0, a0, 8
    jal     ra, print_byte

    # Print the lower byte of the result
    lw      a0, 8(sp)
    jal     ra, print_byte

    # Print the trailng output string
    la      a0, msg2
    jal     ra, puts

    # Restore the return address and sp
```

```
        ld      ra, 0(sp)
        addi    sp, sp, 16

        # Set the exit code to zero and return to caller
        addi    a0, zero, 0
        ret

# Convert the low 4 bits of a0 to an ascii character in a0
nibble2ascii:
        # Reserve stack space and save the return address
        addi    sp, sp, -16
        sd      ra, 0(sp)

        and     a0, a0, 0xF
        sltu    t0, a0, 10
        bne     t0, zero, lt10

        add     a0, a0, ('A' - 10)
        j       done

lt10:
        add     a0, a0, '0'

done:
        ld      ra, 0(sp)
        addi    sp, sp, 16
        ret

# Print a byte in hex
print_byte:
        # Reserve stack space and save the return address
        addi    sp, sp, -16
        sd      ra, 0(sp)

        addi    t1, a0, 0
        srl     a0, t1, 4
        jal     ra, nibble2ascii
```

```
        la      t3, bytes
        sb      a0, 0(t3)

        addi    a0, t1, 0
        jal     nibble2ascii
        sb      a0, 1(t3)

        la      a0, bytes
        jal     ra, puts

        ld      ra, 0(sp)
        addi    sp, sp, 16
        ret

.section .data
msg1:
    .asciz  "[(129 - 66) * (445 + 136)] / 3 = "

bytes:
    .asciz  "??"

msg2:
    .asciz  "h"
```

3. Perform a clean and then rebuild the project (press *Ctrl* + *9* to initiate the clean operation).
4. Select **Debug** under the **Run** menu.
5. Once the debugger starts, open Windows to display the hello.s source file, the **Disassembly** window, and the **Registers** window.
6. Expand the **Registers** tree to display the RISC-V processor registers.
7. Single-step through the program and verify that the text [(129 - 66) * (445 + 136)] / 3 = 2FA9h appears in the console window.

Chapter 12: Processor Virtualization

Exercise 1

Download and install the current version of VirtualBox. Download, install, and bring up Ubuntu Linux as a VM within VirtualBox.

Connect the guest OS to the internet using a bridged network adapter. Configure and enable clipboard sharing and file sharing between the Ubuntu guest and your host operating system.

Answer

Perform the following steps:

1. Download the VirtualBox 6.1 (or later version) installer from https://www.virtualbox.org/wiki/Downloads. Be sure to select the version appropriate for your host operating system.
2. Run the VirtualBox installer and accept the default prompts.
3. Download a VirtualBox image of 64-bit Ubuntu Linux. One source for such an image is https://www.osboxes.org/ubuntu/. If the image is in a compressed format, uncompress it. Use 7-Zip (https://www.7-zip.org/) if the filename ends with .7z. After unzipping, the VirtualBox disk image filename will have the extension .vdi.
4. Start VirtualBox Manager and click the **New** icon. Give the new machine a name, such as Ubuntu, select **Linux** as the type, and select **Ubuntu (64-bit)** as the version. Click **Next**.
5. In the **Memory size** dialog, accept the default memory size (or increase it if you prefer).
6. In the **Hard disk** dialog, select **Use an existing virtual hard disk file**. Click the **Browse** button (it looks like a folder), and then click the **Add** button in the **Hard disk selector** dialog. Navigate to the .vdi file you downloaded and select **Open**. Click **Create** to finish creating the VM.
7. Click the **Settings** icon in VirtualBox. In the **General** section, on the **Advanced** tab, select **Bidirectional** for **Shared Clipboard**.
8. Click **Network**. In the **Adapter 1** tab, select **Bridged Adapter** next to **Attached to:**.
9. Create a folder on the Windows disk named share in your Documents folder. Click **Shared Folders** in the VirtualBox Manager **Settings** dialog for your Ubuntu VM. Click the icon to add a shared folder (it looks like a folder with a plus on it). Select the share folder you just created on the host computer and click **OK**.
10. Click **OK** in the **Settings** dialog to close it.
11. Click the **Start** icon to start the VM. When the Ubuntu system finishes booting, log in with the password osboxes.org.
12. After login has finished, open a terminal window by pressing *Ctrl + Alt + T*.
13. In the VM terminal, install software packages with the following commands. Respond to the prompts to complete the installation:

Appendix

```
sudo apt-get update
sudo apt-get install gcc make perl
sudo apt-get install build-essential linux-headers-'uname -r' dkms
```

14. In the **Devices** menu of the Ubuntu VM window, select **Insert Guest Additions CD Image....** Respond to the prompts to complete the installation. Reboot the VM after installation completes.

15. Log in to the VM and open a terminal window. In the VM terminal, create a directory named share with the following command:

```
mkdir share
```

16. Enter the following command in the VM terminal to mount the shared folder:

```
sudo mount -t vboxsf -o rw,uid=1000,gid=1000 share ~/share
```

17. Create a file in the shared folder on the Ubuntu system:

```
cd ~/share
touch file1.txt
```

18. On the Windows host, verify that file1.txt is now present in your Documents\share directory.

Exercise 2

Within the Ubuntu operating system you installed in *Exercise 1*, install VirtualBox and then install and bring up a virtual machine version of FreeDOS. Verify that DOS commands such as echo Hello World! and mem perform properly in the FreeDOS VM. After completing this exercise, you will have implemented an instance of nested virtualization.

Answer

1. With your Ubuntu VM not running, select the **Settings** icon in the VirtualBox manager for the VM. In the **System** section, under the **Processor** tab, check the box for **Enable Nested VT-x/AMD-V**. You must be running VirtualBox 6.1 or later for this feature to be fully supported. Click **OK** to save the change.

2. Start your Ubuntu VM. Log in to the VM, open a terminal window, and install VirtualBox in the Ubuntu VM with the following command:

```
sudo apt-get install virtualbox
```

3. Install 7-Zip in the Ubuntu VM with this command:

   ```
   sudo apt-get install p7zip-full
   ```

4. Download a VirtualBox virtual disk image for FreeDOS from https://www.osboxes.org/freedos/. Perform the following steps (assuming the downloaded file is in the ~/snap/firefox/common/Downloads directory, and the FreeDOS image filename is 64-bit.7z):

   ```
   cd
   mkdir 'VirtualBox VMs'
   cd 'VirtualBox VMs'
   mv ~/snap/firefox/common/Downloads/64bit.7z .
   7z x 64bit.7z
   ```

5. Start **VirtualBox** with the following command:

   ```
   virtualbox &
   ```

6. Create a new VM in the **VirtualBox** instance running in the Ubuntu VM. Select the following options:

   ```
   Name: FreeDOS
   Type: Other
   Version: DOS
   32MB RAM
   Use an existing virtual hard disk file
   ```

7. Select the VDI file in ~/VirtualBox VMs and complete the VM configuration.

8. Click the **Start** icon in the VirtualBox manager to start the FreeDOS VM.

9. After the VM completes booting, execute these commands in the FreeDOS prompt:

   ```
   echo Hello World!
   mem
   dir
   ```

This screenshot shows the output of the mem command:

Appendix

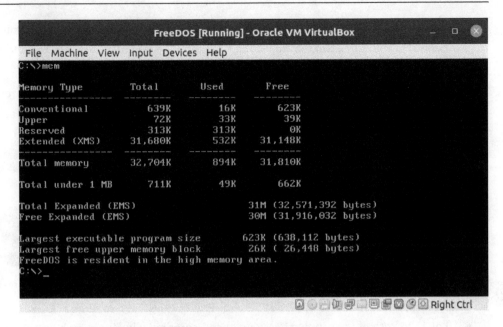

Figure 4: FreeDOS screenshot

10. When you are finished using FreeDOS, close the VM with the following command in the FreeDOS prompt:

```
shutdown
```

Exercise 3

Create two separate copies of your Ubuntu guest machine in your host system's VirtualBox environment. Configure both Ubuntu guests to connect to the VirtualBox *internal* network. Set up the two machines with compatible IP addresses. Verify that each of the machines can receive a response from the other using the `ping` command. By completing this exercise, you have configured a virtual network within your virtualized environment.

Answer

1. In your host system VirtualBox, open the **Settings** dialog for the Ubuntu VM you set up in *Exercise 1* and select the **Network** settings. Set the **Attached to:** network type to Internal and then click **OK**.

2. Right-click on the Ubuntu VM in the VirtualBox manager and select **Clone...** from the context menu. Click **Next** in the **Clone VM** menu. Leave **Full clone** selected and click **Clone**. Wait for the cloning process to complete.

3. Open Command Prompt on your host system and navigate to the installation directory for **VirtualBox**. On Windows, this command takes you to the following default installation location:

```
cd "\Program Files\Oracle\VirtualBox"
```

4. Start a DHCP server for the intnet VirtualBox network with this command:

```
VBoxManage dhcpserver add --netname intnet --ip 192.168.10.1
--netmask 255.255.255.0 --lowerip 192.168.10.100 --upperip
192.168.10.199 --enable
```

5. Start both of the VMs. Based on the DHCP server settings recommended in the previous step, the VMs might be assigned the IP addresses 192.168.10.100 and 192.168.10.101.

6. Log in to both of the running VMs and open a terminal window in each one. Enter the following command in each terminal to display the system IP address:

```
hostname -I
```

7. Ping the other machine. For example, if the current machine's IP address is 192.168.10.100, enter the following command:

```
ping 192.168.10.101
```

You should see a response similar to the following. Press *Ctrl + C* to stop the updates:

```
osboxes@osboxes:~$ ping 192.168.10.101
PING 192.168.10.101 (192.168.10.101) 56(84) bytes of data.
64 bytes from 192.168.10.101: icmp_seq=1 ttl=64 time=0.372 ms
64 bytes from 192.168.10.101: icmp_seq=2 ttl=64 time=0.268 ms
64 bytes from 192.168.10.101: icmp_seq=3 ttl=64 time=0.437 ms
64 bytes from 192.168.10.101: icmp_seq=4 ttl=64 time=0.299 ms
^C
--- 192.168.10.101 ping statistics ---
4 packets transmitted, 4 received, 0% packet loss, time 3054ms
rtt min/avg/max/mdev = 0.268/0.344/0.437/0.065 ms
osboxes@osboxes:~$
```

8. Repeat the ping command on the second machine, switching the target to the IP address of the first machine. Verify that the response is similar to the previous result.

Chapter 13: Domain-Specific Computer Architectures

Exercise 1

Draw a block diagram of the computing architecture for a system to measure and report weather data 24 hours a day at 5-minute intervals using SMS text messages. The system is battery-powered and relies on solar cells to recharge the battery during daylight hours. Assume the weather instrumentation consumes minimal average power, only requiring full power momentarily during each measurement cycle.

Answer

Based on the performance requirements, a processor capable of entering a very low power state for minutes at a time should be able to operate from a moderately sized battery for days at a time. By only powering weather sensors when necessary to take a measurement, and only powering the cellular transceiver when it is time to transmit data, power usage is minimized.

The following diagram represents one possible configuration for this system:

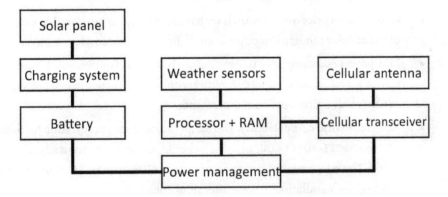

Figure 5: Initial weather data collection system diagram

Exercise 2

For the system of *Exercise 1*, identify a suitable, commercially available processor and list the reasons why that processor is a good choice for this application. Some factors to weigh are cost, processing speed, tolerance for harsh environments, power consumption, and integrated features such as RAM and communication interfaces.

Answer

Perform the following steps:

1. An internet search for `low-power microprocessor` brings up a selection of processors from manufacturers including STM, Analog Devices, Texas Instruments, Microchip Technology, and several others.

2. A second search for `embedded cellular modem` produces a list of cellular modems suitable for this application. Some of these devices are in the form of a **system-on-module (SoM)**, incorporating the RF modem with a programmable processor core in a single module.

3. The MultiTech Dragonfly Nano SoM (`https://www.multitech.com/brands/multiconnect-dragonfly-nano`) appears to be suitable for this application. This device is available for US$103.95 and integrates an ARM Cortex-M4 processor for hosting user applications. The Dragonfly Nano provides I/O interfaces, including a serial UART, USB, I2C, SPI, 9 analog inputs, and up to 29 digital I/O pins. The Cortex-M4 contains 1 MB of flash memory and 128 KB of RAM.

4. The Dragonfly Nano documentation states that when transmitting a small amount of data each day, the device can run for years on two AA-size batteries.

5. The reasons for selecting the Dragonfly Nano for this application are as follows:

    - **Cost**: While a price over $US100 is high for a microprocessor board, the integration of the cellular modem directly accomplishes a key system design goal.

    - **Low power consumption**: Depending on the power requirements for the weather sensors, a small solar panel combined with a small rechargeable battery should easily satisfy system power requirements.

    - **Environmental compatibility**: The temperature range specification for the SoM is -40° to +85°C (-40° to +185°F), which should support operation anywhere in the world. The relative humidity tolerance range (20% to 90% RH, non-condensing) will require installation in a weatherproof enclosure.

    - **Processing power**: The SoM contains an STM32L471QG 32-bit ARM processor operating at 80 MHz. This processor provides a great deal of capability, including an FPU and dynamic voltage scaling. It is possible to perform extensive preprocessing (filtering, sensor fault detection, and so on) on sensor measurements prior to the transmission of data. The flash and RAM within the device should be more than adequate for the application.

- **Certified solution**: The Dragonfly Nano is certified by the FCC and wireless carriers for use on cellular networks.
- **Development support**: Free development tools and online resources are available at https://os.mbed.com/platforms/MTS-Dragonfly-Nano/.

6. The dashed box in the following diagram indicates the portion of the system implemented by the Dragonfly Nano SoM:

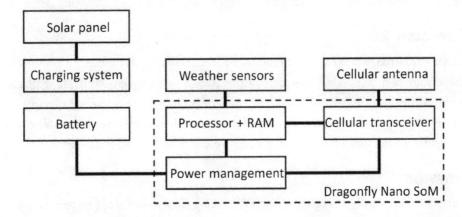

Figure 6: Final weather data collection system diagram

Chapter 14: Cybersecurity and Confidential Computing Architectures

Exercise 1

Where supported, set up two-factor authentication for all your internet-accessible accounts containing data that you care about. This includes bank accounts, email accounts, social media, code repositories (if you are a software developer), medical services, and anything else you value. Ensure at all stages that you are using only information and software applications from trusted sources.

Answer

A comprehensive list of websites and their support (or non-support) for two-factor authentication is available at **2FA Directory** (https://2fa.directory/). 2FA is an abbreviation for two-factor authentication.

The most common method for implementing two-factor authentication is for the site to send an SMS text containing a code to the phone number associated with the account after the user enters a valid username and password.

The code is often a 6-digit number that the user must provide to the website to complete the login process. The two factors used for authentication are the user's knowledge of the account password and the demonstrated access to the phone associated with the account.

Some sites support an app such as the **Duo Mobile app** (https://duo.com/product/multi-factor-authentication-mfa/duo-mobile-app) for two-factor authentication. When accessing a site that uses the app, after entering username and password information, a notification will appear on your phone. With a single tap, you can approve access and finish logging in.

Exercise 2

Create strong passwords for all your internet-accessible accounts containing information of value that cannot be protected by two-factor authentication. A strong password is long (15 characters or more) and includes uppercase, lowercase, numeric, and special characters (for example: ! " # $ % & ' ( ) * +). To keep track of these complicated passwords, install and use a reputable password-safe application. Take care when selecting a password safe and consider its source.

Answer

There are many options in terms of securely storing passwords for use on your computer and on other devices. Most web browsers offer password management, as do most antivirus software packages. Standalone password manager applications are available as well. You can start to narrow your choices by performing an internet search for *password manager*.

When a site asks you to set a password, you can have the password manager generate a long random-looking string of characters as your new password. You won't need to remember the password because it will be stored securely by the password manager.

When selecting a password management solution, you should consider the need to maintain current passwords on all your devices. When you change the password for a site, you do not want to have to update the new password in several places. A browser-based password manager such as **Firefox** (https://www.mozilla.org/en-US/) will take care of this for you as long as you have a Firefox account and you have logged in to it on each device.

Exercise 3

Update the operating system and other applications and services (such as Java) on all computers and other devices under your control. This will ensure that the security updates included in those updates start working to protect you soon after they become available. Set up a plan to continue regularly installing updates as they are released to ensure you are protected in the future.

Answer

1. Go into the update settings for each device and see whether any updates are awaiting installation. If they are, install them.
2. If no updates are waiting, have the device check for updates and install any that are available.
3. Start each application you use and rely on and have it check for updates. Install any that are available.
4. If an application has an option to automatically check for updates, ensure it is turned on. You may want to be notified when updates are available but not have them installed automatically.
5. Set up a repeating reminder in your calendar application to notify you to check for updates for all devices and applications at the shortest interval you believe is reasonable, whether it is weekly, biweekly, or monthly. Don't wait too long between updates because your systems are vulnerable during the period between the identification of a vulnerability and your installation of the update that fixes it.

Chapter 15: Blockchain and Bitcoin Mining Architectures

Exercise 1

Visit the blockchain explorer at https://bitaps.com and locate the list of last blocks on that page. Click on a block number and you will be presented with a display containing the hexadecimal listing of the block header along with its SHA-256 hash. Copy both items and write a program to determine if the hash provided is the correct hash of the header. Remember to perform SHA-256 twice to compute the header hash.

Answer

The Python file Ex__1_compute_block_hash.py contains the block header hashing code:

```
#!/usr/bin/env python

"""Ex__1_compute_block_hash.py: Answer to Ch 15 Ex 1."""

# This is a solution for Bitcoin block 711735
# See https://bitaps.com/711735
```

```python
import binascii
import hashlib

# The block header copied from bitaps.com
header = '00000020505424e0dc22a7fb1598d3a048a31957315f' + \
    '737ec0d00b0000000000000000005f7fbc00ac45edd1f6ca7' + \
    '713f2b048d8a771c95e1afd9140d3a147a063f64a76781ea4' + \
    '61139a0c17f666fc1afdbc08'

# The hash of the header copied from bitaps.com
header_hash = \
    '00000000000000000000bc01913c2e05a5d38d39a9df0c8ba' + \
    '4269abe9777f41f'

# Cut off any extra bytes beyond the 80-byte header
header = header[:160]

# Convert the header to binary
header = binascii.unhexlify(header)

# Compute the header hash (perform SHA-256 twice)
computed_hash = hashlib.sha256(header).digest()
computed_hash = hashlib.sha256(computed_hash).digest()

# Reverse the byte order
computed_hash = computed_hash[::-1]

# Convert the binary header hash to a hexadecimal string
computed_hash = \
    binascii.hexlify(computed_hash).decode("utf-8")

# Print the result
print('Header hash:   ' + header_hash)
print('Computed hash: ' + computed_hash)
```

```
    if header_hash == computed_hash:
        result = 'Hashes match!'
    else:
        result = 'Hashes DO NOT match!'

    print(result)
```

To execute the program, assuming Python is installed and is in your path, execute the command python Ex__1_compute_block_hash.py.

This is the output of a test run:

```
C:\>python Ex__1_compute_block_hash.py
Header hash:
0000000000000000000000bc01913c2e05a5d38d39a9df0c8ba4269abe9777f41f
Computed hash:
0000000000000000000000bc01913c2e05a5d38d39a9df0c8ba4269abe9777f41f
Hashes match!
```

Exercise 2

Set up a full bitcoin peer node and connect it to the bitcoin network. Download the bitcoin core software from https://bitcoin.org/en/download. It is best to have a fast internet connection and at least 200 GB of free disk space.

Answer

1. Download the bitcoin core installer from https://bitcoin.org/en/download and run it.
2. After installation completes, run the bitcoin core application. The application will begin downloading the entire bitcoin blockchain, beginning with the genesis block from 2009 up to the most recently added block. This process will take several hours or days depending on your internet bandwidth.
3. Although the bitcoin core application will consume about 200 GB of disk space during the initial validation process, it will reduce its storage requirements to a disk space limit you select, which defaults to 2 GB.
4. After the blockchain has downloaded, the node will transition into operation as a full network peer. You can display the application's connections to network peers and monitor the addition of freshly created transactions into the pool of transactions awaiting inclusion on a future block to be added to the blockchain.

5. You can also create a bitcoin wallet within the application and use it to conduct your own bitcoin transactions. If you use this application to store a significant quantity of bitcoin, be certain you are enforcing best security practices for all aspects of the host computer operating system and its applications to ensure you don't get hacked and have your coins stolen.

Chapter 16: Self-Driving Vehicle Architectures

Exercise 1

If you do not already have Python installed on your computer, visit https://www.python.org/downloads/ and install the current version. Ensure Python is in your search path by typing python -version at a system command prompt. You should receive a response similar to **Python 3.10.3**. Install TensorFlow (an open source platform for machine learning) with the command (also at the system command prompt) pip install tensorflow. You may need to use the **Run as administrator** option when opening the command prompt to get a successful installation. Install Matplotlib (a library for visualizing data) with the command pip install matplotlib.

Answer

The Windows batch file Ex__1_install_tensorflow.bat contains the commands to install TensorFlow and Matplotlib:

```
REM Ex__1_install_tensorflow.bat: Answer to Ch 16 Ex 1.

REM This batch file installs TensorFlow and Matplotlib in Windows.
REM Python must be installed (see https://www.python.org/downloads/).
REM The Python installation directory must be in the system path.

python --version

pip install tensorflow

pip install matplotlib
```

To run the batch file, assuming Python is installed and is in your path, open an *Administrator* command prompt and execute the command Ex__1_install_tensorflow.bat.

Exercise 2

Create a program using the TensorFlow library that loads the CIFAR-10 dataset and displays a subset of the images along with the label associated with each image. This dataset is a product of the **Canadian Institute for Advanced Research (CIFAR)** and contains 60,000 images, each consisting of 32x32 RGB pixels. The images have been randomly separated into a training set containing 50,000 images and a test set of 10,000 images. Each image has been labeled by humans as representing an item in one of 10 categories: airplane, automobile, bird, cat, deer, dog, frog, horse, ship, or truck. For more information on the CIFAR-10 dataset, see the technical report by Alex Krizhevsky at https://www.cs.toronto.edu/~kriz/learning-features-2009-TR.pdf.

Answer

The Python file Ex__2_load_dataset.py contains the code to load the dataset and display a subset of the images:

```python
#!/usr/bin/env python

"""Ex__2_load_dataset.py: Answer to Ch 16 Ex 2."""

from tensorflow.keras import datasets
import matplotlib.pyplot as plt

def load_dataset():
    (train_images, train_labels), \
        (test_images, test_labels) = \
        datasets.cifar10.load_data()

    # Normalize pixel values to the range 0-1
    train_images = train_images / 255.0
    test_images = test_images / 255.0

    return train_images, train_labels, \
        test_images, test_labels

def plot_samples(train_images, train_labels):
    class_names = ['Airplane', 'Automobile', 'Bird',
```

```
                        'Cat', 'Deer','Dog', 'Frog',
                        'Horse', 'Ship', 'Truck']

    plt.figure(figsize=(14,7))
    for i in range(60):
        plt.subplot(5,12,i + 1)
        plt.xticks([])
        plt.yticks([])
        plt.imshow(train_images[i])
        plt.xlabel(class_names[train_labels[i][0]])

    plt.show()

if __name__ == '__main__':
    train_images, train_labels, \
        test_images, test_labels = load_dataset()
    plot_samples(train_images, train_labels)
```

To execute the program, assuming Python is installed and is in your path, execute the command:

python Ex__2_load_dataset.py.

If you receive an error message stating cudart64_110.dll not found, you can safely ignore the message. This just means you do not have the library installed for running TensorFlow on an Nvidia CUDA GPU. The code will run (more slowly) on your system processor instead.

This is the set of sample images displayed by the code:

Figure 7: Sample CIFAR dataset images

Exercise 3

Create a program using the TensorFlow library that builds a CNN using the structure shown in *Figure 16.1*. Use a 3x3 convolution filter in each convolutional layer. Use 32 filters in the first convolutional layer and 64 filters in the other two convolutional layers. Use 64 neurons in the hidden layer. Provide 10 output neurons representing an image's presence in one of the 10 CIFAR-10 categories.

Answer

This is the CNN structure of *Figure 16.1*:

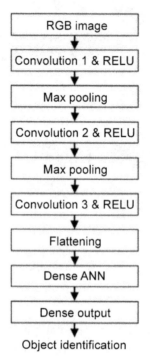

Figure 8: CNN structure for image classification

The Python file Ex__3_create_network.py contains the code to create the CNN model:

```
#!/usr/bin/env python


"""Ex__3_create_network.py: Answer to Ch 16 Ex 3."""

from tensorflow.keras import datasets, layers, models, \
    optimizers, losses

def load_dataset():
    (train_images, train_labels), \
        (test_images, test_labels) = \
        datasets.cifar10.load_data()

    # Normalize pixel values to the range 0-1
```

```python
        train_images = train_images / 255.0
        test_images = test_images / 255.0

        return train_images, train_labels, \
            test_images, test_labels

def create_model():
    # Each image is 32x32 pixels with 3 RGB color planes
    image_shape = (32, 32, 3)

    # The convolutional filter kernel size is 3x3 pixels
    conv_filter_size = (3, 3)

    # Number of convolutional filters in each layer
    filters_layer1 = 32
    filters_layer2 = 64
    filters_layer3 = 64

    # Perform max pooling over 2x2 pixel regions
    pooling_size = (2, 2)

    # Number of neurons in each of the dense layers
    hidden_neurons = 64
    output_neurons = 10

    model = models.Sequential([
        # First convolutional layer followed by max pooling
        layers.Conv2D(filters_layer1, conv_filter_size,
            activation='relu', input_shape=image_shape),
        layers.MaxPooling2D(pooling_size),

        # Second convolutional layer followed by max pooling
        layers.Conv2D(filters_layer2, conv_filter_size,
            activation='relu'),
        layers.MaxPooling2D(pooling_size),

        # Third convolutional layer followed by flattening
```

```python
        layers.Conv2D(filters_layer3, conv_filter_size,
            activation='relu'),
        layers.Flatten(),

        # Dense layer followed by the output layer
        layers.Dense(hidden_neurons, activation='relu'),
        layers.Dense(output_neurons)
    ])

    model.compile(optimizer=optimizers.Adam(),
        loss=losses.SparseCategoricalCrossentropy(
            from_logits=True), metrics=['accuracy'])

    return model

if __name__ == '__main__':
    train_images, train_labels, test_images, \
        test_labels = load_dataset()
    model = create_model()
    model.summary()
```

To execute the program, assuming Python is installed and is in your path, execute the command python Ex__3_create_network.py.

 Note: You can ignore any warning messages about not having a GPU present if your system doesn't have one. The code will execute on the system processor if a GPU is not configured for use with TensorFlow.

This is the output of a test run:

```
C:\>Ex__3_create_network.py
2021-12-12 19:26:07.938984: I tensorflow/core/platform/cpu_feature_guard.
cc:151] This TensorFlow binary is optimized with oneAPI Deep Neural
Network Library (oneDNN) to use the following CPU instructions in
performance-critical operations:  AVX AVX2
To enable them in other operations, rebuild TensorFlow with the
appropriate compiler flags.
2021-12-12 19:26:08.282366: I tensorflow/core/common_runtime/gpu/gpu_
device.cc:1525] Created device /job:localhost/replica:0/task:0/
```

```
device:GPU:0 with 3617 MB memory:    -> device: 0, name: Quadro P2200, pci
bus id: 0000:01:00.0, compute capability: 6.1
Model: "sequential"
_____
 Layer (type)                 Output Shape              Param #
=================================================================
 conv2d (Conv2D)              (None, 30, 30, 32)        896

 max_pooling2d (MaxPooling2D  (None, 15, 15, 32)        0
 )

 conv2d_1 (Conv2D)            (None, 13, 13, 64)        18496

 max_pooling2d_1 (MaxPooling  (None, 6, 6, 64)          0
 2D)

 conv2d_2 (Conv2D)            (None, 4, 4, 64)          36928

 flatten (Flatten)            (None, 1024)              0

 dense (Dense)                (None, 64)                65600

 dense_1 (Dense)              (None, 10)                650

=================================================================
Total params: 122,570
Trainable params: 122,570
Non-trainable params: 0
_____

C:\>
```

Exercise 4

Create a program using the TensorFlow library that trains the CNN developed in *Exercise 3* and test the resulting model using the CIFAR-10 test images. Determine the percentage of test images that the CNN classifies correctly.

Answer

The Python file Ex__4_train_model.py contains the code to create, train, and test the CNN model:

```
#!/usr/bin/env python


"""Ex__4_train_model.py: Answer to Ch 16 Ex 4."""

from tensorflow.keras import datasets, layers, models, optimizers, losses
import matplotlib.pyplot as plt

def load_dataset():
    (train_images, train_labels), \
        (test_images, test_labels) = \
        datasets.cifar10.load_data()

    # Normalize pixel values to the range 0-1
    train_images = train_images / 255.0
    test_images = test_images / 255.0

    return train_images, train_labels, \
           test_images, test_labels

def create_model():
    # Each image is 32x32 pixels with 3 RGB color planes
    image_shape = (32, 32, 3)

    # The convolutional filter kernel size is 3x3 pixels
    conv_filter_size = (3, 3)

    # Number of convolutional filters in each layer
    filters_layer1 = 32
    filters_layer2 = 64
    filters_layer3 = 64

    # Perform max pooling over 2x2 pixel regions
    pooling_size = (2, 2)
```

```python
    # Number of neurons in each of the dense layers
    hidden_neurons = 64
    output_neurons = 10

    model = models.Sequential([
        # First convolutional layer followed by max pooling
        layers.Conv2D(filters_layer1, conv_filter_size,
            activation='relu', input_shape=image_shape),
        layers.MaxPooling2D(pooling_size),

        # Second convolutional layer followed by max pooling
        layers.Conv2D(filters_layer2, conv_filter_size,
            activation='relu'),
        layers.MaxPooling2D(pooling_size),

        # Third convolutional layer followed by flattening
        layers.Conv2D(filters_layer3, conv_filter_size,
            activation='relu'),
        layers.Flatten(),

        # Dense layer followed by the output layer
        layers.Dense(hidden_neurons, activation='relu'),
        layers.Dense(output_neurons)
    ])

    model.compile(optimizer=optimizers.Adam(),
        loss=losses.SparseCategoricalCrossentropy(
            from_logits=True), metrics=['accuracy'])

    return model

def train_model(train_images, train_labels, \
            test_images, test_labels, model):
    history = model.fit(train_images, train_labels,
        epochs=10, validation_data=(test_images, test_labels))

    test_loss, test_acc = model.evaluate(test_images,
```

```
        test_labels, verbose=2)

    return history, test_acc

def plot_model_accuracy(history):
    plt.figure()
    plt.plot(history.history['accuracy'], label='Accuracy')
    plt.plot(history.history['val_accuracy'],
        label = 'Validation Accuracy')
    plt.xlabel('Epoch')
    plt.ylabel('Accuracy')
    plt.ylim([0.5, 1])
    plt.legend(loc='upper left')
    plt.grid()
    plt.show()

if __name__ == '__main__':
    train_images, train_labels, test_images, \
        test_labels = load_dataset()
    model = create_model()
    history, test_acc = train_model(train_images, \
        train_labels, test_images, test_labels, model)
    print()
    print('='*31)
    print('| Validation accuracy: {:.2f}% |'.
        format(100*test_acc))
    print('='*31)
    plot_model_accuracy(history)
```

To execute the program, assuming Python is installed and is in your path, execute the command python Ex__4_train_model.py.

Note: You can ignore any warning messages about not having a GPU present if your system doesn't have one. The code will execute on the system processor if a GPU is not configured for use with TensorFlow.

Your results should indicate an accuracy of approximately 70%. For such a simple CNN, this is a tremendous improvement over the accuracy of random guessing, which would be 10%.

This is the output of a test run:

```
2021-12-12 17:55:19.402677: I tensorflow/core/platform/cpu_feature_guard.
cc:151] This TensorFlow binary is optimized with oneAPI Deep Neural
Network Library (oneDNN) to use the following CPU instructions in
performance-critical operations:   AVX AVX2
To enable them in other operations, rebuild TensorFlow with the
appropriate compiler flags.
2021-12-12 17:55:19.802026: I tensorflow/core/common_runtime/gpu/
gpu_device.cc:1525] Created device /job:localhost/replica:0/task:0/
device:GPU:0 with 3617 MB memory:  -> device: 0, name: Quadro P2200, pci
bus id: 0000:01:00.0, compute capability: 6.1
Epoch 1/10
2021-12-12 17:55:21.475358: I tensorflow/stream_executor/cuda/cuda_dnn.
cc:366] Loaded cuDNN version 8301
1563/1563 [==============================] - 9s 5ms/step - loss: 1.5032 -
accuracy: 0.4521 - val_loss: 1.2326 - val_accuracy: 0.5559
Epoch 2/10
1563/1563 [==============================] - 7s 5ms/step - loss: 1.1306 -
accuracy: 0.5996 - val_loss: 1.0361 - val_accuracy: 0.6318
Epoch 3/10
1563/1563 [==============================] - 8s 5ms/step - loss: 0.9704 -
accuracy: 0.6589 - val_loss: 1.0053 - val_accuracy: 0.6517
Epoch 4/10
1563/1563 [==============================] - 7s 5ms/step - loss: 0.8831 -
accuracy: 0.6904 - val_loss: 0.8999 - val_accuracy: 0.6883
Epoch 5/10
1563/1563 [==============================] - 7s 5ms/step - loss: 0.8036 -
accuracy: 0.7177 - val_loss: 0.8924 - val_accuracy: 0.6956
Epoch 6/10
1563/1563 [==============================] - 7s 5ms/step - loss: 0.7514 -
accuracy: 0.7374 - val_loss: 0.9180 - val_accuracy: 0.6903
Epoch 7/10
1563/1563 [==============================] - 7s 5ms/step - loss: 0.7020 -
accuracy: 0.7548 - val_loss: 0.8755 - val_accuracy: 0.7074
Epoch 8/10
```

```
1563/1563 [==============================] - 7s 5ms/step - loss: 0.6599 -
accuracy: 0.7694 - val_loss: 0.8505 - val_accuracy: 0.7116
Epoch 9/10
1563/1563 [==============================] - 8s 5ms/step - loss: 0.6180 -
accuracy: 0.7842 - val_loss: 0.8850 - val_accuracy: 0.7058
Epoch 10/10
1563/1563 [==============================] - 8s 5ms/step - loss: 0.5825 -
accuracy: 0.7943 - val_loss: 0.8740 - val_accuracy: 0.7128
313/313 - 1s - loss: 0.8740 - accuracy: 0.7128 - 648ms/epoch - 2ms/step


================================
| Validation accuracy: 71.28% |
================================
```

This figure displays the classification accuracy of the CNN on the training images (**Accuracy**) and on the test images (**Validation Accuracy**) after each of the 10 training epochs:

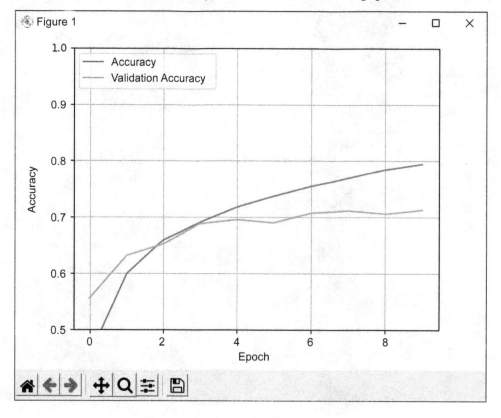

Figure 10: CNN image classification accuracy

Chapter 17: Future Directions in Computer Architectures

Exercise 1

Install the Qiskit quantum processor software development framework by following the instructions at https://qiskit.org/documentation/getting_started.html. The instructions suggest installation of the Anaconda (https://www.anaconda.com/) data science and machine learning toolset. After installing Anaconda, create a Conda virtual environment named qiskitenv to contain your work on quantum code and install Qiskit in this environment with the command pip install qiskit. Be sure to install the optional visualization dependencies with the command pip install qiskit-terra[visualization].

Answer

1. Download the Anaconda installer from https://www.anaconda.com/distribution/. Select the current version and the appropriate 32-bit or 64-bit variant for your host computer.

2. Run the Anaconda installer and accept the defaults. Close the installer after it completes.

3. Start Anaconda from the Windows search box by typing anaconda and clicking on **Anaconda Prompt** when it appears in the search list. A console window will appear.

4. In the Anaconda prompt, create and activate a virtual environment named qiskitenv with the following commands. Install any recommended packages:

    ```
    conda create -n qiskitenv python=3.8
    conda activate qiskitenv
    ```

5. Install Qiskit and the visualization dependencies with the following commands:

    ```
    pip install qiskit
    pip install qiskit-terra[visualization]
    ```

6. This completes the installation.

Exercise 2

Create a free IBM Quantum account at https://quantum-computing.ibm.com/. Locate your IBM Quantum Services API token at https://quantum-computing.ibm.com/account and install it in your local environment using the instructions at https://qiskit.org/documentation/getting_started.html.

Answer

1. Visit https://quantum-computing.ibm.com/. If you don't already have an account, click the **Create an IBMid account** link to get started.
2. Once you are logged in, locate **API token** on the screen. Click the button to copy your API token to the clipboard.
3. Return to the Anaconda prompt for the `qiskitenv` environment you created in *Exercise 1*.
4. Enter the following commands at the Anaconda prompt to set up your API token. You will need to replace `MY_TOKEN` with the token you copied to the clipboard in *step 2*:

```
python
import qiskit
from qiskit import IBMQ
IBMQ.save_account('MY_TOKEN')
```

Exercise 3

Work through the example quantum program at https://qiskit.org/documentation/tutorials/circuits/1_getting_started_with_qiskit.html. This example creates a quantum circuit containing 3 qubits that implements a **Greenberger–Horne–Zeilinger (GHZ)** state. The GHZ state exhibits key properties of quantum entanglement. Execute the code in a simulation environment on your computer.

Answer

1. Start an Anaconda prompt console. Type anaconda in the Windows search box and click on **Anaconda prompt** when it appears in the search list. A console window will appear.
2. Enter the `qiskitenv` environment with this command:

```
conda activate qiskitenv
```

3. Enter the following commands at the Anaconda prompt:

```
python
import numpy as np
from qiskit import *
```

4. Create a quantum circuit containing a 3-qubit GHZ state and add measurements for each qubit:

```
circ = QuantumCircuit(3)
```

```python
# Add an H gate to qubit 0, creating superposition
circ.h(0)
# Add a CX (CNOT) gate. Qubit 0 is control and qubit 1 is target
circ.cx(0,1)
# Add a CX (CNOT) gate. Qubit 0 is control and qubit 2 is target
circ.cx(0,2)

# Add a measurement to each of the qubits
meas = QuantumCircuit(3, 3)
meas.barrier(range(3))
meas.measure(range(3),range(3))

# Combine the two circuits
circ.add_register(meas.cregs[0])
qc = circ.compose(meas)
```

5. Display the circuit on screen:

```
qc.draw()
```

The output of this command should appear as follows:

```
>>> qc.draw()
          ┌───┐               ┌─┐
q_0: ─────┤ H ├──■─────■──────┤M├──────
          └───┘┌─┴─┐   │      └╥┘┌─┐
q_1: ──────────┤ X ├───┼───────╫─┤M├───
               └───┘ ┌─┴─┐     ║ └╥┘┌─┐
q_2: ────────────────┤ X ├─────╫──╫─┤M├
                     └───┘     ║  ║ └╥┘
c: 3/══════════════════════════╩══╩══╩═
                                0  1  2
>>>
```

6. Run the circuit on your computer using the `qasm_simulator` simulator. The `shots` parameter provides a count of the number of times the circuit will be executed to collect statistical results:

```
backend_sim = Aer.get_backend('qasm_simulator')
job_sim = backend_sim.run(transpile(qc, backend_sim), shots=1024)
```

7. Retrieve and display the count of the number of times each bit pattern resulted from a simulation run:

```
result_sim = job_sim.result()
counts_sim = result_sim.get_counts(qc)
counts_sim
```

8. You should see results similar (but not identical) to these:

```
>>> counts_sim
{'111': 506, '000': 518}
>>>
```

Exercise 4

Execute the code from *Exercise 3* on an IBM quantum computer.

Answer

1. Repeat *steps 1-5* from *Exercise 3* to create the quantum circuit.
2. Import your IBMQ account information and list the available quantum computing providers:

```
from qiskit import IBMQ
IBMQ.load_account()
provider = IBMQ.get_provider(group='open')
provider.backends()
```

3. If you visit the IBM Quantum home page at https://quantum-computing.ibm.com/, you will be able to see the length of the job queues for the available quantum computers. Select a system with sufficient qubits for your circuit and a short job queue. This example assumes that you choose the ibmq_bogota computer.

4. Add your job to the queue and monitor its status with these commands. The shots parameter provides a count of the number of times the circuit will be executed to collect statistical results:

```
backend = provider.get_backend('ibmq_bogota')
from qiskit.tools.monitor import job_monitor
job_exp = execute(qc, backend=backend, shots=1024)
job_monitor(job_exp)
```

Appendix

After the run completes, you will see the following output line:

```
Job Status: job has successfully run
```

5. After the job completes, retrieve the results with this command:

```
result_exp = job_exp.result()
```

6. Retrieve and display the count of the number of times each bit pattern resulted from a quantum computer run:

```
counts_exp = result_exp.get_counts(qc)
counts_exp
```

Approximately 50% of the time, the output bit string for this circuit should be 000, and the other 50% of the time it should be 111. However, these systems are **noisy, intermediate-scale quantum (NISQ)** computers. You should see results similar (but not identical) to these:

```
>>> counts_exp
{'000': 467, '001': 15, '010': 23, '011': 17, '100': 21, '101': 127, '110': 16, '111': 338}
>>>
```

Join our community Discord space

Join the book's Discord workspace for a monthly *Ask me Anything* session with the author:
https://discord.gg/7h8aNRhRuY

packt.com

Subscribe to our online digital library for full access to over 7,000 books and videos, as well as industry leading tools to help you plan your personal development and advance your career. For more information, please visit our website.

Why subscribe?

- Spend less time learning and more time coding with practical eBooks and Videos from over 4,000 industry professionals
- Improve your learning with Skill Plans built especially for you
- Get a free eBook or video every month
- Fully searchable for easy access to vital information
- Copy and paste, print, and bookmark content

At www.packt.com, you can also read a collection of free technical articles, sign up for a range of free newsletters, and receive exclusive discounts and offers on Packt books and eBooks.

Other Books You May Enjoy

If you enjoyed this book, you may be interested in these other books by Packt:

The Art of Writing Efficient Programs

Fedor G. Pikus

ISBN: 9781800208117

- Discover how to use the hardware computing resources in your programs effectively
- Understand the relationship between memory order and memory barriers
- Familiarize yourself with the performance implications of different data structures and organizations
- Assess the performance impact of concurrent memory accessed and how to minimize it
- Discover when to use and when not to use lock-free programming techniques
- Explore different ways to improve the effectiveness of compiler optimizations
- Design APIs for concurrent data structures and high-performance data structures to avoid inefficiencies

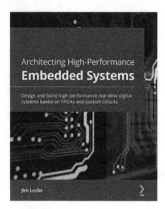

Architecting High-Performance Embedded Systems

Jim Ledin

ISBN: 9781789955965

- Understand the fundamentals of real-time embedded systems and sensors
- Discover the capabilities of FPGAs and how to use FPGA development tools
- Learn the principles of digital circuit design and PCB layout with KiCad
- Construct high-speed circuit board prototypes at low cost
- Design and develop high-performance algorithms for FPGAs
- Develop robust, reliable, and efficient firmware in C
- Thoroughly test and debug embedded device hardware and firmware

Packt is searching for authors like you

If you're interested in becoming an author for Packt, please visit `authors.packtpub.com` and apply today. We have worked with thousands of developers and tech professionals, just like you, to help them share their insight with the global tech community. You can make a general application, apply for a specific hot topic that we are recruiting an author for, or submit your own idea.

Share your thoughts

Now you've finished *Modern Computer Architecture and Organization, Second Edition*, we'd love to hear your thoughts! Scan the QR code below to go straight to the Amazon review page for this book and share your feedback or leave a review on the site that you purchased it from.

https://packt.link/r/1803234512

Your review is important to us and the tech community and will help us make sure we're delivering excellent quality content.

Index

Symbols

32-bit ARM architecture 273-275
  32-bit ARM assembly language 282-284
  ARM addressing modes 277
  ARM instruction categories 279
  ARM register set 275, 276
  data types 274
  distinct execution privilege levels 274

64-bit ARM architecture 284
  64-bit ARM assembly language 286, 287
  instruction set 285
  limitations 285

64-bit RISC-V 310, 311

256-bit Secure Hashing Algorithm (SHA-256) 130

6502 addressing modes 66
  absolute addressing mode 67, 68
  absolute indexed addressing mode 68-70
  immediate addressing mode 66
  indirect indexed addressing mode 70, 71

6502 architecture
  instruction categories, implementing 71

6502 architecture, instruction categories
  arithmetic instructions 73
  branching instructions 74
  interrupt-related instructions 75
  logical instructions 74
  memory load instructions 72
  memory store instructions 72
  NOP instruction 76
  processor flag instructions 75
  register-to-register data transfer instructions 72
  stack instructions 72, 73
  subroutine call and return instructions 75

6502 flags 20, 21

6502 instruction set 22-24

6502 microprocessor 18-22

6502 processor 54, 55
  Arithmetic Logic Unit (ALU) 54, 58-63
  control unit 54-57
  functional units 54
  interrupt handling 76
  registers 64, 65
  register set 54

6502 processor, interrupt input
  BRK instruction processing 79, 80
  IRQ processing 76-78
  NMI processing 78

8087 floating-point coprocessor 237, 238

80386 processor
  features 250, 251

A

A64 284
AArch64 284
abort exception 325
absolute addressing mode 67, 68
absolute indexed addressing mode 68-70
accelerometer 428
access violation exception 181

Index

accumulator addressing mode 74
activation function 165
Ada programming language 47
Adaptive Cruise Control (ACC) 423
adders 41
addressing mode 22
address space layout randomization (ASLR) 176
adiabatic quantum computation 460, 461
Advanced Micro Devices (AMD) 250
Advanced RISC Machine (ARM) 126
Advanced Schottky Transistor-Transistor Logic (AS) TTL 44
Advanced Vector Extensions (AVX) 222
air-gapped computing 379
aliasing 204
Alienware Aurora Ryzen Edition R10 gaming desktop 347
  Nvidia GeForce RTX 3090 GPU 348, 349
  Ryzen 9 5950X branch prediction 348
  subsystems 350, 351
altcoins 417
AMD64 268
AMD virtualization (AMD-V) 327
amperes (amps) 28
analog 152
analog-to-digital converter (ADC) 152-155
AND gate 33
Android Debug Bridge (adb) tool 284
Android Studio integrated development environment (IDE) 282
Android Studio Native Development Kit (NDK) 283
Antminer S19 Pro 414

AntPool
  reference link 413
Apple Neural Engine 344
application binary interface (ABI) 296
application software
  security, ensuring 390
application virtualization 322
architectural level
  security, designing 386
architecture-independent device drivers 126
arithmetic instructions 73
Arithmetic Logic Unit (ALU) 54, 58-63
ARM addressing modes 277
  double register indirect 279
  double register indirect, with scaling 279
  immediate 277
  register direct 278
  register indirect 278
  register indirect, with offset 278
  register indirect with offset, post-incremented 278
  register indirect with offset, pre-incremented 278
ARM architectures 273
ARM instruction categories
  arithmetic and logic 280
  breakpoint 281
  comparisons 280
  conditional execution 281, 282
  control flow 281
  load/store 279
  other instruction categories 282
  register movement 280
  stack manipulation 279
  supervisor mode 281

Index

ARM processor virtualization 334
ARMv8-A 275
Ars Technica
  URL 465
artificial intelligence (AI) 165
artificial neural networks (ANNs) 165, 435
ASICs
  used, for bitcoin mining 413-416
assembly language 22
associative memory 189
atomic 149
atomic memory operation (AMO) 307
automated computing devices
  Charles Babbage's Analytical Engine 2, 3
  Electronic Numerical Integrator and Computer (ENIAC) 4
  evolution 2
  IBM PC 5
  iPhone 9, 10
autonomous driving stages
  edge cases, handling 425, 426
  obstacle avoidance 425
  road following 425
  traffic rules, obeying 425
autonomous vehicle
  components and processes 444
  computing architecture 443, 444
  state and surroundings, sensing 427
Avalanche 344

B

backing store 196
bandpass filter 159
bandstop filter 159
bare metal hypervisor 321
Base Pointer (BP) 6
B extension 309
bfloat16 format 365
bi-endianness 67
big data 164
binary coded decimal (BCD) 21, 238
binary number 14-18
binary semaphore 150
binary translation 327
BIOS 124, 125
BIOS boot 128
bipolar 89
bipolar junction transistor (BJT) 89
bit 15
bitcoin 396-399
  full node 403
  genesis block 399
  miners 397
  transaction fees 407
bitcoin block header fields
  bitcoin version number 410
  bits 410
  Merkle root 410
  nonce 410
  previous block hash 410
  time 410
Bitcoin cash 418
bitcoin core software 403
  download link 403
bitcoin mining
  economics 416
  with ASICs 413-416
  with CPU 408
  with FPGAs 411, 412
  with GPU 408, 409

bitcoin mining computer
    architectures 409, 410
bitcoin mining pool 405-408
bitcoin mining process 404, 405
bitcoin mining systems 397
bitcoin network
  block, updating with new transactions 404
  data, modifying in transaction header 405
  timestamp, altering within block 404
bitline 94
Bitmain 413
  reference link 413
Blizzard 344
blockchain 396-399
boot configuration data (BCD) 129
boot loader 128
boot priority order
  versus multi-booting 128
boot process 127, 128
  BIOS boot 128
  embedded devices 131
  trusted boot 130
  UEFI boot 128, 129
bot-herder 372
botnet 372
branching instructions 74, 213
branch prediction 219
Break flag 21
BRK instruction processing 79, 80
brute force algorithm 401
brute force password attack 373
BTC.com
  reference link 413
buffer overflow 175, 390, 391

business desktop computer 168
bypass 216
byte 15

C

cache coherence problem 208
cache hit 196
cache lines 201
cache memory 196-198
  examples 196, 197
cache miss 196
capacitance 93
capacitor 92, 93
carbon nanotube field-effect transistor
    (CNTFET) 462, 463
  advantages 463
  barriers 463
Carry flag 21
Central Processing Unit (CPU) 54
  used, for bitcoin mining 408
Charles Babbage's Analytical Engine 2, 3
chiplets 454
chipset 100
Chisel 315
classical physics 456
cleared
  versus set 17
client node 403
clock multiplier 45
clock signal 44, 45
closed system 10
cloud application, WSC
  deploying 359-362
cloud compute server 167

Index

cloud computing
   versus virtualization 338
   virtualization, using 338
code segment (CS) 332
combinational logic 35
command-line interface 126
command shell 391
Common Weakness Enumeration (CWE)
   reference link 393
comparator 153
compiler 22
complementary MOS (CMOS) 91
complex instruction set
      computer (CISC) 64, 215
computational instructions 298
   functions 298, 299
computer architect
   considerations 342, 343
computer architecture 14
   6502 instruction set 22-24
   6502 microprocessor 18-22
   binary number 14-18
   evolution 450, 451
   hexadecimal number 14-18
   numbers, representing with
      voltage levels 14
conditional branching 218, 219
confidential computing 383-385
continuous learning 464, 465
control and status registers (CSRs) 300
control flow instructions 299
control unit 54-57
   instruction, executing 57, 58
convolution 157

Convolutional Neural
      Network (CNN) 431, 432
   example implementation 433-435
   using, in autonomous
      driving applications 435-440
cooperative multithreading 137
Coordinated Universal Time (UTC) 410
counter type ADC 154
counting semaphore 150
covert channel attacks 379
critical path 43
critical section 151
cross-site scripting attack 391
cryptocurrency 417-419
cryptographic hash function 130
current-generation motor vehicle operations
   accelerator 422
   brake 422
   gear selection 422
   steering 422
current program status register (CPSR) 276
cyberattack techniques
   botnet 372
   password attacks 372, 373
   phishing 372
   vulnerability exploitation 373
cybersecurity threat categories
   criminal groups 371
   hacktivists 371
   industrial spies 371
   insiders 372
   national governments 370
   terrorist organizations 371
   white-hat hacking 371
cybersecurity threats 370
   post-exploitation actions 376, 377

D

data at rest 383
data block
  SHA-256 hash, computing 402
data caches 197
data compression 161
data execution prevention (DEP) 176
dataflow processing 218
data in transit 383
data in use 383
data parallelism 163
data processors
  GPUs as 164
DDR5 SDRAM 96-98
  reference link 97
deadlock 150
Decimal mode flag 21
deep learning 165-167
Denial-of-Service (DoS) attack 372, 375
depletion mode 89
DeskPro 9
Destination Index (DI) 6
device driver 118, 119
  parallel port 119, 120
  PCIe device drivers 121
  structure 122, 123
device memory protection 245
device specialization 454
digital filtering 158, 159
digital hold input signal 153
digital signal processor (DSP) 152, 177
digital-to-analog converters (DACs) 152-155
Digital Visual Interface (DVI) 108
digital wallets 396
direct addressing mode 67
direct-mapped cache 201-205
Direct Memory Access (DMA) 83, 84, 122
dirty bit 208
discrete cosine transform (DCT) 161
discrete Fourier transform (DFT) 159
disk drive caches 196
DisplayPort 109
Dogecoin 418
doping 30
dot product 155
double data rate (DDR) 96
drain terminal 89
DRAM bit cell 94, 95
DRAM circuits
  capacitor 93
  constructing, with MOSFET 92
  DDR5 SDRAM 96-98
  DRAM bit cell 94-96
  Graphics DDR (GDDR) 99
  prefetching 99
driving task fallback 423
DSP hardware features 155-157
dual inline memory module (DIMM) 97
DVI variants
  DVI-A 108
  DVI-D 108
  DVI-I 108
dynamic RAM (DRAM) 92, 196
dynamic translation 328
dynamic voltage and frequency scaling (DVFS) 241, 242, 301

E

eager execution 219
easy 6502
  reference link 24
edge detection 436
edge-sensitive device 38
electrical circuits 28, 29
electrically erasable programmable read-only memory (EEPROM) 178
electrical power consumption 339
electromagnetic emissions 380
Electron Device Engineering Council (JEDEC) 97
Electronic Numerical Integrator and Computer (ENIAC) 4
embedded devices 131
enclaves 384
encryption of data at rest 381
encryption of data in transit 381
Engadget
  URL 465
enhancement mode 89
error correcting code (ECC) 97
Ether 418
Ethereum 418
Ethernet 110, 111
exahash 406
exception
  abort 325
  fault 325
  handling 226-229
  in programming languages 229
  trap 325
exponent 235

extrapolation 426

F

fair scheduling 138
farad 93
Fast Fourier transform (FFT) 159-162
  applications in signal processing 160, 161
fault exception 325
Federal Information Processing Standards (FIPS) 400
feedforward network 166
field-programmable gate arrays (FPGAs) 47
  RISC-V, implementing 313-317
  used, for bitcoin mining 411, 412
filter bank 161
Fin Field-Effect Transistor (FinFET) 445
finite state machines 39
first come, first served (FCFS) 137
fixed-priority preemptive scheduling 137
flag 20
flashing the BIOS 124
flattening 438
Flexible Intermediate Representation for RTL (FIRRTL) 315
flip-flops 38, 39
floating-point arithmetic 234-237
floating-point instructions 213
floating-point unit (FPU) 213, 237
flushing the pipeline 219
frames 111
frequency bin 160
frequency selective filter 158
full adder 42
full-duplex 101

full virtualization 324
fully associative cache 206, 207
future-tolerant skill set
  building 464
  college education 466, 467
  conferences and literature 467, 468
  continuous learning 464, 465

G

gate array 46
gated D latch 36
gate terminal 89
general-purpose I/O (GPIO) 314
generation low-power double data rate RAM (LP-DDR4x) 344
Gizmodo
  URL 465
globally unique identifier (GUID) 127
global minimum 460
Global Positioning System (GPS) 428
Global System for Mobile communications (GSM) 10
Google Compute Engine
  reference link 322
GRand Unified Bootloader (GRUB) 126
graphic equalizer 161
Graphics DDR (GDDR) 99
graphics display 106, 107
  Digital Visual Interface (DVI) 108
  DisplayPort 109
  High-Definition Media Interface (HDMI) 109
  Video Graphics Array (VGA) 108
graphics processing unit (GPU) 13, 107
  as data processors 164
  processing 162, 163
  used, for bitcoin mining 409
guests 321
GUID partition table (GPT) 127

H

half adder 42
half-duplex 101
half-precision floating-point 252
hard fault 185
Hardware 3.0 (HW3) 444
hardware debugger 244
hardware description languages 46
hardware emulation 328
hardware, warehouse-scale computer (WSC)
  fault management 358
  implementing 352, 353
Harvard architecture 156, 176, 177
hash algorithm 247
hash collision 401
hashrate 406
headless mode 354
hexadecimal number 14-18
hidden layers 166
high-bandwidth memory (HBM) 366
High-Definition Media Interface (HDMI) 109
high-end smartphone 168
highpass filter 158
High-Performance Computing (HPC) 310
high-performance gaming computer 168
High-Speed Input Output (HSIO) lanes 102
holes 89
host 321
hosted hypervisor 321

Hot plugging 121
hyperparameter 439
hypervisor 321
  type-1 hypervisor 321
  type-2 hypervisor 321
  versus virtual machine monitor 321
hypervisor-extended supervisor mode (HS) 335

I

IBM PC 5
  Intel 8088 microprocessor 6, 7
  Intel 80286 microprocessor 8, 9
  Intel 80386 microprocessor 8, 9
IBM Quantum
  reference link 462
IEEE 754 floating-point standard 239
imaginary number 160
immediate addressing mode 22, 66
implied addressing mode 72
impulse response 158
indirect indexed addressing mode 70, 71
inertial sensors 428
infrastructure attacks 375
input layer 166
input/output operations
  Direct Memory Access (DMA) 83, 84
  fundamentals 80, 81
  interrupt-driven I/O 82, 83
  programmed I/O 82
Institute of Electrical and Electronic Engineers (IEEE) 110, 239
instruction-level parallelism (ILP) 214
instruction pipelining 211-214
Instruction Pointer (IP) 6
instruction set architecture (ISA) 294
instructions per clock (IPC) 212, 347
integer instructions 213
integrated circuit
  design approach 453
integrated graphics 107
Intel 64 268
Intel Nervana neural network processor 363
Intel virtualization technology (VT-x) 327
internal encryption engine with key storage 244
internet of things (IoT) 310
interpolation 426
interrupt descriptor table register (IDTR) 332
Interrupt disable flag 21
interrupt-driven I/O 82, 83
interrupt-related instructions 75
Interrupt Request (IRQ) 73, 82
interrupts
  handling 226-229
interrupt service routine (ISR) 227
interrupt vector table (IVT) 227
inverse Fourier transform 159
I/O subsystem 100
  M.2 105
  parallel data bus 100-102
  PCI Express (PCIe) 103, 104
  Serial AT Attachment (SATA) 104
  serial data bus 100-102
  Thunderbolt 106
  Universal Serial Bus (USB) 105, 106
iPhone 9, 10

iPhone 13 Pro Max 343
  components 345
  subsystems 345-347
IRQ processing 76-78

J

J extension 309
junk conferences 467

K

Kalman filter 429
kernel 118, 231
kernel-based virtual machine (KVM) 337
kernel-mode drivers 118
keyboard 112
key logger 375
key page frame states
  active page 188
  bad page 188
  free page 188
  modified page 188
  standby page 188
  zeroed page 188

L

L1 cache 200, 201
L1 D-cache 201
L1 I-cache 201
Lane-Keeping Assistance (LKA) 423
Last-In, First-Out (LIFO) data structure 21
latch 35-37
least-recently used (LRU) 206
level 2 processor caches 209, 210
level 3 processor caches 209, 210
level-sensitive device 37
L extension 309
lidar-based driving systems
  advantages 440, 441
  disadvantages 441
light detection and ranging (lidar) 430
  localization 440, 441
  sensors 440
Litecoin 418
little-endian processor 67
load/store architecture 274
load value injection (LVI) 385
local area network (LAN) 110
locality of reference 197
local minimum 460
lockings 4
logical instructions 74
logic gates 30-35
long mode 268
lookup tables (LUTs) 313
lossless compression 161
lossy compression 161
lowpass filter 158
Low-Power DDR4 (LPDDR4) 445
Lunar Atmosphere Dust and Environment Explorer (LADEE) 459

M

M.2 105
malware (malicious software) 373
  denial of service (DoS) attack 375
  infrastructure attacks 375
  key logger 375
  man in the middle (MITM) attack 375

ransomware 374
spyware 374
SQL injection 375
viruses 374
worms 374
man in the middle (MITM) attack 375
mantissa 235
maskable interrupt 77
master boot record (MBR) 127
max pooling 437
memory access instructions 299
memory load and store instructions 72
memory management unit (MMU) 189, 190
  functions 190, 191
memory pools 187, 188
memory protection extensions (MPX) 263
memory protection unit (MPU) 245
memory subsystem 88, 89
message 402
message queue 151
metal-oxide-semiconductor field-effect transistor (MOSFET) 89-92
  used, for constructing DRAM circuits 92
mezzanine card 363
micro-operations 217, 218
mill 4
modern computer motherboard
  specification 114-116
modern processors
  virtualizing 331
modified Harvard architecture 156, 177
monitor 79
Moore's law 11-14, 46, 452
mouse 112-114

multi-booting
  versus boot priority order 128
multi-core processors 141
multilevel feedback queue 138
multilevel processor caches 198, 199
multipath 111
multiple-input-multiple-output (MIMO) 111
multiple instruction, multiple data (MIMD) 141, 222
multiple-issue processing 214
multiplexer 34
multiply-accumulate (MAC) 155, 178, 222, 366
multiprocessing computer 141, 142
multiprogramming 179
multitasking 148
mutual exclusion (mutex) 149
mutual exclusion (mutex) issues
  deadlock 150
  priority inversion 149
  thread preemption 149

N

natural boundary, of data type 252
natural language processing (NLP) 364
near-field communication (NFC) 169
Negative sign flag 20
Nervana neural network processor for inference (NNP-I) 366
  form factors 367
Nervana neural network processor for training (NNP-T) 363, 366
  NNP-T1000 363
  NNP-T1300 363, 364
  NNP-T1300, versus NNP-T1400 366

NNP-T1400 363
nested interrupt 77
nested virtualization 322
network interface 110
  Ethernet 110
  Wi-Fi 111, 112
network virtualization 323
neural networks and machine learning architectures 363
  Intel Nervana neural network processor 363
neuron 165
N extension 309
nibble 16, 282
NMI processing 78
nonce 398
Non-Maskable Interrupt (NMI) 73
non-paged pool 187
non-preemptive scheduling 135
non-real-time operating systems 132
  versus real-time operating system (RTOS) 132
nonvolatile memory (NVM) 125
NOP instruction 76
NOT gate 32
NPM transistor 30
null pointer exception 191
numbers
  representing, with voltage levels 14
Nvidia GeForce RTX 3090 GPU 348
  features 349

O

object tracking 441
Ohm's law 28

OnlineU, online college reviews
  reference link 467
opcode 55
Open Compute Project (OCP) accelerator module (OAM) 363
operands 58
operating system 131, 132
operating system virtualization 321
operation code/opcode 22
OR gate 33
out-of-order instruction execution (OoO) 216
output layer 166
overclocking 114
Overflow flag 20

P

page 180
Page Directory Base Register (PDBR) 183
paged pool 188
paged virtual memory 182-186
page fault 184
page frame 182
page frame number (PFN) 188
page status bits 186, 187
page swapping 180
page table 182
page table directory 182
parallel data bus 100-102
  limitations 100
parallel port 119, 120
paravirtualization 327
passband 159

Index

password attacks 372, 373
password-protected hardware debugger interface 244
path traversal vulnerability 392, 393
PC AT (Advanced Technology) 8
PCIe device drivers 121
PCI Express (PCIe) 103, 104
perceptron 348
Peripheral Component Interconnect (PCI) 102, 119
personal computer architecture 347
   Alienware Aurora Ryzen Edition R10 desktop PC 347
petabytes (PB) 164
P extension 309
Phase-locked loop (PLL) 45
phishing 372
physical memory 178, 179
piezoelectric effect 45
pipeline bubble 216
pipeline hazards 215-217
point cloud 431
pointer 191
polyvinyl chloride (PVC) 28
pooling 437
Portable Operating System Interface (POSIX) 322
positive edge-triggered D flip-flop 38
post-exploitation actions 376, 377
post-quantum cryptography 460
power consumption fluctuations 380
power management 240
   techniques 240
Power-On Self-Test (POST) 124

predication 281
preemptive scheduling 135
prefetching 99
principle of least privilege 388
priority inheritance 150
priority inversion 149
private key 246
privileged processor modes 226
privilege escalation 376
privilege levels 304, 305
process control block (PCB) 135
processes 133-136
process identifier (PID) 135
processor cache write policies 207, 208
   write-back 208
   write-through 208
processor context 135
processor flag instructions 75
processor instruction caches 197
processor instruction set 65
processor virtualization methods
   binary translation 327
   hardware emulation 328
   paravirtualization 327
   trap-and-emulate virtualization 324-327
process priority 136-141
process states
   blocked 134
   initializing 134
   running 134
   waiting 134
Program Counter (PC) 55
programmed I/O 82
proof of stake 418

proof of work 404
propagation delay 43, 44
protection rings 230-232
pseudo-instructions 302, 303
public key 246
public-key cryptography 246
pulse's time of flight 430, 440

Q

Q extension 309
Qisket 462
  URL 462
quantum bits/qubits 13, 458
quantum code-breaking 459, 460
quantum computing 458-462
quantum conferences
  reference link 467
quantum decoherence 461
quantum entanglement 459
quantum error correction 461
quantum physics 455, 456
quantum superposition 458
quantum supremacy 461
queue 151
quick emulator (QEMU) 337
  URL 329

R

R-2R ladder 152
rack-based servers, WSC 354-357
radio detection and ranging (radar) 430
radio frequency (RF) transmissions 152
random-access memory (RAM) 6, 55, 92, 178

ransomware 374
rate-monotonic scheduling (RMS) 137
raytracing 348
read-only memory (ROM) 6, 124, 178
real mode 251
real-time computing 146-148
real-time operating system
  (RTOS) 131, 148-151
  versus non-real-time operating systems 132
Rectified Linear Unit (RELU) 437
recurrent network 166
reduced instruction set
  computer (RISC) 64, 215
register load and store instructions
  offset 277
  register 277
  scaled register 277
register renaming 217, 218
registers 19, 40, 41, 64, 65
register set 54
register-to-register data transfer
  instructions 72
register-transfer level (RTL) 315
relative addressing mode 75
remote computing 384
Resource Monitor tool 135
reverse engineering 386
RGB 432
ring counter 39
ripple carry adder 42
RISC-V
  implementing, in FPGA 313-317
RISC-V 32-bit integer instruction
  set (RV32I) 305

RISC-V A extension 306, 307
RISC-V applications 294-298
RISC-V architecture 294-298
   extensions 308
   variants 309, 310
RISC-V assembly language 312
RISC-V base instruction set 298
   computational instructions 298, 299
   control flow instructions 299
   memory access instructions 299
   privilege levels 304, 305
   pseudo-instructions 302, 303
   system instructions 300, 301
RISC-V C extension 307, 308
RISC-V configurations 311
RISC-V D extension 308
RISC-V extensions 305, 306
RISC-V F extension 308
RISC-V M extension 306
RISC-V processor virtualization 334, 335
round-robin scheduling 137
routers 110
row hammering 379, 380
RV32E architecture 308
Ryzen 9 5950X processor
   branch prediction 348
   features 347

S

safer mode extensions (SMX) 263
sample-and-hold circuit 153
sandbox 320
Scala 315
Scalable Link Interface (SLI) 115

scheduling algorithms 136, 138-141
sector 105
Secure Guard Extensions (SGX) 384
secure hardware, features
   attacks, anticipating 378, 379
   data, identifying to be protected against unauthorized disclosure 378
secure system design
   comprehensive secure design 387
   principle of least privilege 388
   zero trust architecture 388-390
secure system design, features
   cryptographically secure key generation 382
   encryption, of data at rest 381
   encryption, of data in transit 381
   secure boot procedure 382
   secure key storage 381
   tamper-resistant hardware design 383
security
   avoiding, through obscurity 386, 387
   designing, at architectural level 386
   ensuring in application software 390
   ensuring in system software 390
security technologies
   device memory protection 245
   internal encryption engine with key storage 244
   password-protected hardware debugger interface 244
self-driving vehicles
   autonomy levels 422-424
   hardware and software requirements 426
   overview 422
   safety concerns 424-426
self-driving vehicles, autonomy levels
   conditional driving automation 423
   driver assistance 423

full driving automation 424
high driving automation 423
no driving automation 422
partial driving automation 423

self-modifying code 175

semaphore 150

semiconductor 29

sensor data
converting, into driving decisions 431

sensor data processing 442
lane-keeping task 442
objects, avoiding 443
road rules, complying with 442
vehicle path, planning 443

sensor types, used in autonomous vehicle designs
Global Positioning System (GPS) 428
inertial sensors 428
light detection and ranging (lidar) sensor 430
radio detection and ranging (radar) 430
sound navigation and ranging (sonar) 431
speedometer 428
video cameras 429

sequential logic 45

Serial AT Attachment (SATA) 104

serial data bus 100-102

Serial Peripheral Interface (SPI) 310

set
versus cleared 17

set associative cache 205, 206

Set-Reset (SR) 35

SHA-256 hash algorithm 400, 401
computing, of data block 402

shadow page tables 330

shellcode 391

shift register 39

Shor's algorithm 460

signal processing algorithm 157
convolution 157
digital filtering 158, 159
Fast Fourier transform (FFT) 159-162

signal-to-noise ratio (SNR) 157

silicon n-channel transistors 453

simultaneous multithreading 220, 221

single data rate (SDR) 96

single instruction, multiple data (SIMD) 10, 141, 309
processing 221-223

single processor instruction
execution, steps 211

small outline DIMM (SODIMM) 97

Smart contracts 418

smartphone architecture 343
iPhone 13 Pro Max 343

Society of Automotive Engineers (SAE) 422

SoC integration 454

soft fault 185

soft real-time system 146

software attestation procedure 384

software weaknesses 390
buffer overflow 390, 391
cross-site scripting 391
path traversal vulnerability 392, 393
SQL injection 391, 392

solid-state drive (SSD) controller 345

solid-state drives (SSDs) 105

solo mining 408

sound navigation and ranging (sonar) 431

source code security scans 393
Source Index (SI) 6
source terminal 89
spatial locality 197
spectral analysis 160
spectrum analyzers 160
speedometer 428
spin quantum number 457
spintronics 456-458
spin valve 457
split cache 201
spyware 374
SQL injection 375, 391, 392
square wave signal 44
stack instructions 72, 73
Stack Pointer (SP) 6, 21
Static RAM (SRAM) 199, 200
static translation 328
Status Flags (FLAGS) 6
stopband 159
stopband suppression 158
storage virtualization 323
   centralized management 323
   data migration 324
   replication 324
Streaming SIMD Extensions (SSE) 222
subroutine call and return instructions 75
successive approximation 154
superpipelining 214, 215
superscalar processors 221
supervisor mode 232, 233
swap file 180
symmetric multiprocessing systems 141

synapses 165
synchronous circuit 40
synchronous DRAM (SDRAM) 96
synchronous logic 46
system calls 233, 234
system instructions 300, 301
system-on-chip (SoC) 273, 309
system security management 243, 244
system software
   security, ensuring 390

T

T32 instruction set 273
tamper-resistant hardware design 383
TechCrunch
   URL 465
temporal locality 197
tensor 365
TensorFlow 349
tensor processor clusters (TPCs) 366
terahash 406
terminate and stay resident (TSR) 179
Tesla HW3 Autopilot 444, 445
   features 445, 446
test-and-set instruction 149
T extension 309
thermal fluctuations 380
thread control block (TCB) 135
thread identifier (TID) 135
thread preemption 149
threads 133-136
Thunderbolt 106
thwarting cyberattackers 246, 247

tile-based deferred rendering (TBDR) 344

time-slicing 133

TNW
  URL 465

Tom's Hardware
  URL 465

Top 25 Most Dangerous Software Weaknesses
  reference link 393

training set 364

transactional synchronization extensions (TSX) 263

transistor 29, 30

Transistor-Transistor Logic (TTL) 119

translation lookaside buffer (TLB) 189, 196

trap-and-emulate virtualization 324-327

trap exception 325

trillion operations per second (TOPS) 366

Trusted Platform Module (TPM) 130, 245
  authentication services 246
  cryptographic key generation 245
  cryptographic key storage 246
  random number generation 245
  system health monitoring 246
  system integrity verification 246

truth table 32

two-input multiplexer 34

two's complement 17, 18

U

Unified Extensible Firmware Interface (UEFI)
  applications 126
  boot 128, 129

UEFI features
  architecture-independent
    device drivers 126
  faster booting 127
  secure boot 127

Unified Extensible Firmware Interface (UEFI) 126, 127

unipolar transistor 89

Universal Serial Bus (USB) 105, 106

unsafe instructions 330, 332

unshielded twisted-pair (UTP) 110

user mode 232, 233

U.S. News & World Report, online college
  reference link 467

V

V bit 335

vector 155

velocity 427

Verilog 47

very high-speed integrated circuit (VHSIC) 47

V extension 309

VHSIC hardware description language (VHDL) 47-51

video cameras
  challenges 429
  using 429

Video Graphics Array (VGA) 108

Virtual Address Extension (VAX) 181

VirtualBox 335

virtualization 320
  using, in cloud computing 338
  versus cloud computing 338

virtualization challenges 329
  security 331
  shadow page tables 330

unsafe instructions 330
virtualization tools 335
  kernel-based virtual machine (KVM) 337
  QEMU 337
  VirtualBox 335
  VMware ESXi 336
  VMware Workstation 336
  Xen 337
virtualization types
  application virtualization 322
  network virtualization 323
  operating system virtualization 321
  storage virtualization 323
virtual local area networks (VLANs) 323
virtual machine 321
virtual machine extensions (VMX) 263
virtual machine monitor 321
  versus hypervisor 321
virtual memory 182
  benefits 180, 181
  usage 180
Virtual Memory System (VMS) 181
virtual supervisor (VS) mode 335
virtual user (VU) mode 335
viruses 374
VMware ESXi 336
VMware Workstation 336
voltage levels
  used, for representing numbers 14
von Neumann architecture 5, 156, 174
  bottleneck 156, 175
  security considerations 175
vulnerability exploitation 373

W

warehouse-scale computer (WSC) architecture 351
  as multilevel information cache 358
  cloud application, deploying 359-362
  electrical power consumption 358
  hardware 352, 353
  hardware fault management 357
  rack-based servers 354-357
water hammer arrestors function 93
web browser caches 197
web crawlers 357
white-hat hacking 371
wide area network (WAN) 110
Wi-Fi 111, 112
Wi-Fi Protected Access 2 (WPA2) 112
Windows boot loader 129
Windows Boot Manager 129
Windows x64 application programming interface (API) 270
Wine
  URL 322
Wired
  URL 465
word length 19
wordline 94
working set 188
workloads 336
worms 374

X

x64 architecture 268
  features 268
  instruction set 268

x64 assembly language 270-272
x64 instruction categories 270
x64 instruction formats 270
x64 register set 269

x86-64 268

x86 addressing modes 255
  based indexed addressing 257
  based indexed addressing, with scaling 257, 258
  direct memory addressing 256
  immediate addressing 256
  implied addressing 256
  indexed addressing 257
  register addressing 256
  register indirect addressing 256

x86 architecture 250, 252
  addressing modes 255
  instruction set 250, 251
  x86 assembly language 264-267
  x86 instruction categories 258
  x86 instruction formats 264
  x86 register set 252-255

X86 current privilege level (CPL) 332

x86 hardware virtualization 333

x86 instruction categories 258
  AES instructions 263
  arithmetic and logic instructions 259, 260
  common instruction patterns 263
  control flow instructions 261
  conversion instructions 261
  data movement instructions 258
  flag manipulation instructions 262
  floating-point instructions 263
  input/output instructions 262
  miscellaneous instructions 262
  MPX instructions 263
  other instruction categories 263
  protected mode instructions 262
  SIMD instructions 263
  SMX instructions 263
  stack manipulation instructions 259
  string manipulation instructions 261
  TSX instructions 263
  VMX instructions 263

x86 instruction formats
  address displacement bytes 264
  immediate value bytes 264
  ModR/M byte 264
  opcode bytes 264
  prefix bytes 264

x86 processor virtualization 331-333

Xen 337

XOR gate 34

Z

zero-day vulnerability 373

Zero flag 21

zero-page addressing mode 71

zero trust architecture 388-390

Zicsr extension 309

Zifencei extension 309